The Cambridge Illustrated Glossary of Botanical Terms

This comprehensive and beautifully illustrated glossary comprises over 2400 terms commonly used to describe vascular plants. The majority are structural terms referring to parts of plants visible with the naked eye or with a x10 hand lens, but some elementary microscopical and physiological terms are also included, as appropriate. Each term is defined accurately and concisely, and whenever possible, cross referenced to clearly labelled line drawings made mainly from living material. The illustrations are presented together in a section comprising 127 large format pages, within which they are grouped according to specific features, such as leaf shape or flower structure, so allowing comparison of different forms at a glance. In addition to supporting the definitions, the illustrations therefore also provide a unique compilation of information that can be referred to independently of the definitions. This makes the glossary a particularly versatile reference work for all those needing a guide to botanical terminology, equally useful to beginners as well as to those more advanced in their botanical knowledge.

MICHAEL HICKEY trained in botany and horticulture, including a period at the University Botanic Garden, Cambridge, before embarking on a teaching career that has spanned over 40 years, ranging from teaching biology to school children through to instructing adults in basic botany and the art of botanical illustration. Awarded the Royal Horticultural Society's Silver Gilt and Silver Lindley Medals for his pen and ink drawings, he has exhibited in the UK and the USA and has contributed illustrations to numerous publications. He is author of *Plant Names: A Guide for Botanical Artists* (1993), *Drawing Plants in Pen and Ink* (1994), and *Botany for Beginners* (1999). With Clive King, he has co-authored *100 Familes of Flowering Plants* (1981, 1988) and *Co,nmon Families of Flowering Plants* (1997).

CLIVE KING spent over thirty years at the University Botanic Garden, Cambridge as Assistant Taxonomist and Librarian. During that time he was involved in the identification of plants belonging to a wide range of families, the botanical instruction of the annual intake of students, and the day-to-day running of the specialist library housed there. He has translated *Collins Guide to Tropical Plants* (1983) and *Collins Photoguide to the Wild Flowers of the Mediterranean* (1990), and has co-translated *Collins Photographic Key to the Trees of Britain and Northern Europe* (1988), all from the original editions in German. With Michael Hickey he has co-authored *100 Families of Flowering Plants* (1981, 1988) and *Common Families of Flowering Plants* (1997).

The Cambridge Illustrated Glossary of Botanical Terms

MICHAEL HICKEY

and CLIVE KING

PUBLISHED BY THE PRESS SYNDICATE OF THE UNIVERSITY OF CAMBRIDGE
The Pitt Building, Trumpington Street, Cambridge, United Kingdom

CAMBRIDGE UNIVERSITY PRESS
The Edinburgh Building, Cambridge, CB2 2RU, UK
40 West 20th Street, New York, NY 10011-4211, USA
10 Stamford Road, Oakleigh, VIC 3166, Australia
Ruiz de Alarcón 13, 28014 Madrid, Spain
Dock House, The Waterfront, Cape Town 8001, South Africa

http://www.cambridge.org

First published 2000

Printed in the United Kingdom at the University Press, Cambridge

Set in Times by the authors

A catalogue record for this book is available from the British Library

Library of Congress Cataloguing in Publication data applied for

ISBN 0 521 79080 8 hardback
ISBN 0 521 79401 3 paperback

Contents

Foreword

The collaboration between Michael Hickey and Clive King, authors of this new Glossary, has become a very familiar facet of the life of the Cambridge University Botanic Garden, and the small part I have been able to play over the years in furthering this productive partnership has given me continuing pleasure.

This entirely new book represents a logical further step in the concern of the authors to supply the increasing educated public interested in the naming and classification of vascular plants with the tools of their trade. Technical terms, derived almost exclusively from the classical languages of Latin and Greek, are indispensable if we wish to proceed beyond a very superficial knowledge of the great variety of plants in gardens or in nature, and I feel very confident that this carefully constructed illustrated Glossary will meet all our needs – including my own, as I continue, in old age, to enjoy my expert hobby as much as I ever did.

I commend the book unreservedly to you, the next generation, for whom it is designed.

S.M. Walters,
Formerly Director,
University Botanic Garden,
Cambridge

Acknowledgements

We would like to express our appreciation and thanks to Dr S.M. Walters for his encouragement and helpful comments, and for consenting to write the Foreword to this book. We would also like to thank Professor John Parker, Director of the University Botanic Garden, Cambridge, for enabling us to use the facilities within the Botanic Garden and Cory Library, as well as for spending his valuable time looking through the manuscript. We are also grateful for the advice generously given by Dr James Cullen, Director of the Stanley Smith Horticultural Trust, and by Caroline Poole, teacher of biology at Cheltenham Ladies' College, and for the help provided by the staff of Cheltenham and Gloucester College of Higher Education.

In particular we would like to thank Robert King for his expert use of computer equipment which resulted in the production of the camera-ready copy, and also Diane Hudman, who had the complex job of labelling all the illustrations.

Finally, we are indebted to Dr Maria Murphy of Cambridge University Press for her much appreciated advice and assistance with the layout of this book.

Preface

Most glossaries are found on a few pages at the end of text-books, floras, monographs, and other botanical works, and these are often only partially illustrated, and sometimes not at all. We have tried to fill this gap by producing an independent and well illustrated glossary that includes all the terms most commonly used in describing vascular plants, as well as some that are found in more specialised works.
In addition to morphological terms, referring to parts of plants visible with the naked eye or with a hand lens of x10 magnification, we have included some elementary histological, cytological, and genetical terms found in the more general books on botany. Chemical terms, however, have been largely excluded, as it was felt that these were outside the scope of this book.
The definition of each term has been kept brief, but this has been supplemented wherever possible by a line drawing.
The illustrations are for the most part original, and have been drawn from living material. Many of these have a naturalistic rendering, though some, by necessity, are diagrammatic. In a few cases, we have had to make adaptations of existing drawings, either because suitable specimens could not be obtained, or because a particular illustration could not be improved upon.
Although the terms are listed in the traditional alphabetical order, the illustrations have been grouped according to their content in order to make the arrangement more 'user-friendly'. It is hoped that its large format will make the book clear and easy to use, and that readers at all levels of understanding, both amateur and professional, will find it helpful in their chosen area of study, especially plant sciences, horticulture, field studies and botanical illustration.

Notes to Readers

1. Symbols, prefixes, and suffixes precede the glossary, but abbreviations are found under their appropriate letter.

2. In some cases, an alternative term (in round brackets) occurs after the term listed. This may either be followed by an explanation of the term or it may indicate where the definition may be found.

3. The number [in square brackets] after the definition of a term, or the relevant part of it, shows the page on which there is an illustration of the term concerned.

4. Where a term is used in more than one context, or the authors feel that additional drawings would be helpful, several page references are given.

5. As far as possible illustrations referring to a particular part of a plant have been placed on adjacent pages, beginning with roots and other underground parts and moving upwards through stems and leaves to flowers and fruits.

6. Certain pages have been devoted to plant families or classes in which vegetative or floral structure requires the use of special terms.

7. An arrow has been used to indicate a particular feature in illustrations where possible confusion might arise.

8. Practical studies will often need the use of a hand lens with x10 magnification. In a few cases, e.g. when observing cell tissue, spores, or pollen grains, access to a microscope will be necessary (see illustrations below).

Using a hand lens Hold the lens close to the eye, and in good light move the specimen towards the lens

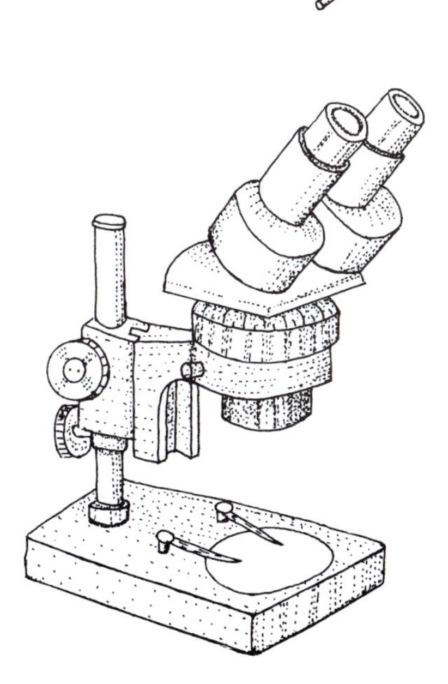

Zoom stereo microscope
Approximate magnification range: x 7 to x 40
Note: Cheaper models without a zoom facility are also available

Compound optical microscope
Approximate magnification range: x 40 to x 60. Note: Advanced models magnify up to x 1000+

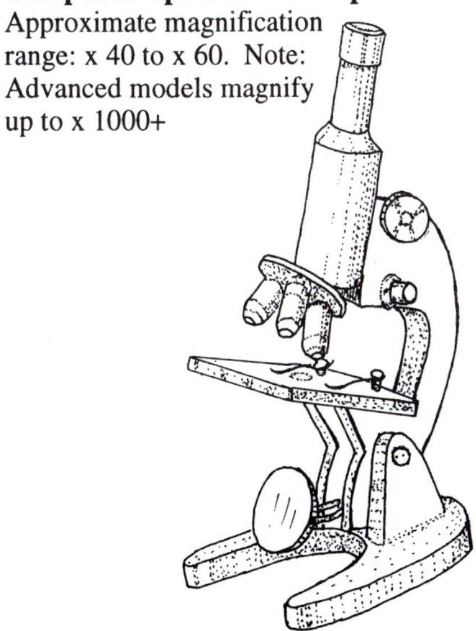

Symbols

+	(before a botanical plant name) Graft hybrid.
+	(between numbers) Indicates that the plant parts concerned are in two groups.
-	(between numbers) Indicates the range of variation.
±	More or less.
>	More than.
<	Less than.
≥	At least, not less than.
≤	At most, not more than.
( )	(enclosing a number) Connate.
⌢	(above two letters) Adnate.
∞	Indefinite number.
♂	Male.
♀	Female.
⚥	Bisexual, hermaphrodite.
⊕	Actinomorphic, regular.
·|·	Zygomorphic, irregular.
§	Section, or other divisions of a genus.

Prefixes (General)

a-	Not, without
acro-	Uppermost, farthest
amphi-	On both sides, around
andro-	Male
aniso-	Unequal
anti-	Against, opposite to
apo-	Separate
bathy-	Lowest
circum-	Around
chloro-	Green
chori-	Separate
chromo-	Colour(ed)
crassi-	Thick, dense
dialy-	Separate
e-	Without
eco-	Ecological
endo-	Within
epi-	Upon, above, over
ex-	Without
exo-	Outside
extra-	Outside
gamo-	United
gymno-	Naked, exposed
gyno-	Female
haplo-	Single
hetero-	Different
holo-	Whole
homo-, homoio-	Same, alike
hypo-	Beneath, below
infra-	Below
inter-	Between, among
intra-	Within, inside
iso-	Equal, same
macro-	Large
medi-	Middle
mega-	Large
meso-	Middle
micro-	Small, very small
multi-	Many
n-, notho-	Hybrid
ob-	Inverse(ly)
oligo-	Few
ortho-	Straight
palaeo-, paleo-	Ancient
pauci-	Few
peri-	Around, enclosing
pleio-	More
pluri-	Several
poly-	Many
post-	After
pro-	Before, in front of, earlier
proto-	First
pseudo-	False
schizo-	Split
sub-	Beneath, less than, approximately
super-, supra-	Above
sym-, syn-	With, together
tenui-	Thin

Prefixes (Numerical)

	LATIN	GREEK
½	semi-	hemi-
1	uni-	mono-
2	bi-	di-
3	tri-	tri-
4	quadri-	tetra-
5	quinque-	penta-
6	sex-	hexa-
7	septem-	hepta-
8	octo-	octo-
9	novem-	ennea-
10	decem-	deca-
11	undecim-	endeca-
12	duodecim-	dodeca-
20	viginti-	icosa-
100	centi-	hecato-
1000	milli-	chilio-

Suffixes (details of taxonomic ranks are given in the Glossary)

-aceae	Family, a taxonomic rank.
-ales	Order, a taxonomic rank.
-ara	Added to a personal name to indicate an intergeneric orchid hybrid.
-eae	Tribe, a taxonomic rank.
-ferous	Bearing, producing.
-fid (leaves)	Divided up to half-way towards the rachis or base.
-foliolate (leaves)	Consisting of a certain number of leaflets.
-idae	Subclass, a taxonomic rank.
-inae	Subtribe, a taxonomic rank.
-ineae	Suborder, a taxonomic rank.
-merous (flowers)	Having the parts of the flower of a particular number, or a multiple of that number.
-oid	Resembling.
-oideae	Subfamily, a taxonomic rank.
-opsida	Class, a taxonomic rank.
-partite (leaves)	Divided from half to two-thirds of the way towards the rachis or base.
-phyta	Division, a taxonomic rank.
-phytina	Subdivision, a taxonomic rank.
-ploid (genetics)	The number of sets of chromosomes in each cell.
-sect (leaves)	Divided almost to the rachis or base.

Measurements

c. (circa)	About, approximately
m	Metre(s)
cm	Centimetre(s)
mm	Millimetre(s)

Taxonomic Ranks

The following list of taxonomic ranks is arranged in descending order. Details of the ranks are given in the Glossary.
Principal ranks are shown in bold type, e.g. **kingdom**, secondary ranks in roman type, e.g. tribe, and additional ranks (all with the prefix 'sub-') are in italics.

kingdom

subkingdom

division

subdivision

class

subclass

order

suborder

family

subfamily

tribe

subtribe

genus

subgenus

section

subsection

series

subseries

species

subspecies

variety

subvariety

form

subform

Glossary

Latin plurals: The following examples show how the plural is formed for most of the Latin terms listed below.
Less common Latin plurals and Greek plurals, where required, are given after the terms concerned.

SINGULAR	PLURAL
locul-us	**locul-i**
pinn-a	**pinn-ae**
haustori-um	**haustori-a**

A (in a floral formula) Androecium, e.g. A10 indicates an androecium composed of 10 stamens.
abaxial The side of an organ away from the axis, dorsal. **[103]**
aberrant Not typical, differing from the normal form.
abortion Non-formation or incompletion of a part. **[178]**
abortive Imperfectly developed.
abscission The shedding of parts of a plant, e.g. leaves, flowers, etc. by means of an abscission layer, either naturally from old age or prematurely from stress. **[85]**
abscission layer A layer of cells that develops across the base of a petiole or pedicel and then weakens, causing the leaf or flower to fall off. **[85]**
acanthophyll A spine, often large, derived from a leaflet. **[167]**
acaulescent Without a stem, or apparently so. **[74, 166]**
accrescent Increasing in size with age, as the calyx of some plants after flowering, e.g. *Physalis alkekengi* (Chinese Lantern). **[127]**
accumbent Of cotyledons, having the edges adjacent to the radicle. **[68]**
achene A small, dry, one-seeded, indehiscent fruit, strictly of one carpel, as in the genera *Ranunculus* and *Rosa*. **[177]**
achlamydeous Without a perianth, as the flowers of *Salix* (willow). **[128]**
acicle A stiff bristle or slender prickle, sometimes with a gland at its apex. **[94]**
acicular Needle-shaped. **[105]**
acid soils Soils with a pH value of 6.5 or below.
acidophile A plant that thrives in an acid soil.
acidophilic, acidophilous Adapted to or thriving in an acid soil.
acorn The fruit of the genus *Quercus* (oak) in the Fagaceae. **[98]**
acrogen A flowerless plant, e.g. a fern, in which growth occurs only at the apex of the stem.
acrogenous Growing only at the apex of the stem.
acropetal Produced, developing, or opening in succession from base to apex.
acrophyll One of the mature fronds of a climbing fern that occur in the upper part of the plant. **[201]**
acroscopic On the side towards the apex. [**202]**
acrospire The first sprout of a germinating seed.
acrostichoid Resembling *Acrostichum*, one of several genera of ferns in which the sporangia are distributed over the lower surface of the fertile lamina. **[203]**
acrotonic A type of branching in which the shoots nearest the apex of the stem show the greatest development.
actinomorphic (**regular**, **radially symmetric**) Divisible through the centre of the flower in several or many longitudinal planes, the halves of the flower being mirror images in every case. **[126]**
actinostele A type of protostele, in which the xylem forms a star-shaped structure with phloem between its rays. **[204]**
aculeate Bearing prickles. **[94]**
acuminate Narrowing gradually to a point. **[109, 110]**
acute Sharply pointed. [**109, 110]**
acyclic Arranged spirally rather than in whorls.
adaxial The side of an organ towards the axis, ventral. **[103]**
adherent In close contact with a different part, but not fused with it.
adnate United with a different part, as stipules to a petiole **[92]**, or a bract to a peduncle. **[98]**
adpressed (see **appressed**)
adventitious bud A bud that arises from any part of a plant other than the axil of a leaf.
adventitious root A root that arises from any part of a plant other than the primary root system. **[51, 53, 56, 58, 66, 79]**
adventive Growing spontaneously in a particular region but not native there.
aerenchyma Tissue with well-developed air spaces between the cells, characteristic of the roots and stems of water plants. **[165]**
aerial root An adventitious root that does not grow down into the soil, e.g. the roots of epiphytic orchids that absorb water from the surrounding air. **[52]**

aerotaxis Movement in response to the source of oxygen.
aerotropic Turning towards or away from the source of oxygen.
aerotropism The growth movement of a plant in response to oxygen.
aestival Occurring in early summer.
aestivation The arrangement of the calyx or corolla in a flower bud. (see illustrations of **vernation** for terms used)
aff. (**affinis**) Having affinity with, near to. Usually precedes the name of a species to indicate a plant not conforming exactly with the description of that species but clearly related to it.
agamospermy (**seed apomixis**) A form of apomixis in which seed is set, but without sexual fusion. Offspring produced in this way have the genetic constitution of the parent plant. Genera in which this process occurs include *Taraxacum* (dandelion), *Hieracium* (hawkweed), and *Rubus* (blackberry, etc.).
agg. (**aggregatum**) Aggregate, added to the name of a species to signify the inclusion of other taxa in a closely related group.
aggregate fruit A fruit formed by the joining of several carpels that were separate in the flower, as in the genus *Rubus* (Rosaceae). **[177]**
aggregate species (**collective species**) A group of two or more closely related species which for convenience have been given a shared name, e.g. *Rubus fruticosus* (blackberry).
ala A wing; one of the two lateral petals in the flowers of plants of subfamily Papilionoideae in the Leguminosae. **[148]**
alar (see **axillary**)
alar flower A flower borne in the fork between the two branches of a dichasium, e.g. as in some genera of the Caryophyllaceae. **[121]**
alate (**winged**) Having a wing or wings. **[67, 85, 91, 178]**
albumen Nutritive material stored within the seed.
albuminous Possessing albumen.
alien Not native to the region concerned.
alkaline soils Soils with a pH value above 7.5.
allele, allelomorph Any one of the alternative forms of a particular gene.
allogamy (**cross-fertilisation**) Fertilisation of the ovules of a flower by pollen from a different flower. (see **geitonogamy** and **xenogamy**)
allopatric Of plant species or populations, not growing in the same geographical area.
allopolyploid A polyploid of hybrid origin, containing sets of chromosomes from two or more different species.
alternate Placed singly along the stem or axis, not opposite or whorled. **[101]**
alternation of generations In the life cycle of ferns and fern allies, the alternation of a haploid gametophyte generation, reproducing sexually, with a diploid sporophyte generation, reproducing asexually. **[199]**
ament (**amentum, catkin**) A spicate, often pendulous inflorescence of unisexual, apetalous flowers. **[121]**
amphicarpic, amphicarpous Producing two kinds of fruit, differing in one or more characters.
amphidiploid An allopolyploid containing a diploid set of chromosomes from each of two different species.
amphiphloic siphonostele (**solenostele**) A type of stele in which a central core of pith is surrounded first by a ring of phloem, then by a ring of xylem, followed by a second ring of phloem. **[204]**
amphistomatal, amphistomatic With stomata on both upper and lower surfaces of the leaf.
amphitropous (**hemitropous**) Curved, so that both ends of the ovule are brought near to each other. **[140]**
amplexicaul Clasping the stem, but not completely encircling it. **[102]**
amyloplast A form of leucoplast occurring in storage organs that can convert sugar into starch.
anastomosing Having the veins branched, the vein branches sometimes meeting only at or near the margin of the leaf. **[104]**
anastomosis A cross-connection of veins in a leaf, producing a somewhat denser network of veins towards the margin. **[104]**
anatomy The science or study of the structure of plants, based on dissection.
anatropous With the body of the ovule inverted so that it lies alongside the funicle. **[140]**
ancipitous Having two edges and being flattened, as the pseudobulbs of *Laelia rubescens*.
androdioecious A species in which individual plants bear only male flowers or only bisexual flowers.
androecium The male sex organs (stamens) collectively. **[131-135]**
androgynophore A stalk bearing both androecium and gynoecium, as in the flowers of many members of the Passifloraceae. **[146]**
andromonoecious Having male and bisexual flowers on the same plant.
androphore A stalk bearing the androecium, as in the flowers of some members of the Tiliaceae.
anemochore A plant whose seeds or fruits are dispersed by the wind.
anemochorous Of seeds or fruits, dispersed by the wind.
anemochory The dispersal of seeds or fruits by the wind.

anemophilous Depending on the wind to convey pollen for fertilisation.

anemophily Pollination by means of the wind.

Angiospermae (**Anthophyta**, **Magnoliophyta**) The angiosperms, flowering plants whose ovules are enclosed in an ovary. **[130]**

angustiseptate Having the partition (septum) across the narrowest diameter of the fruit, as in *Capsella bursa-pastoris* (Shepherd's Purse). **[184]**

anisophyllous With two leaves of a pair differing in shape or size.

anisophylly The condition of being anisophyllous.

annual A plant that completes its life cycle within a single year.

annual ring A growth ring, formed in the course of a year in the stem or root of a woody plant, that consists of a band of large xylem cells produced in the spring (see **spring wood**), followed by progressively smaller cells produced in the late summer and autumn (see **autumn wood**). **[82]**

annular Ring-like.

annulate Composed of rings or having that appearance. **[53]**

annulus The specialised ring of cells on a sporangium which is involved in the release of the spores. **[199]**

antenna One of the pair of slender structures on the pollinarium of the genus *Catasetum* that, when touched by an insect, cause the pollinia to be forcibly ejected. **[171]**

antepetalous (see **antipetalous**)

antesepalous (see **antisepalous**)

anterior Front, away from the axis. **[144]**

anthela The panicle in some species of *Juncus* (rush), in which the upper branches are overtopped by the lower ones.

anther The part of the stamen that produces pollen. **[129, 133]**

anther cell (see **theca**)

antheridium The male sex organ in ferns that produces antherozoids. **[199]**

antheriferous Bearing anthers.

antherozoid (**spermatozoid**) A male sex cell with sets of flagella that enable it to move in water. **[194]**

anthesis (**efflorescence**) Flowering time. **[163]**

anthocarp A structure comprising a fruit enclosed in a persistent perianth, as in the Nyctaginaceae. **[125]**

anthocyanins The pigments present in solution in the vacuoles of plant cells that are responsible for the red, blue, or purple colouring in flowers, fruits, and other parts of flowering plants. (see also **carotenoids**)

anthophore An extension of the receptacle above the calyx that appears as a short stalk bearing the corolla, stamens, and ovary. **[125]**

Anthophyta (see **Angiospermae**)

antidromous Having the stipules joined by their outer margins, as in *Alchemilla mollis* (Rosaceae). **[92]**

antipetalous Of stamens, situated on the same radii as the corolla segments, as distinct from alternating with them. **[131]**

antipodal cell One of the group of usually three cells, typically haploid, that lie in the embryo sac at the opposite end to the micropyle. **[130]**

antisepalous Of stamens, situated on the same radii as the calyx segments, as distinct from alternating with them. **[131]**

antrorse Pointing forwards or upwards.

aperturate Having pores. **[129]**

aperture An opening, often circular. **[129]**

apetalous Without petals.

apex (plural **apices**) The tip of an organ. **[57, 110]**

apical At the apex of an organ. **[85]**

apical placentation (**pendulous placentation**) The arrangement in which the placenta is situated at the top of the ovary and the ovule or ovules hang down from it. **[140]**

apiculate With an apiculus. [**110**]

apiculus A short sharp point. **[110]**

apocarpous (**dialycarpic**) Having free carpels. **[136]**

apocarpy The condition of being apocarpous. **[136]**

apogamy Asexual reproduction in ferns, in which a sporophyte is produced directly from a prothallus without the union of gametes.

apomictic Reproducing either by seeds produced asexually, as many species of *Taraxacum* (dandelion) and *Hieracium* (hawkweed), or vegetatively, as when a part of a plant may become detached and develop into a separate plant without any sexual reproduction having taken place.

apomixis Reproduction without fertilisation, either vegetatively (see **vegetative apomixis**) or by seed (see **agamospermy**).

apopetalous (**choripetalous**, **dialypetalous**, **polypetalous**) With a corolla of separate petals, as *Geranium*. **[150]**

apophysis The part of a cone-scale that remains exposed when the cone is closed. **[192]**

aposepalous (**chorisepalous**, **dialysepalous**, **polysepalous**) With a calyx of separate sepals, as *Geranium*. **[150]**

apospory The development of a gametophyte from a sporophyte without the production of spores. **[201]**

appendage An attached subsidiary part.
appressed (**adpressed**) Lying flat against. **[81]**
aquatic Living in water or a waterlogged environment. **[87]**
arachnoid hairs Fine, interlaced hairs resembling a spider's web, like those on the leaves of *Sempervivum arachnoideum* (Houseleek). **[118]**
arboreous, arborescent Tree-like in growth or general appearance.
archegonium The female sex organ in ferns that produces the egg. **[199]**
arching, arcuate Bending over, curved.
areole One of the small areas surrounded by veins in a leaf with reticulate venation **[202]**; one of the small, spine-bearing areas on the stem of a cactus. **[159]**
aril An outgrowth of the funicle, forming an appendage or outer covering of a seed, e.g. the fleshy, scarlet outer covering of the seeds of *Taxus* (yew). **[67]**
arillate With an aril. **[67, 176]**
arilloid Resembling an aril.
arista An awn or stiff bristle. **[110]**
aristate With an arista. **[110]**
aroid A member of the Araceae, as *Arum maculatum* (Lords-and-Ladies). **[124]**
articulate, articulated Jointed. **[202]**
arundinaceous Reed-like.
ascending Sloping or curving upwards. **[71, 73]**
ascidiate Bearing pitcher-shaped structures, as the leaves of *Nepenthes*. **[95]**
ascidium A little pitcher.
asepalous Without sepals.
aseptate Without partitions.
asexual Not sexual, i.e. not involving the fusion of male and female cells.
assimilatory Capable of converting inorganic substances into the constituents of the plant system.
assurgent Rising upwards.
asymmetric With one side of the leaf larger then the other **[109]**; having flowers not divisible into equal halves, as the flowers of the genus *Canna*. **[154]**
atactostele A type of stele characteristic of monocotyledons, in which the individual vascular bundles are distributed throughout the ground tissue. **[79]**
atropous (see **orthotropous**)
attenuate Drawn out and gradually narrowing. **[109]**
atypical Not conforming to type.
auct. (**auctorum**) Of authors, used after a botanical plant name to indicate that that particular name has been accepted by various authors, but not by the original one.
auct. non A phrase placed after a botanical plant name to signify that the name has been misapplied. It is followed by the name of the original author.
auricle A small lobe or ear-like appendage. **[160]**
auriculate With one or more auricles. **[109]**
authority The author(s) of a plant name, i.e. the person(s) responsible for giving a name to a particular taxon. These personal names are normally used in scientific descriptions of plants, on plant labels, etc. for added precision. They are often abbreviated, as in Papaveraceae Juss. (= Jussieu), *Papaver* L. (= Linnaeus), and *Papaver glaucum* Boiss. & Hausskn. (= Boissier and Haussknecht). Certain changes in the classification of a plant may result in a name with a 'double authority' as in *Lobularia maritima* (L.) Desv. In this case, the plant (Sweet Alison or Sweet Alyssum) was originally named *Alyssum maritimum* by Linnaeus, but was subsequently transferred to the genus *Lobularia* by Desvaux.
autochory Dispersal of seeds by the plant itself, e.g. by means of an explosive or ejective mechanism. **[150]**
autogamy (**self-fertilisation**) Fertilisation of the ovules of a flower by pollen from the same flower.
autopolyploid A polyploid containing three or more sets of chromosomes, all from the same species.
autumnal Occurring in autumn.
autumn wood (**late wood, summer wood**) Wood, darker in colour and with smaller xylem cells than spring wood, that is produced in late summer and autumn. **[82]**
auxin One of a group of hormone-like substances, formed in the actively growing parts of plants, that control the growth and development of the plant.
awl-shaped (see **subulate**)
awn A bristle-like appendage, often occurring on the glumes or lemmas of grasses **[162]**; one of the linear structures projecting from the base of the anther in some species of the genus *Erica* **[133]**; the strip of tissue attached to a mericarp in the Geraniaceae. **[150]**
axil The angle formed by the upper side of a leaf and the stem. **[103]**
axile placentation The arrangement in which the placentas are situated on the central axis of the ovary in the angles formed by the septa. **[139]**
axillary (**alar**) In the axil. **[85, 92]**
axis (plural **axes**) A central line of symmetry in a plant or part of a plant, e.g. stem, root, or rachis. **[189, 192]**

baccate Berry-like.
bacciferous Bearing berries.
back-cross A cross between a hybrid plant and one of its parents.
baculate Of pollen grains, covered with rods that are higher than wide and not constricted at their bases.
baculiform Rod-like.
balausta A many-celled, many-seeded, indehiscent fruit with a tough pericarp, as *Punica granatum* (Pomegranate). **[182]**
band (see **retinaculum**)
banner (see **vexillum**)
barb A hooked hair. **[178]**
barbed With hooked hairs. **[178]**
bark A collective term for all the tissues outside the cambium of a woody stem. **[82]**
basal At the base of an organ. **[79, 109]**
basal placentation The arrangement in which the placenta is situated at the bottom of the ovary. **[140]**
basal plate The 'disc' or reduced stem at the base of a bulb. **[57]**
base The part of attachment of any organ. **[133, 162]**
basifixed With the anther attached by the base to the filament. **[133]**
basipetal Produced, developed, or opening from apex to base.
basiscopic On the side towards the base.
basitonic A type of branching in which the shoots nearest the base of the stem show the greatest development.
bast (see **phloem**)
bathyphyll One of the first or basal fronds of a climbing fern. **[201]**
beak The slender projection from the apex of certain fruits. (see also **rostrum**) **[153, 175, 184]**
beard The line of dense hairs at the base of the outer perianth segments ('falls') in flowers of the genus *Iris*. **[156]**
Beltian body A structure formed at the end of a leaflet in certain plants, e.g. *Acacia cornigera* (Bull's-horn Acacia), that is used as food by ants. **[95]**
berry A fleshy, indehiscent fruit with the seed or seeds immersed in pulp. **[124]**
betalain A red or yellow pigment found only in plants of the order Caryophyllales, e.g. *Beta vulgaris* (Beet).
biauriculate Having two auricles.
bicarinate Having two keels.
biciliate With two cilia.
bicollateral bundle A vascular bundle having phloem on two sides of the xylem.
bicoloured Of two colours.
biconvex (see **lenticular**)
bidentate Having two teeth; (leaves) with the margin composed of larger and smaller teeth. **[108]**
biennial A plant that completes its life cycle within two years, producing only vegetative growth in the first year, and flowering in the second.
bifid Divided up to about half-way into two parts. **[117]**
bifurcate Forked, divided into two more or less equal branches. **[137]**
bigeneric Composed of two different genera, as the orchid x *Laeliocattleya*, a hybrid genus produced by crossing a species of *Laelia* with one of *Cattleya*.
bijugate Of a compound leaf, having two pairs of leaflets. **[167]**
bilabiate With two lips, as the corolla in many members of the Labiatae.
bilaterally symmetric (see **zygomorphic**)
bilobed With two lobes. **[125, 193]**
bilocular Having two loculi or compartments.
binomial The botanical name of a plant, comprising the name of a genus followed by the name of the species.
binomial or **binominal nomenclature** The system devised by the Swedish botanist Linnaeus and published in his Species Plantarum (1753), in which plants are distinguished by a two-word name, the first word being the name of the genus (generic name) and the second the name of the species (specific epithet).
biochemical Involved with a chemical process in a living organism.
biovulate Having two ovules.
bipartite Divided almost to the base into two parts.
bipinnate Pinnate, with the primary leaflets again pinnate. **[107]**
bipinnatifid Pinnately lobed, with the lobes themselves similarly divided. **[200]**
biseriate In two series, rows or whorls. **[184]**
biserrate With a saw-toothed margin composed of larger and smaller teeth. **[108]**
bisexual (**hermaphrodite**, **monoclinous**) Having both stamens and carpels in the same flower.
bitegmic Of an ovule, having two integuments.
biternate Consisting of three parts, each part again divided into three. **[107]**
bivalved With two valves. **[125]**
blade (see **lamina**)
blind Of a flower bud, failing to develop into a flower.
bloom The waxy, often bluish green covering on some leaves and fruits; a flower.
blossom A flower, especially one on a fruit tree; the mass of flowers on a fruit tree.
bole The trunk of a tree. **[71]**

boll The spherical or ovoid capsule of *Gossypium* (cotton). **[68]**
bostryx (helicoid cyme) A spiral inflorescence, with axes on different planes, branching always in the same direction. **[123]**
bough One of the main branches of a tree.
bowed Curved, arched. **[104]**
bract A much-reduced leaf, especially the small or scale-like leaves associated with a flower or flower cluster. **[98, 124]**
bracteate Bearing bracts.
bracteolar Relating to bracteoles.
bracteolate Bearing small bracts.
bracteole A small bract, especially when borne on the pedicel of a flower, usually one in monocotyledons, and two (often opposite but sometimes staggered) in dicotyledons. In some species of the genus *Erica* there are three bracteoles, whorled or almost so. **[98]**
bractlet A small bract.
branch A division or subdivision of an axis.
branchlet A small branch or twig. **[195]**
breathing root (see **pneumatophore**)
bristle A stiff hair. **[150]**
bromeliad A member of the Bromeliaceae.
bud A young shoot, protected by scale leaves, from which either leaves or flowers may develop. **[56, 57, 81]**
bud break The stage in the development of a bud when leaves become visible at its apex. **[81]**
bud scale One of the scales that enclose a bud. **[81]**
bud sport A branch, inflorescence, or flower that differs genetically from the remainder of the plant, the differences persisting when the plant is vegetatively propagated from the part concerned.
bulb A usually underground organ, consisting of a short disc-like stem bearing fleshy scale leaves and one or more buds, often enclosed in protective scales. **[57, 79]**
bulbous Having or resembling a bulb.
bulbil A small bulb, often one that arises from the axil of a leaf or the inflorescence. **[61]**
bulblet A small bulb.
bullate With blister-like swellings on the surface. **[115]**
bulliform cell (hinge cell) One of the cells that lie in the grooves on the upper surface of the leaves of grasses, and participate in the folding and unfolding of the leaf. **[160]**
bur, **burr** A rough or prickly fruit of a plant, aiding dispersal of its seeds by animals; the excrescence on the trunk of a tree formed by the bases of epicormic shoots, as in *Tilia* (lime). **[74]**
bursicle The flap-like or pouch-like base of the rostellum in some members of the Orchidaceae. **[170]**
buttress root An adventitious root that grows out from the lower part of the trunk of a tree and remains connected to it down to ground-level. **[53]**

C (in a floral formula) Corolla, e.g. C5 indicates a corolla composed of 5 petals.
c. (circa) About, approximately.
$CaCO_3$ (see **calcium carbonate**)
caducous Falling off early.
caespitose (cespitose) Tufted. **[74]**
calathidium The head of flowers in members of the Compositae, or, in a narrower sense, the involucre only.
calcarate Spurred. **[144]**
calcicole Growing on soils containing lime.
calcifuge Growing on lime-free soils.
calciphile A plant that prefers soils containing lime.
calcium carbonate ($CaCO_3$) Lime, chalk.
caliciform Resembling a calyx.
calloused Having a callus. **[85]**
callus A hard or tough tissue that develops over a wound after an injury **[85]**, or occurs naturally on the labellum of some orchids. **[172]**
calycanthemous Having sepals wholly or partially converted into petals.
calycle, **calyculus** A small calyx, or a calyx-like structure often composed of bracts.
calyptra A cap-like structure covering some flowers and fruits, as the calyx of *Eschscholzia californica* (Californian Poppy). **[149]**
calyptrate Bearing or resembling a calyptra. **[149]**
calyx (plural **calyces**) The outer perianth, composed of free or united sepals. **[67, 127]**
calyx lobe One of the free parts that is joined to the tube of a gamosepalous calyx. **[147]**
calyx segment (see **sepal**)
calyx tube The tube of a gamosepalous calyx. **[68]**
cambial Relating to cambium.
cambium (see **vascular cambium** and **phellogen**)
campanulate Bell-shaped. **[128]**
campylotropous With the body of the ovule curved so that it appears to be attached by its side to the funicle. **[140]**
canaliculate Channelled.
candelabriform Candelabra-like, with tiered whorls of radiating branches, as the inflorescence in some species of *Primula*. **[126]**
cane The hollow, jointed, woody stem of certain grasses, e.g. *Saccharum officinarum* (Sugar Cane) and species of bamboo; the solid stem of certain species of palm, e.g. *Calamus*, used for making rattan and Malacca canes; the stem of species of *Rubus* (Raspberry, Blackberry, etc.).

canescent Densely covered with short, greyish white hairs. **[117]**
capillary Hair-like.
capitate Pin-headed, as the stigma in the genus Primula **[138]**; growing in heads, as the flowers of the Compositae.
capitellate Diminutive of capitate.
capitular Relating to a capitulum.
capitulum A head of sessile or almost sessile flowers surrounded by an involucre, the inflorescence especially characteristic of the Compositae and Dipsacaceae. **[122]**
capsular Relating to, or in the form of a capsule.
capsule A dry, dehiscent fruit formed from a syncarpous ovary. **[124, 125, 136, 164, 176]**
capsuliferous Bearing capsules.
carbohydrate A compound based on carbon, hydrogen and oxygen, e.g. sugar, starch, and cellulose.
carbon dioxide (**CO_2**) A colourless, odourless gas that is necessary for photosynthesis to take place.
carina (**keel**) The structure formed by the two more or less united lower petals in the flowers of plants of subfamily Papilionoideae in the Leguminosae. **[148]**
carinate Keel-shaped. **[103]**
carnivorous plant (**insectivorous plant**) One of some 400 species in several different families that live in habitats poor in nutrients, and have developed insect traps of various kinds in order to obtain the nourishment required for their continued existence. **[95]**
carnose, carnous Fleshy, pulpy.
carotenes A group of red or orange pigments, belonging to the carotenoids, that occur in the chromoplasts of plant cells. They are found in the roots of carrots, and in some flowers and fruits.
carotenoids Red, orange, or yellow pigments, including carotenes and xanthophylls, that occur in the chromoplasts of plant cells and often act as accessory photosynthetic pigments. (see also **anthocyanins**)
carpel One of the units forming the gynoecium, usually consisting of ovary, style, and stigma. **[124, 136]**
carpellate (see **pistillate**)
carpophore The stalk which bears the fruit in the genus *Silene* and some other members of subfamily Silenoideae in the Caryophyllaceae **[125]**, also in the family Umbelliferae. **[177]**
cartilaginous Gristly.
caruncle A protuberance near the hilum of a seed. **[65]**
carunculate With a caruncle. **[65]**
carunculoid Resembling a caruncle.
caryopsis (plural **caryopses**) A dry, one-seeded, indehiscent fruit, characteristic of grasses, having the pericarp united to the seed; the grain of a cereal grass. **[66, 163]**
casual An alien plant that has not become naturalised.
catalyst A substance that increases the rate of a reaction without itself being changed, e.g. an enzyme.
cataphyll A reduced leaf, e.g. a bract, bracteole, bud scale, or one of the papery, sheathing leaves which enclose the whole of the newly developing aerial shoot in the genus *Crocus*. **[56]**
catkin (see **ament**)
caudate With a tail-like appendage. **[110]**
caudex The stem of a plant, especially a fern or a woody monocotyledon, e.g. a palm. **[166, 204]**
caudicle The flexible, stalk-like group of threads connecting a pollinium with the viscidium in the Orchidaceae. **[171]**
caulescent Having an obvious stem. **[74]**
cauliflorous Exhibiting cauliflory. **[125]**
cauliflory The production of flowers on the trunk and branches of trees rather than at the ends of twigs, as in *Cercis siliquastrum* (Judas Tree) and *Theobroma cacao* (Cocoa Tree). **[125]** (see also **ramiflory**)
cauline Borne on the stem. **[79, 80]**
cell The smallest unit of plant tissue that can function independently, containing a nucleus surrounded by cytoplasm **[85, 111]**; a compartment of an anther or an ovary. (see **loculus** and **theca**)
cellular Relating to a cell.
censer mechanism A form of seed dispersal occurring in genera such as *Papaver* (poppy) and *Antirrhinum* in which the capsules open in such a way that seeds are scattered only when the capsule is swung from side to side, as in a strong wind.
centrifugal Developing from the centre towards the margin.
centripetal developing from the margin towards the centre.
cephalium A woody enlargement, bearing a dense mass of hairs, at or near the top of the stem in certain cacti. **[159]**
cereal Any grass that produces an edible grain (caryopsis).
cespitose (see **caespitose**)
cf., cfr. (**confer**) Compare, used in a similar way to the abbreviation 'aff.'.
chaff Thin, dry, membranous bracts or scales, especially the bracts at the base of the florets in members of the Compositae. **[152]**
chaffy Chaff-like.

chalaza The basal portion of the nucellus of an ovule. **[140]**
chalazal Relating to a chalaza.
chalazogamous With the pollen tube entering the ovule through the chalaza.
chalazogamy The entry of the pollen tube into the ovule through the chalaza.
chamaephyte A small, woody or herbaceous perennial, having resting buds not more than 25 cm above soil level. **[75]**
chartaceous Paper-like.
chasmogamy The production of flowers which open in the normal way to expose the reproductive organs.
chasmogamous With flowers opening normally for reproduction to take place.
chasmophyte A plant that grows in rock crevices.
chemonasty A nastic movement in response to a chemical stimulus.
chimaera, chimera A plant or part of a plant composed of cells of two genetically different types, either by a mutation or by the grafting together of parts from two different individuals. (see **graft chimaera**)
chiropterophilous Pollinated by bats.
chiropterophily Pollination by bats.
chlorenchyma Parenchymatous tissue containing chloroplasts, e.g. the mesophyll in a leaf. **[111, 190]**
chlorophyll The green pigment in plants which allows photosynthesis to take place.
chlorophyllous Containing chlorophyll.
chloroplast A plastid containing the green pigment chlorophyll necessary for photosynthesis. **[111]**
chlorosis An unhealthy condition due to a deficiency of chlorophyll that causes the green parts of the plant to become yellowish.
choripetalous (see **apopetalous**)
chorisepalous (see **aposepalous**)
chromoplast A plastid containing a pigment, especially the red, orange or yellow pigments known as carotenoids.
chromosome One of the pairs of strands in the nucleus of a cell that bears genes in a linear order. The number of chromosomes in a cell will vary according to the species, cultivar, hybrid etc. concerned.
ciliate Fringed with long hairs. **[108]**
ciliolate Fringed with short hairs.
cilium One of the fine hairs, resembling eyelashes, that arise from the margin of an organ. **[108]**
cincinnus (scorpioid cyme, spiralled cyme) A cylindrical inflorescence, with axes on different planes, branching alternately to one side and the other. **[123]**
circinate Coiled in a flat spiral, like a young fern frond. **[103]**
circumfloral Around the flower.
circumscissile Dehiscing by a line round the fruit or anther, the top coming off as a lid. **[176]**
circumscissile capsule (see **pyxis**)
cirrhiferous Bearing tendrils. **[93]**
cirrhose, cirrhous, cirrose Ending in a long, coiled tip, tendril-like. **[110]**
cl. Clone.
clade A group of plants believed to have evolved from a common ancestor.
cladistic Relating to a clade.
cladistics A method of classification based on the assumed divergence of groups of plants from a common ancestor.
cladode (cladophyll, phylloclade) A branch taking on the form and functions of a leaf, as in some members of the Euphorbiaceae and the genus *Ruscus* in Liliaceae. **[95, 159, 190]**
cladogenesis The formation of a species by evolutionary divergence from an ancestral species.
cladogram A tree diagram representing the relationships between species or groups of species based on the cladistic method of classification..
cladophyll (see **cladode**)
class A taxonomic rank standing between division and order. Names of classes end in '-opsida'.
classification The arrangement of plants in increasingly specialised categories because of similarities in their structure.
clathrate Resembling lattice-work.
clavate Club-shaped, thickened towards the apex. **[169]**
clavellate Diminutive of clavate.
claw The narrowed base of some petals. **[145]**
cleft Deeply divided into two lobes at the apex. **[110]**
cleistogamous With self-pollination occurring within the unopened flower. **[125]**
cleistogamy The production of often inconspicuous flowers that do not open, allowing only self-pollination to take place, as in some members of the Violaceae. **[125]**
climber A plant that grows upwards by twining round nearby plants and other supports, or by clinging to them with tendrils. **[71, 91]**
clinandrium Of orchids, the tissue at the apex of the column lying beneath the anther.
cline A gradual variation in one or more characters within a species or population.
clip (see **corpusculum**)
clockwise twining Looked at from above, the tip

of the climbing plant grows in the direction of the hands of a clock, forming an apparently left-hand spiral, as in *Lonicera* and a few other plants. **[71]**
clone (cl.) A group of plants that have arisen by vegetative reproduction from a single parent and which are therefore genetically identical.
cluster An indeterminate inflorescence containing several flowers. **[125]**
clustered Closely arranged in a group. **[81, 101]**
cm Centimetre(s).
CO_2 (see **carbon dioxide**)
coalescence Union.
coalescent Having grown together.
cob The infructescence of *Zea mays* (Maize, Sweet Corn) consisting of a woody stem to which the grains are attached. **[66, 163]**
cochleate Spirally twisted, like a snail shell.
coconut fibre (see **coir**)
coenocarpium A fruit formed from an entire inflorescence, such as *Ficus* (fig) or *Ananas* (pineapple). **[182]**
coenocarpous Bearing a coenocarpium.
coensorus An extended sorus or a group of sori that have combined so as to appear as one. **[203]**
co-florescence A lateral branch of a synflorescence.
coherent In close contact with a similar part, but not fused with it. **[134]**
coir A fibrous material, processed from the husks of *Cocos nucifera* (Coconut), that is used in the manufacture of mats, ropes and brushes, and also horticulturally as a substitute for peat.
coleoptile The sheath that surrounds the plumule in grasses, and protects it as it grows to the surface of the soil. **[66]**
coleorhiza The sheath that surrounds the radicle in grasses. **[66]**
collateral Lying or standing side by side.
collective fruit (see **multiple fruit**)
collective species (see **aggregate species**)
collenchyma A specialised form of parenchyma, with thickening in the cell walls, that acts as supporting tissue in leaf veins, petioles, and the cortex of stems.
colleter One of the multicellular, glandular hairs that are found on the adaxial surface of sepals, stipules, the base of petioles, and the adjacent surface of stems in e.g. Loganiaceae and Rubiaceae. **[118]**
colpate Furrowed. **[129]**
colporate Having a composite aperture, consisting of a furrow and a pore. **[129]**
colpus An elongated aperture that appears as a furrow on the surface of a pollen grain. **[129]**
columella The persistent central axis round which the carpels of some fruits are arranged.
column The combination of stamens and styles into a single structure, as in Asclepiadaceae and Orchidaceae. **[151, 170]** (see also **rod**)
columnar Growing in the shape of a vertical cylinder, as some conifers. **[73]**
coma The tuft of hairs at the end of some seeds, as those of *Asclepias* (milkweed) and *Gossypium* (cotton) **[67, 68]**; the apical crown of leaves on the fruit of *Ananas comosus* (Pineapple). **[182]**
commensalism A form of symbiosis in which two different organisms co-exist, only one benefiting, but the other not being harmed.
commissure The surface along which adjoining carpels are appressed. **[177]**
comose Having a tuft of hairs.
compatible Of any two plants, able to fertilise each other.
composite A member of the Compositae.
compound Composed of two or more similar parts. **[91, 107, 121, 122]**
compressed Flattened.
compression wood A kind of reaction wood found on the lower sides of branches and inclined trunks of softwood trees and characterised by rounded tracheids with intervening intercellular spaces.
concave Appearing as if hollowed out. **[191]**
concolorous Having the same colour throughout.
conduplicate Of leaves, folded once lengthwise **[103]**; of a style, grooved down one side, giving the appearance of being folded. **[137]**
cone The conical expansion at the base of the beak of the fruit of *Taraxacum* (dandelion). **[153]** (see also **strobile**)
confluent Merging together.
congeneric Belonging to the same genus.
conical Cone-shaped. **[73]**
conifer A plant bearing cones, as *Pinus sylvestris* (Scots Pine). **[189-192]**
Coniferae (see **Pinopsida**)
conifer allies The Cycadopsida, Ginkgoopsida, and Gnetopsida. **[193-195]**
coniferous Bearing cones.
conker The seed of *Aesculus hippocastanum* (Horse-chestnut). **[176]**
connate United with a similar part as stipules **[92]**, bracts **[124]**, stamens **[134]**, or styles. **[137]**
connate-perfoliate With the bases of two opposite leaves joined together so that the stem appears to pass through them. **[102]**
connective The part of the stamen, a continuation of the filament, that joins together the two pairs of anther cells. **[133, 144]**
connivent (convergent) Gradually approaching each other and meeting at the tips.
conoid, conoidal Cone-shaped.

conspecific Belonging to the same species.
contiguous Touching, but not united with.
contorted Twisted **[72]**; of sepals or petals, overlapping at one margin and overlapped at the other.
contractile root A root that can shorten in order to keep a bulb, corm or rhizome at a particular level. **[56-58]**
convar. Convariety.
convariety (**convarietas**, plural **convarietates**) A concept developed by the German botanist F. Alefeld (1820-72) to group together closely related varieties of cultivated plants, e.g. *Brassica oleracea* ssp. *oleracea* convar. *fruticosa*, which comprises var.*ramosa* (Thousand-headed Kale) and var. *gemmifera* (Brussel Sprout).
convars. Convarieties.
convergent (see **connivent**)
convex Having a rounded surface.
convolute In vernation, having one leaf rolled inside another. **[104]**
cordate Heart-shaped. **[105, 109]**
core The central part, containing the seeds, of the fruit of *Malus* (apple), *Pyrus* (pear) and similar fruits. **[181]**
coriaceous Leathery.
cork (see **phellem**)
cork cambium (see **phellogen**)
corky Having the outer surface cork-like. **[85]**
corm A swollen underground stem, somewhat bulb-like in appearance, but solid and not composed of fleshy scale leaves. **[56]**
cormel (**cormlet**) A small corm.
corniculate Bearing a small horn or horn-like outgrowth.
corolla The inner perianth, composed of free or united petals. **[125, 144]**
corolla lobe One of the free parts that is joined to the tube of a gamopetalous corolla. **[147]**
corolla segment (see **petal**)
corolla tube The tube of a gamopetalous corolla. **[147]**
corona A structure occurring between (and sometimes united with) the stamens and the corolla, as the cup-shaped or trumpet-shaped outgrowth in the genus *Narcissus*, the horned cuculli in *Asclepias curassavica*, and the ring of filaments in the genus *Passiflora* (passion flower). **[146, 151]**
coronal scale One of the ring of scales on the inner surface of the corolla, as at the junction of the limb and the claw in *Silene dioica* (Red Campion). **[145]**
corolloid Corolla-like.
corpus The inner layers of cells in an apical meristem, which divide to produce the inner tissues of the shoot.
corpusculum The clip connecting the two bands (retinacula) which are attached to the pollinia in the Asclepiadaceae. **[151]**
cortex (plural **cortices**) The outer part of an organ; the tissue in a stem or root between the epidermis and the vascular tissue. **[54, 79]**
cortical Relating to the cortex.
corticate Having bark, or a bark-like covering.
corymb A racemose inflorescence with pedicels of different lengths, causing the flower cluster to be flat-topped. **[122]**
corymbiform In the form of a corymb.
corymbose In the form of a corymb; bearing corymbs.
costa A rib, the mid-vein of a simple leaf or the rachis of a compound leaf.
costapalmate With the petiole extending into the blade of a palmate leaf, as in some palms. **[168]**
costate Ribbed.
costule The midrib of a pinnule or of the lobe of a pinna.
cotyledon (**seed leaf**) One of the first leaves of the embryo of a seed plant, typically one in monocotyledons, two in dicotyledons, and two or more in gymnosperms (up to eighteen in the Pinaceae). **[65-68, 189]**
coumarin An aromatic substance, the smell of newly mown hay, especially of the grass *Anthoxanthum odoratum*.
counter-clockwise twining Looked at from above, the tip of the climbing plant grows in the opposite direction to the hands of a clock, forming an apparently right-hand spiral. Most climbers are in this category. **[71]**
crassinucellate Having a nucellus of considerable bulk, because of the numerous cells lying between the epidermis and the embryo sac.
creeper A plant that grows along the ground, over fences, or up walls, e.g. *Hedera helix* (Ivy).
cremocarp A dry fruit, composed of two one-seeded carpels, that at maturity separate into mericarps, e.g. *Heracleum* (hogweed) and other members of the Umbelliferae. **[177]**
crenate With rounded teeth. **[108]**
crenulate With small, rounded teeth. **[108]**
crest An irregular or toothed ridge on the upper part of an organ **[138, 151]**; in the genus *Iris*, one of the two lobes at the end of each style branch. **[156]**
crested Bearing a crest.
crispate With the margin curled or crumpled. **[108]**
cristate Crested.
croceate Saffron-coloured.
cross-fertilisation (see **allogamy**)

cross-pollination The transfer of pollen from one plant to another.
crown The upper, branched part of a tree above the bole **[71, 72]**; the rootstock of e.g. *Rheum* x *cultorum* (Rhubarb).
crownshaft The cylinder formed by tubular leaf sheaths at the apex of the stem in some species of palm. **[166]**
crozier (**fiddlehead**) A coiled young fern frond. **[199]**
cruciate (see **cruciform**)
crucifer A member of the Cruciferae, a family in which the flowers typically have four petals, arranged in the form of a cross, and tetradynamous stamens. **[132]**
cruciferous Resembling a crucifer.
cruciform (**cruciate**) In the form of a cross. **[128]**
cryptobiosis Extended dormancy, as when dehydrated seeds and pollen-grains, also certain flowering plants and ferns in dry habitats remain dormant for long periods but rapidly become active again with the arrival of rain.
cryptobiotic Capable of remaining dormant for a long time in the absence of water.
cryptogam (**flowerless plant**) A plant belonging to the Cryptogamia, a division in former classifications that included all the plants that do not produce seeds, as ferns, fern allies, mosses, fungi, and algae.
cryptophyte A plant with resting buds lying either beneath the surface of the ground, as on rhizomes or in bulbs, or submerged in water. **[75]**
cucullate Hood-shaped.
cucullus A corona hood in the Asclepiadaceae. **[151]**
culm The jointed stem, especially the flowering stem of grasses. **[160]**
cultigen One of many plants found only in cultivation, including cultivars, many hybrids, also a number of ornamental and crop plants which have been grown and developed by man for so long that their wild origins are now uncertain or have been lost entirely.
cultivar (**cv., cvar.**) A cultivated variety, not necessarily attached to a single botanical species, that has been selected for a particular attribute or combination of attributes. Its name is printed in Roman type with a capital initial letter, and enclosed in single quotation marks, e.g. *Prunus avium* 'Plena', the double-flowered form of *Prunus avium* (Wild Cherry).
cuneate Wedge-shaped. **[106, 109]**
cup (see **cupule**)
cupular Relating to or shaped like a cupule. **[98, 167]**
cupulate Bearing a cupule. **[98]**
cupule A cup-shaped structure composed of coalescent bracts, as in the Fagaceae. In e.g. *Quercus* (oak) the cupule (acorn) is indehiscent, but in *Castanea* (chestnut) the cupule splits into several valves. **[98]**
cupuliferous Having a cup-shaped structure surrounding the seed.
cupuliform Shaped like a cupule.
cushion The part of the prothallus of a fern that bears the archegonia, often thicker than the surrounding area. **[199]**
cuspidate Ending rather abruptly in a sharp point. **[110]**
cuticle The waxy layer of cutin that covers the outer surface of the epidermis and restricts the passage of water and gases into and out of the plant. **[111, 190]**
cutin A mixture of fatty substances that comprise the cuticle.
cv., **cvar.** Cultivar.
cvs., **cvars.** Cultivars.
cyathiform Cup or beaker-shaped.
cyathium The inflorescence of the genus *Euphorbia*. **[124]**
Cycadopsida The class that comprises the cycads. **[194]** (see **Gymnospermae**)
cyclic Arranged in whorls.
cylindric, cylindrical Shaped like a straight tube, but completely solid.
cymbiform (see **navicular**)
cyme A branching, determinate inflorescence, with a flower at the end of each branch. **[121]**
cymose In the form of a cyme; bearing cymes.
cymule A small cyme or portion of one.
cypsela A small, dry, one-seeded, indehiscent fruit formed from two united carpels, as in the Compositae. Sometimes treated as a form of achene. **[153]**
cyst A sac or bladder-like structure, containing a liquid secretion.
cystolith A stone-like mass, usually of calcium carbonate, that forms within the epidermal cells of certain plants. **[112]**
cytoplasm The protoplasm of a plant cell excluding the nucleus. **[129]**

daughter bulb A small bulb that is produced by a mature (mother) bulb. **[79]**
daughter cell One of the two cells produced when a cell divides.
decandrous Having ten stamens.
decaploid (**10n**) Having ten sets of chromosomes in each cell.
deciduous Falling off, as leaves in autumn or petals after flowering time. **[85]**

declinate Arching downwards, then turning up towards the apex.
decumbent Lying along the ground, but with the tip ascending. **[71]**
decurrent Running down, as when the base of a leaf is prolonged down the stem as a wing. **[102, 137]**
decussate In opposite pairs, each pair at right angles to the next. **[101]**
deflexed Bent abruptly downwards.
degenerate To deteriorate, to lose normal qualities. **[85]**
dehisce Open spontaneously when ripe.
dehiscence The process of splitting open at maturity, usually applied to an anther shedding pollen or a fruit releasing seeds. **[135, 136]**
dehiscent Opening naturally.
deliquescent Rapidly becoming semi-liquid.
deltoid Triangular. **[105]**
deme A group of plants in a particular locality forming a subpopulation of a species, capable of interbreeding freely amongst themselves but genetically distinct from neighbouring plants of the same species.
dendriform Tree-like.
dendritic Of hairs, branched like a tree. **[118]**
dendrochronology The study of the annual rings present in trees to determine their age and also to ascertain variations in the climatic conditions during their period of growth; often used in the case of ancient timber to date past events.
dendrogram A tree diagram reflecting the relationships of species or groups of plants, e.g. cladogram.
dendrology The study of trees.
dentate Toothed. **[108]**
denticulate Minutely toothed. **[108]**
depauperate In an impoverished condition through lack of the essential elements for growth.
depressed-globose (see **subglobose**)
dermal Relating to the outer covering of an organ.
dermatogen The outermost layer of the apical meristem which develops into the epidermis.
determinate A form of inflorescence in which the terminal flower opens first and prevents further growth of the stem or branch, e.g. a cyme. **[121]**
diadelphous With two groups of stamens, as in some of the Leguminosae. **[132]**
dialycarpic See **apocarpous**).
dialypetalous (see **apopetalous**)
dialysepalous (see **aposepalous**)
diandrous Having two stamens.
diarch root A root with two protoxylem strands in the stele. **[55]**
diatropic Growing at right angles to the source of the stimulus.
diatropism The growth movement of part of a plant at right angles to the source of the stimulus, as horizontal rhizomes to gravity.
dichasial cyme, dichasium A cyme with lateral branches on both sides of the main axis. **[121]**
dichlamydeous Having a perianth of two whorls, i.e. both calyx and corolla are present.
dichogamous With stamens and stigmas of a flower not ripening at the same time, i.e. either protogynous or protandrous.
dichogamy The state of being dichogamous.
dichotomous Forked, having the axis divided into two equal branches. **[166]**
dichotomous key A device for the identification of plants, composed of couplets of opposing statements (leads), whereby the user successively rejects one of the two statements until the final lead is reached, resulting in the identification of the plant concerned.
dichotomy Division into two parts.
diclinous (see **unisexual**)
dicliny The state of being diclinous.
dicotyledons (**Dicotyledonae, Magnoliopsida**) Flowering plants having two cotyledons. **[80]**
dicotyledonous Having two cotyledons.
dictyostele A type of stele, derived from a siphonostele, which is broken up into a network of meristeles. **[204]**
didymous In pairs, as the mericarps forming the fruit of plants in the Umbelliferae. **[177]**
didynamous Having one pair of stamens longer than the other pair, as in many of the Labiatae. **[132]**
diffuse Widely or loosely spreading.
diffuse-parietal With the ovules scattered over the inner surface of a carpel.
digestive zone The area within the pitcher-like leaf of a carnivorous plant where the trapped insects decompose and become the source of nutrients. **[95]**
digitate Palmate with narrow leaflets. **[107]**
dilated Broadened, enlarged.
dimerous Having the parts of the flower in twos.
dimorphic Occurring in two forms.
dioecious Having male and female flowers on different plants of the same species. **[145]**
dioecism, dioecy The state of being dioecious. **[145]**
diploid (**2n**) Having two sets of chromosomes in each cell.
diplostemonous Having the stamens in two whorls, the inner whorl opposite the petals and the outer whorl opposite the sepals. **[131]**
diporate Of a pollen grain, having two pores.
dipterous Two-winged.
disc A development of the receptacle within the perianth, often bearing nectar glands; the central

part of the capitulum in members of the Compositae **[152]**; the basal plate or reduced stem at the base of a bulb. **[57]**

disc floret One of the central, regular flowers in the flower heads of some members of the Compositae. **[87, 152]**

disciferous Bearing a disc, often with nectar glands.

discoid, discoidal Resembling a disc, i.e. thin, flat, and circular. **[138]**

discrete Separate.

disjunct Discontinuous, with the distribution split into two or more areas.

disk (see **disc**)

dispersal The process of scattering seeds away from the parent plant, by means of the wind, birds, animals, etc.

disperse To scatter, especially in regard to seeds.

dissected Deeply divided into segments.

distal Situated away from the point of attachment. **[101]**

distichous Arranged in two vertical ranks. **[101]**

distinct Having similar parts completely separate from each other, e.g. as one petal from another.

distribution The occurrence of a plant species from a geographical point of view.

distyly Dimorphic heterostyly, the occurrence of two different lengths of style in flowers of the same plant species, as *Primula veris* (Cowslip). **[147]**

disulcate Of a pollen-grain, having two grooves or furrows.

disymmetric Divisible through the centre of the flower in only two longitudinal planes for the halves of the flower to be mirror images, e.g. *Dicentra* (Fumariaceae).

dithecous With two anther cells. **[133]**

diurnal Occurring in the daytime.

divaricate Widely spreading or greatly divergent.

divarication Wide-angled branching and subsequent interlacing of the branches of shrubs growing in dry, wind-swept habitats.

divergent Spreading away from each other. **[72]**

divided With lobes extending almost or entirely to the midrib or the base of the leaf.

division (phylum) The highest principal taxonomic rank into which the plant kingdom is divided. Names of divisions end in '-phyta'.

dolabriform Shaped like the head of an axe or a hatchet.

domatium A cavity or tuft of hairs on a plant that acts as a shelter for insects, mites, or other similar organisms. **[96]**

dormancy The state of being inactive.

dormant Resting, inactive. **[56, 81]**

dorsal The side of an organ facing away from the axis, abaxial. **[103]**

dorsifixed With the anther attached by the back to the filament. **[133]**

dorsiventral Having the dorsal and ventral surfaces of the leaf different from each other.

double flower A flower having petals or petal-like structures (petaloids) additional to those in the typical form. Where there is a large number of petals in the double form, the intermediate state may be described as semi-double. **[127]**

downy (see **pubescent**)

drepanium A sickle-shaped cyme, with axes only on one plane, branching always to the same side. **[123]**

drip tip The long, narrow apex of the leaves of many plants living in wet, tropical regions that enables water to be rapidly shed from the upper surface. **[95]**

drooping (see **nutant**)

dropper A shoot growing downwards from a bulb or corm and producing a bulb or corm at its apex. **[60]**

drupaceous Resembling a drupe.

drupe A fleshy fruit containing one or more seeds, each enclosed within a stony endocarp, as in species of the genus *Prunus* (Rosaceae). **[179]**

drupelet One of the small drupes which together form an aggregate fruit, as in the genus *Rubus* (Rosaceae). **[177]**

duct Any of the tubular structures forming the vascular tissue which allow the passage of air or water throughout the plant.

duramen (see **heartwood**)

dwarf A plant of much smaller size than the average of its kind; a part of a plant that is undersized in comparison with the average for that kind of part e.g. dwarf shoots. **[189]**

ear A spike or head of a cereal grass, e.g. *Triticum* (wheat) or *Hordeum* (barley), that contains the grains. **[160]**

early wood (see **spring wood**)

ebracteate Without bracts.

ebracteolate Without bracteoles.

ecalcarate Without spurs.

eccentric Off-centre, having the axis not centrally placed. **[137]**

ecdemic Not native.

echinate Bearing spines. **[176]**

echinulate Bearing small spines or bristles.

ecoclimate The climate of a particular habitat.

ecodeme A deme occupying a particular ecological habitat.

ecological Relating to ecology.

ecology The study of the plants and animals that occupy a particular habitat, and the interaction between them and their environment.

ecosystem The concept of a community of plants and animals that occupy a particular habitat, with special emphasis on the interaction between them and their environment. The living part of the system is usually divided into 'producers' (green plants), 'consumers' (animals), and 'decomposers' (bacteria and fungi), and the whole process can be represented as a flow diagram.

ecotype The variation within a species, often treated as a subspecies, that occurs in a particular habitat.

ectophloic siphonostele A type of stele in which a central core of pith is surrounded first by a ring of xylem and then by a ring of phloem. **[204]**

edaphic Relating to the soil.

efflorescence (see **anthesis**)

e.g. (exempli gratia) For example.

egg (see **ovum**)

eglandular Without glands.

elaiosome An appendage on the seeds of some plants (e.g. *Helleborus*, *Viola*) that contains oily substances attractive to ants, which assist in the disperal of the seeds by carrying them away from the parent plant. **[68]**

elater In species of *Equisetum* (horsetail), an elongated appendage attached to a spore which uncurls as it dries and assists in its dispersal. **[207]**

ellipse A figure shaped like a flattened circle. **[105]**

ellipsoid Elliptical in outline with a three-dimensional body.

elliptic, elliptical In the form of an ellipse. **[105]**

emarginate Distinctly notched at the apex. **[110]**

embryo A young plant developed sexually or asexually from the ovum. In spermatophytes it is contained within the seed. **[65-67, 140]**

embryonic Relating to an embryo or any other part in a rudimentary state of development.

embryo sac The cell inside the ovule of an angiosperm in which fertilisation occurs, and which develops into the female gametophyte. **[130]**

em., emend. (emendavit) An abbreviation that indicates an emendation or correction. It is followed by the name of the author responsible for the alteration.

endemic Restricted to a particular country or region.

endocarp The innermost layer of the pericarp. **[179]**

endodermal Relating to the endodermis.

endodermis The innermost layer of cells of the cortex in stems and roots. **[54, 55, 190]**

endogenous Originating or developing from inside the plant.

endophyllous Formed from within a sheathing leaf.

endophyte A plant that lives inside another plant.

endorhizal With the radicle, instead of lengthening, giving rise to secondary rootlets, as in monocotyledons. **[66]**

endosperm The albumen of a seed, particularly that deposited within the embryo sac. **[65-67, 180]**

endospermic or **endospermous seed** A seed in which the nutritive tissue is absorbed more slowly into the developing embryo so that part of it remains at least until germination, e.g. *Ricinus communis* (Castor-oil Plant) and *Zea mays* (Maize, Sweet Corn). **[65, 66]**

endozoic Having seeds or fruits dispersed by animals, e.g. edible fruits, whose seeds are later expelled with the waste material.

endozoochory Dispersal of seeds or fruits by passing unharmed through the digestive system of an animal.

ensiform Sword-shaped. **[105]**

entire With an unbroken margin, i.e. not toothed or lobed. **[108]**

entomophilous Depending on insects to convey pollen for fertilisation.

entomophily Pollination by insects.

enucleate Without a nucleus.

environment The sum total of external influences acting on a plant.

enzyme A protein produced by a living cell which acts as a catalyst for a biochemical reaction.

ephemeral Lasting for a relatively short period.

epiblem, epiblema (see **piliferous layer**)

epicalyx (plural **epicalyces**) A whorl of sepal-like organs just below the true sepals. **[127]**

epicarp (exocarp) The outermost layer of the pericarp. **[179]**

epicaulic (see **epicormic**)

epichile The apical portion of the labellum in an orchid. **[172]**

epicormic Of shoots, growing out from the trunk of a tree or from a substantial branch. **[74]**

epicotyl The part of a seedling above the cotyledon(s) that gives rise to the stem and leaves. **[65]**

epidermal Relating to the epidermis.

epidermis The outermost layer of cells on plant organs, forming a protective covering to roots, stems, leaves, flowers and fruits. **[54, 79, 80, 82, 111]**

epigeal A type of germination in which the cotyledons emerge above the ground. **[65]**

epigynous With the sepals, petals and stamens inserted near the top of the ovary. **[135]**

epinasty The tendency in part of a plant to grow more rapidly on the upper side so that it curves downwards.

epipetalous Borne upon the petals or corolla-segments, as the stamens in many species. **[131]**

epiphyllous Borne upon the leaves, as the flowers of *Helwingia* (Cornaceae).

epiphylly Growth upon a leaf, as flowers or vegetative buds, e.g. *Kalanchoe daigremontiana* (Crassulaceae). **[61]**

epiphyte A plant that grows on another plant but does not derive nourishment from it, as many species of tropical orchids, bromeliads and ferns. **[52, 96, 201]**

epiphytic Growing on another plant, but not deriving nourishment from it. **[52, 96, 201]**

epistomatal, epistomatic With stomata only on the upper surface of the leaf.

epitepalous Borne upon the tepals.

epithelium A layer of cells, often secretory, that line internal cavities in plants, e.g. resin and gum canals. **[66, 190]**

epithem A group of water-secreting cells in the mesophyll of the leaves of some plants.

epizoic Having seeds or fruits dispersed by animals, e.g. by being furnished with hooks that become entangled in an animal's fur or hair.

epizoochory Dispersal of seeds or fruits by being carried away on the coats of animals.

equator The surface area of a pollen grain midway between the poles. **[129]**

equitant Overlapping in two ranks, as the leaves in the genus *Iris*. **[101, 104]**

erect Upright. **[72]**

ericaceous Relating to the Ericaceae.

ericetum A heath plant-association dominated by the genus *Erica*.

ericoid Resembling the genus *Erica*.

erose Having an eroded or jagged margin.

escape A plant, previously growing in a garden, that has reproduced and become established outside it.

estipulate Without stipules.

etaerio An aggregate fruit composed of achenes, follicles, berries, or drupes. **[177]**

et al. (**et alii**) And others, a phrase used after the name of an author to indicate that other authors also were involved in the naming of a particular plant.

etiolated Grown pale and weak through lack of light. **[86]**

etiolation The condition of a green plant growing in insufficient light, resulting in weak stems, long internodes, and small, yellowish or whitish leaves. **[86]**

evanescent Short-lived, not permanent.

even-pinnate (see **paripinnate**)

evergreen Retaining most of its leaves throughout the year.

exalbuminous Lacking endosperm.

exarillate Without an aril.

exceeding Longer than (another organ).

exfoliating Peeling off in layers, flakes or scales, as the bark of *Platanus* (plane).

exfoliation The process of shedding pieces of bark.

exine (**extine**) The outer part of the wall of a pollen grain, composed of the base, rods or columns, tectum, and ornamentation. **[129]**

exocarp (see **epicarp**)

exodermal Relating to the exodermis.

exodermis The outermost layer of cells in the cortex of a root. **[54]**

exogenous Originating from outside the plant.

exotic Not native; originating in a foreign country, especially one in the tropics.

explicative (see **replicate**)

explosive mechanism A form of seed dispersal in which the ripe fruit suddenly opens, scattering its seeds away from the parent plant. Examples include *Impatiens* (the 5 valves of the capsule roll up), *Ulex* and other members of the Leguminosae (the 2 valves of the capsule separate and twist), and *Ecballium* (seeds ejected in a watery fluid through a hole at the end of the capsule).

exserted Protruding beyond the surrounding parts.

exstipulate Without stipules.

extine (see **exine**)

extrafloral Outside the flower, sometimes used to describe nectaries forming a ring round the reproductive organs, in other cases used to describe nectaries situated on vegetative parts of the plant.

extranuptial nectary A nectary situated away from the reproductive organs of the flower.

extrastaminal Situated outside the whorl of stamens.

extrorse With the anthers facing and opening outwards, away from the centre of the flower. **[135]**

eye One of the pits on a potato tuber containing a bud **[58]**; one of the three roundish areas at the base of a coconut. Two of these represent aborted carpels, the third (larger and softer than the others) indicates the place from which the shoot will emerge after germination. **[180]**

f. (**fide**) According to.

f., fil. (**filius**) Son, as in e.g. Linnaeus fil. or L.f., son of Linnaeus, a means of distinguishing the author of a plant name.

f. (forma) Form.
F_1 The first filial generation, i.e. the hybrids produced from the first crossing of two parent plants, usually for commercial purposes.
F_2 The second filial generation, i.e. the hybrids produced by allowing certain plants of the F_1 generation to interbreed.
falcate Sickle-shaped. **[105]**
fall One of the outer perianth segments in flowers of the genus *Iris*, usually narrow at the base but expanding into a broad, pendulous blade. **[156]**
false fruit (see **pseudocarp**)
false indusium A protective covering of the sporangia formed by the reflexed margin of the frond, as in the genus *Adiantum*. **[203]**
false septum (see **replum**).
false whorl (see **verticillaster**)
family A group of genera resembling one another in general appearance and technical characters. The names of families normally end in '-aceae', but eight long-established family names not ending in '-aceae' are still in use. These are given here with their modern equivalents: Compositae (Asteraceae), Cruciferae (Brassicaceae), Gramineae (Poaceae), Guttiferae (Clusiaceae), Labiatae (Lamiaceae), Leguminosae (Fabaceae), Palmae (Arecaceae), and Umbelliferae (Apiaceae).
fan-shaped (see **flabellate**)
farina The powdery or flour-like covering found on some leaves, stems or floral parts.
farinaceous Having a powdery appearance.
farinose Covered with a fine powder. **[115]**
fasciation The abnormally broad and flattened growth of a stem caused by meristem damage, bacterial infection or mutation so that it resembles several stems placed side by side. The condition, which also affects the inflorescence, occurs commonly in *Taraxacum* (dandelion) and *Plantago* (plantain). The pot-plant *Celosia cristata* (Cockscomb) is grown for its curious inflorescence induced by mutation. **[97]**
fascicle A close cluster or bundle. **[134, 191]**
fascicled (see **fasciculate**)
fascicular cambium The layer of meristematic cells lying between xylem and phloem in a vascular bundle. **[82]**
fasciculate (fascicled) Arranged in clusters or bundles. **[101]**
fastigiate With the branches more or less erect and appressed, as in *Populus nigra* 'Italica' (Lombardy Poplar). **[73]**
feathered With feather-like markings, as those on the perianth segments of some species of *Crocus*; formed like a feather, as the bract in *Euphorbia helioscopia* (Sun Spurge). **[124]**
felted Densely matted, with intertwined hairs.
female flower Having a functional gynoecium, but only rudimentary or sterile stamens. **[124, 145, 183]**
fenestrate, fenestrated Having a fenestration. **[95, 147]**
fenestration (window, window pane) A translucent area in a flower or leaf. In the corolla tube of some gentians and in the lip of the orchid *Cypripedium* it is probably a guide to pollinating insects, while in the domed top of the leaf of *Darlingtonia*, a carnivorous plant, it may assist the plant to catch its insect prey. **[95, 147]**
fern A member of the Filicopsida, a class of flowerless plants with usually large leaves (megaphylls), often called fronds. **[199-204]**
fern ally A member of the Psilotopsida (fork ferns), Lycopsida (club mosses and quillworts) or Equisetopsida (horsetails), classes of flowerless plants with usually small leaves (microphylls). **[204-208]**
ferruginous Rust-coloured.
fertile Able to reproduce sexually. **[124, 134]**
fertilisation The fusion of a male and female reproductive cell resulting in a zygote. **[130]**
festucoid Resembling the grass *Festuca* (fescue).
fetid (see **foetid**)
Fibonacci series A sequence of numbers named after the Italian mathematician Leonardo Fibonacci (c.1170 - c.1250) in which each number is the sum of the preceding two numbers. It is relevant to botany in certain spiral arrangements, e.g. alternate leaves on a stem, florets in the heads of flowers in the Compositae, scales in the cones of the Gymnospermae, and the individual fruits which comprise the multiple fruit in *Ananas* (pineapple). **[87]**
fibres Elongated cells, pointed at both ends, that together with the sclereids, compose the sclerenchyma.
fibril A small fibre.
fibrous Thread-like.
fibrous root A thread-like root, particularly one of the adventitious roots arising from the base of grass stems. **[51]**
fiddlehead (see **crozier**)
fil. (see **f., fil.**)
filament A fine, elongated, thread-like structure, especially the stalk of an anther. **[133, 146]**
filiform Thread-like. **[92, 105, 137]**
fimbriate, fimbriated Fringed, having the margin cut into long, slender lobes. **[108, 138, 151]**
fish-tail leaflet A leaflet shaped more or less like the tail of a fish, as in some palms, e.g. *Caryota*. **[168]**
fish-tail nectary The forked structure, covered in

nectar glands, at the mouth of the pitcher-like trap formed by the leaf of *Darlingtonia*. **[95]**
fissure A usually long, narrow opening, caused by the separation of the parts of an organ.
fistulose Cylindrical and hollow, like a pipe.
fl. (flos) Flower.
flabellate, flabelliform Fan-shaped, as the leaves of *Ginkgo* and fan palms. **[106, 137]**
flaccid Weak, flabby, limp.
flagellum One of the whip-like hairs on an antherozoid. **[194]**
flexuous Wavy. **[74]**
floccose With tufts of woolly hairs. **[117]**
flora The plant population of a particular region; a book listing and describing the plants found in a given area.
floral Relating to the flower.
floral cup (see **hypanthium**)
floral diagram A stylised drawing of the cross-section of a flower, showing the number and relative position of the various parts. **[148]**
floral envelope The perianth of a flower.
floral formula A system for representing the structure of a flower, using the capital letters K (calyx), C (corolla), A (androecium) and G (gynoecium) in that order. The letter P (perianth) is used instead of K and C in cases where the calyx and corolla are not clearly differentiated. Each letter is followed by a figure to indicate the number of parts of which each whorl is composed. E.g. C5 = 5 separate petals, C(5) = 5 petals joined together, A5+5 = 2 whorls of stamens with 5 in each whorl, A12-20 = the range of variation in the number of stamens. Sometimes a curved line above two adjacent letters is used to indicate that these whorls are joined together.
flore pleno With a double flower. **[127]**
florescence The part of a synflorescence that includes the main axis.
floret A small flower, as in the Gramineae and Compositae. **[152, 163, 183]**
floribunda A type of garden rose that bears abundant flowers in dense clusters.
floriferous Bearing numerous flowers.
floristic Relating to a flora.
floury Powdery.
flower The structure in angiosperms concerned with sexual reproduction. **[79, 80, 143-156]**
flowerless plant A cryptogam.
fl. pl. Flore pleno.
foetid (fetid) Having a highly unpleasant smell.
foliaceous Leaf-like; leafy.
foliage The leaves of a plant, considered as a whole.
foliar Relating to leaves.
foliose Leafy.
follicle A dry fruit formed from a single carpel, containing more than one seed, and splitting open along the suture. **[67, 176]**
follicular In the form of a follicle.
food body (see **elaiosome**)
forb Any herbaceous plant other than a grass.
form, forma (f.) A rank used to indicate a minor variant of a species, subspecies, or variety, such as a randomly occurring white-flowered plant in a population with typically coloured flowers.
fornix A small, arched scale, as those in the throat of the tubular corolla of the genus *Myosotis* (Forget-me-not).
foveate Pitted.
foveolate With small pits or depessions.
fr. (fructus) Fruit.
free Having adjoining, but different, parts completely separate from each other, e.g. as stamens from petals. **[136]**
free-central placentation The arrangement in which the placentas are situated on a central column that arises from the base of an ovary that is unilocular and not divided by septa. **[139]**
frond The leaf of a palm **[166]**, cycad **[194]** or fern. **[199]**
frondescent Breaking into leaf.
fructiferous Bearing fruit.
fructification The process of bearing fruit; the fruit of a flowering plant; the spore-bearing structures in a cryptogam.
fruit A mature ovary with its enclosed seeds and sometimes with attached external structures. **[66, 67, 80, 95, 98, 175-185]**
fruitlet One of the individual parts that compose an aggregate fruit. **[177]**
frutescent Becoming shrubby.
fruticose Shrubby.
fruticulose Somewhat shrubby.
fugacious Soon falling off or withering.
fulvous Tawny.
funicle (funiculus) The stalk connecting an ovule to its placenta. **[140]**
funicular Relating to a funicle.
funiculus (see **funicle**)
funnelform, funnel-shaped With the limb of the corolla widening gradually from a short tube. **[128]**
furcate Forked.
furfuraceous Scaly, scurfy.
furrowed Having a channel or channels along the part concerned, often broader and deeper than grooved. **[85]**
fuscous Dark brown or grey.
fused United completely. **[66, 98]**
fusiform Spindle-shaped. **[169]**
fusion Complete union of parts.

G (in a floral formula) Gynoecium, e.g. G3 indicates a gynoecium composed of 3 carpels. A line above the 'G' indicates that the ovary is inferior; a line below the 'G' indicates that the ovary is superior.
g. (**gen.**) Genus.
galbulus The fruit of *Juniperus* (juniper), a modified cone that becomes fleshy and berry-like as it matures. **[192]**
galea A hood or helmet-shaped structure formed by the perianth segments of certain flowers. **[125, 128]**
galeate With a galea. **[128]**
gall Abnormal growth of plant tissue in response to an attack by insects, fungi, bacteria, mites etc., the characteristic form of the gall often revealing the cause of the abnormality. **[96, 183]**
gamete One of the male or female sex cells (usually haploid) that unite at fertilisation to form a zygote.
gametophyte The sexual stage in the life cycle of a plant when the chromosomes in each cell are reduced to half the usual number, typically diploid reduced to haploid. **[199]**
gamodeme A deme forming a relatively isolated community.
gamopetalous (**sympetalous**) Having petals that are united at least at the base, as *Primula veris*. **[147]**
gamophyllous Having united petals and sepals.
gamosepalous (**synsepalous**) Having sepals that are united at least at the base, as *Primula veris*. **[147]**
geitonogamy A form of allogamy, in which the ovules of a flower are fertilised by pollen from another flower on the same plant.
gelatinous Jelly-like.
geminate (**paired**) Arranged in pairs. **[81, 98, 144, 189, 193]**
gemma An adventitious bud arising on a fern frond that can develop into a plantlet. **[206]**
gemmiferous, **gemmiparous** Bearing gemmae.
gen. (**g.**) Genus.
gene One of the units of heredity occupying a fixed position (locus) on a chromosome, that either by itself or in combination with other genes is responsible for a particular characteristic, e.g. height, flower colour, etc.
generative cell The smaller of the two cells into which the nucleus of the pollen grain divides while still in the pollen sac. This cell subsequently divides, sometimes before pollination, and gives rise to two male nuclei (sperm cells). **[130]**
genet One or more individuals produced by asexual reproduction from a single zygote.
genetic Relating to genes.
genetics The study of variation and heredity.
geniculate Abruptly bent or 'knee-like'. **[137, 160]**
genus (plural **genera**) A botanical rank, comprising one or more similar species. The names of genera are written with a capital initial letter.
geocarpic With fruits ripening below ground, as *Arachis hypogaea* (Groundnut). **[86]**
geocarpy The ripening of fruits below ground from flowers borne above the ground, the young fruits being pushed into the soil by the curving action of the stalk. **[86]**
geophyte An herbaceous plant that perennates by means of underground buds, e.g. bulbs, corms, etc.
geotropic (**gravitropic**) Responding to gravity, either positively geotropic as roots that grow downwards, or negatively geotropic as stems that grow upwards. **[86]**
geotropism (**gravitropism**) The growth movement of plants in response to gravity. **[86]**
germ cell A cell specialised for reproductive purposes, which gives rise to male or female gametes.
germination The development of a spore into a prothallus or a seed into a seedling. **[65, 130]**
gibbose, gibbous With a pouch-like swelling. **[128]**
Ginkgoopsida The class that includes *Ginkgo*. **[193]** (see **Gymnospermae**)
girdle scar The scar left by the terminal bud of the previous year. **[81]**
glabrescent Becoming hairless.
glabrate Almost hairless.
glabrous Without hairs.
gland A organ producing a secretion. **[91, 124]**
glandiferous Bearing glands.
glandular Possessing glands.
glandular hair A hair with a gland at its apex. **[80, 117]**
glaucescent Becoming glaucous, but sometimes incorrectly used to mean slightly glaucous.
glaucous With a waxy, greyish blue bloom.
globose, globular Spherical or globe-shaped, as the flower of *Trollius europaeus* (Globe Flower). **[127, 192]**
glochid (see **glochidium**)
glochidiate Bearing barbed hairs or bristles.
glochidium (**glochid**) A barbed hair or bristle. **[159, 205]**
glomerule A compact cluster of cells or spores; a condensed cyme of almost sessile flowers.
glossy (see **lustrous**)
glume One of the pair of bracts at the base of a spikelet in the Gramineae **[162]**; the single bract subtending the flower in the Cyperaceae. **[165]**
glutinous Sticky.

Gnetopsida The class containing the families Ephedraceae, Gnetaceae, and Welwitschiaceae. **[195]** (see **Gymnospermae**)
graft A portion of a plant inserted into and uniting with a larger part of another plant, as a scion into a stock.
graft chimaera, graft hybrid A plant that has originated by grafting rather than by sexual reproduction, e.g. + *Laburnocytisus adamii*, derived from *Laburnum anagyroides* and *Cytisus purpureus*.
grain (see **caryopsis**)
graminaceous, gramineous Relating to the Gramineae.
graminoid Grass-like.
graniferous Bearing grain, or seed resembling grain.
granular, granulose Having a slightly rough surface.
grex (plural **greges** or **grexes**) A collective name covering all the progeny of two parent plants, and now used only for orchid hybrids, e.g. *Paphiopedilum* Maudiae. It is printed in Roman type with a capital initial letter.
grooved Having a channel or channels along the part concerned, often narrower and shallower than furrowed. **[85]**
ground cover The lowest layer of vegetation in a wood or forest, consisting of herbaceous plants sometimes specially planted.
ground tissue Tissue other than vascular tissue, e.g. pith, cortex, etc.
growth ring (see **annual ring**)
grp. Group.
guard cell One of the pair of specialised cells of a stoma that control the size of the aperture and regulate the flow of gases in and out of the plant. **[111, 190]**
guttation The secretion of droplets of water from a plant, typically from hydathodes at the tips or on the margins of leaves.
Gymnospermae, Gymnospermophyta The gymnosperms, a group comprising Cycadopsida (cycads), Ginkgoopsida, Gnetopsida, and Pinopsida (conifers), plants whose ovules are naked, i.e. not enclosed in an ovary. **[189-195]**
gynaecium (see **gynoecium**)
gynandrous Having the stamens and style(s) united, as members of the Orchidaceae.
gynobasic With the style arising from below the ovary and between the carpels, as in many genera of the Boraginaceae and Labiatae. **[137]**
gynodioecious A species in which individual plants bear only female flowers or only bisexual flowers.
gynoecium (gynaecium) The female sex organs (carpels) collectively. In some species, e.g. *Vicia faba* (Broad Bean) the gynoecium consists only of a single carpel, but in most plants the gynoecium is composed of several carpels. **[130, 136]** (see **apocarpous** and **syncarpous**)
gynomonoecious Having female and bisexual flowers on the same plant.
gynophore The stalk bearing a carpel or gynoecium.
gynostegium The staminal crown in the flowers of some members of the Asclepiadaceae. **[151]**
gynostemium The column formed by the androecium and gynoecium combined, as in the flowers of some members of the Aristolochiaceae. **[143]**

habit The general appearance of a plant.
habitat The location in which a plant normally grows, determined by type of soil, amount of water, temperature, and other environmental factors.
haft The lower, usually narrower, part of the falls or standards in flowers of the genus *Iris*. **[156]**
half-epigynous, half-inferior (see **semi-inferior**)
half-parasite (see **hemi-parasite**)
halophyte A plant that is adapted to grow in saline soils.
halophytic Growing in saline soils.
hamate Hooked.
hamulate Bearing small hooks.
hapaxanthic (see **monocarpic**)
haplochlamydeous (see **monochlamydeous**)
haploid Having half the usual number of chromosomes in each cell, typically a single set.
haplostele A type of protostele consisting of a central core of xylem surrounded by a ring of phloem. **[204]**
haplostemonous With a single series of stamens in one whorl.
haptotropic (see **thigmotropic**)
hardiness The ability of a plant to withstand unfavourable conditions, especially cold.
hardwood Wood obtained from broad-leaved, dicotyledonous trees.
hastate Spearhead-shaped, with basal lobes directed outwards. **[106, 109]**
hastula A flap of tissue, borne at the junction of the petiole and the lamina in some palm leaves. **[167]**
haulm The stem of various herbaceous plants, e.g. peas, beans, potatoes, hops, and grasses.
haustorium An outgrowth from a parasitic plant that enables it to absorb nutrients from its host. **[53]**
head A short, dense spike of flowers; the capitulum in the Compositae. **[122, 152, 153]**
heartwood (duramen) The inner, older layers of wood in the trunk or branch of a tree or in the

stems of a shrub, usually denser and darker than the surrounding sapwood, and no longer able to conduct sap. **[82]**

heel The small piece of tissue that is pulled away from the parent stem when a cutting is taken from a plant.

heliciform Coiled like a snail shell.

helicoid cyme (see **bostryx**)

heliophilic Sun-loving or light-loving.

heliophobic Shade-loving.

heliophyte A plant adapted to living in high light intensities.

heliotropic Turning towards the sunlight (positively heliotropic) or away from it (negatively heliotropic). **[86]**

heliotropism The growth response of a plant to the stimulus of sunlight. **[86]**

helophyte A marsh plant, with resting buds below the surface of the marsh.

hemicryptophyte A plant with resting buds at or near the level of the soil. **[75]**

hemi-parasite (**half-parasite**, **partial parasite**, **semi-parasite**) A plant that derives some of its nourishment from a host plant. **[54]**

hemi-parasitic Partially parasitic, as members of the Loranthaceae, e.g. *Viscum album* (Mistletoe), and some genera of subfamily Rhinanthoideae in the Scrophulariaceae. **[54]**

hemitropous (see **amphitropous**)

heptaploid (**7n**) Having seven sets of chromosomes in each cell.

herb A non-woody plant, or one that is woody only at the base.

herbaceous Composed of soft, non-woody tissue. **[80]**

herbarium A collection of dried plants or parts of plants, usually mounted on sheets of thick paper of a uniform size, and kept for purposes of reference or research; the building in which such a collection is housed.

herbarium label A label affixed to an herbarium sheet giving information about the plant which cannot be obtained from study of the specimen alone, e.g. the collector, place and date of collection, dimensions (if a large plant), etc., also details which may change in the course of time, e.g. colour of flowers.

hermaphrodite (see **bisexual**)

hesperidium A berry in which the fleshy part is divided into segments and the outer skin is a tough, leathery rind, e.g. *Citrus* (orange, lemon etc.). **[180]**

heterobrochate Of pollen grains, having reticulate sculpturing, but with the meshes of the network differing in size.

heterochlamydeous Having a perianth that is clearly divided into calyx and corolla.

heterogamous With each flower-head composed of two or more kinds of flowers, as in many species of Compositae. **[152]**

heterogeneous Composed of dissimilar parts.

heteromorphic, heteromorphous Having two or more different forms.

heterophyllous Having more than one kind of leaf on the same plant. **[87]**

heterophylly The condition of being heterophyllous. **[87]**

heterosis (see **hybrid vigour**)

heterosporous Producing different kinds of spores, typically microspores and megaspores.

heterostylous Having variation in the length of the style (and stamens) in different flowers of the same species. **[147]**

heterostyly Variation in the length of the style (and stamens) in different flowers of the same species, as in some genera of the Primulaceae and Lythraceae. **[147]**

hexamerous Having the parts of the flower in sixes.

hexandrous Having six stamens.

hexaploid (**6n**) Having six sets of chromosomes in each cell.

hexarch root A root with six protoxylem strands in the stele.

hibernaculum A winter bud, formed when the plant dies down, and from which it regenerates.

hibernal Occurring in winter.

hilum The scar left on a seed where it was previously attached to the funicle. **[65]**

hinge cell (see **bulliform cell**)

hip The false fruit or pseudocarp in the genus *Rosa* (rose), developed from the fleshy, hollow hypanthium and containing achenes. **[181]**

hippocrepiform Horseshoe-shaped.

hirsute Covered in rough, coarse hairs. **[116]**

hirsutullous Slightly hirsute. **[117]**

hirtellous Minutely hirsute. **[117]**

hispid Having stiff, bristly hairs. **[116]**

hispidulous Having small, stiff, bristly hairs.

histogen One of the three layers (dermatogen, periblem, and plerome) considered to be present in an apical meristem, more recently challenged by the concept of tunica and corpus.

hoary Covered with small, whitish hairs, giving the surface a frosted appearance.

homochlamydeous (**homoiochlamydeous**) Having a perianth composed of similar segments, and therefore not clearly divided into calyx and corolla. **[79]** (see **tepal**)

homogamous With only one kind of flower; with anthers and stigmas maturing simultaneously.

homogeneous Of uniform structure, composed of similar or identical parts.

homoiochlamydeous (see **homochlamydeous**)
homologous Similar in structure and origin.
homosporous Producing only one kind of spore.
homostylous Having styles of the same length.
homostyly The state of being homostylous.
honey gland (see **nectary**)
honey guide (see **nectar guide**)
honey leaf A nectary, as in many members of the Ranunculaceae. **[144]**
hood The rounded lid that forms a canopy over the mouth of the pitcher-shaped leaves in e.g. *Darlingtonia* and *Sarracenia*. **[95]**
horizontal With the branches growing at right-angles to the trunk. **[72, 73]**
hormone A substance formed in a plant which has a specific effect on its growth or development. **[85]**
hose-in-hose The unusual arrangement of flowers in some forms of *Primula vulgaris* (Primrose) and *P. veris* (Cowslip), in which the flowers are in pairs, one growing from the centre of the other. **[97]**
host A plant from which a parasite obtains nutrients. **[53, 54]**
husk The external, membranous covering of certain seeds; one of the bracts that surround and protect the female inflorescence of *Zea mays* (Maize, Sweet Corn). **[163]**
hyaline Thin, colourless, and translucent.
hybrid A plant resulting from a cross between two or more plants, genetically unlike and belonging to different taxa, e.g. *Geum* x *intermedium*, a cross between two species in the same genus (*G. rivale* and *G. urbanum*), or x *Pyronia veitchii*, a cross between two species in different genera (*Cydonia oblonga* and *Pyrus communis*). **[127]**
hybrid vigour An increase in desirable characteristics, e.g. growth rate, yield, etc., exhibited by hybrids in comparison with their parents.
hydathode A pore or gland that exudes water.
hydrochoric, hydrochorous Having seeds that are dispersed by water.
hydrochory Dispersal of seeds by water.
hydromorphic Exhibiting hydromorphy.
hydromorphy The specialised structure present in the submerged stems and leaves of aquatic plants.
hydrophilous Depending on water to convey pollen for fertilisation.
hydrophily Pollination by water.
hydrophyte An aquatic plant, one that grows in water or needs a waterlogged habitat. **[87]**
hydrophytic Growing in a wet environment. **[87]**
hydrotropic Turning towards the source of water. **[86]**
hydrotropism The growth of a plant in response to the stimulus of water, as when the root of a plant turns towards the source of moisture. **[86]**
hygrochastic A type of plant movement resulting from the absorption of water, as capsules that open in moist air.
hygromorphic Exhibiting hygromorphy.
hygromorphy The specialised structure present in land plants growing in very damp habitats which promotes transpiration.
hygrophilous Requiring abundant moisture.
hygrophyte A land plant adapted to a perpetually damp habitat.
hygroscopic Extending or shrinking according to changes in moisture content.
hypanthial Relating to a hypanthium.
hypanthium (floral cup) A cup-shaped or tubular enlargement of the receptacle or of the bases of the floral parts. **[146, 181]**
hypochile The basal portion of the labellum in an orchid. **[172]**
hypocotyl The part of a seedling below the cotyledons which gives rise to the root. **[59, 65, 67]**
hypocrateriform (salverform, salver-shaped) Having a slender tube that expands abruptly into a flat or saucer-shaped limb. **[128]**
hypodermis The layer of cells immediately below the epidermis. **[190]**
hypogeal A type of germination in which the cotyledons remain underground. **[65, 66]**
hypogynous With the sepals, petals and stamens attached to the receptacle or axis below the ovary. **[135]**
hyponasty The tendency in part of a plant to grow more rapidly on the lower side so that it curves upwards.
hypopodium The lower portion of a sylleptic shoot from the adjoining stem to the first leaf or leaves.
hypostomatal, hypostomatic With stomata only on the lower surface of the leaf.
hysteranthous Describes leaves which are produced after the plant has flowered.

I.C.B.N. The International Code of Botanical Nomenclature, a set of internationally accepted rules, first published in 1952, that govern the naming of botanical taxa. These regulations and recommendations are amended when necessary at the International Botanical Congresses that are held in different countries every five or six years.
i.e. (id est) That is.
imbricate Overlapping, like fish scales or roof tiles. Some kinds of overlapping have been given

separate names, e.g. contorted, quincuncial. **[81, 98, 101, 104, 109]**

imbricate-ascending Having the vexillum within the lateral petals in bud, as in members of subfamily Caesalpinioideae in the Leguminosae. **[175]**

imbricate-descending (**vexillary**) Having the vexillum outside the lateral petals in bud, as in most members of subfamily Papilionoideae in the Leguminosae. **[175]**

immaculate Without any spots or markings.

imparipinnate (**odd-pinnate**) Pinnate, with a terminal leaflet. **[107]**

imperfect flower A flower in which only the androecium or the gynoecium is functional.

implexed Entangled, as the hairs on some species of *Stachys* (Labiatae).

inaperturate Without openings or pores.

incised Cut deeply and sharply into narrow, angular divisions. **[108]**

included Not projecting, contained within another organ.

incompatible Unable to produce viable offspring despite the presence of fertile gametes, e.g. pollen, although functional, may not grow down the style of a particular flower and will therefore fail to fertilse the ovule.

inconspicuous Not easily seen, blending with the surrounding parts.

incumbent Lying close along a surface; of cotyledons, having dorsal sides parallel to the radicle. **[68]**

incurved Curved inwards.

indefinite Of a large enough number to make a precise count difficult.

indehiscent Remaining closed at maturity.

indeterminate A form of inflorescence in which the outer or lower flowers open first and the stem or branch continues to grow, e.g. a spike or raceme. **[121]**

indigenous (**native**) Occurring naturally in the region concerned.

indumentum The covering of hairs or scales.

induplicate Folded inwards or upwards. **[167]**

indurate Hardened.

indusiate Having an indusium. **[199]**

indusium A protective covering of the sporangia formed by an outgrowth from the frond. **[199]**

ined. (**ineditus**) Unpublished.

inferior Below, as when the ovary appears embedded in the pedicel below the other floral parts. **[135, 143]**

infertile Not fertile.

inflated Enlarged, as the calyx of *Physalis alkekengi* (Chinese Lantern). **[127, 175]**

inflexed Abruptly bent inwards.

inflorescence The arrangement of flowers on the floral axis; a flower cluster. **[80, 97, 98, 121-126]**

infrafoliar Borne below the leaves. **[166]**

infrapetiolar Borne below the petiole.

infraspecific Below the rank of species.

infructescence A cluster of fruits, derived from an inflorescence. **[80]**

infructuous Not bearing fruit.

infundibular, infundibuliform Funnel-shaped. **[128]**

initial A cell in a meristem that divides into two daughter cells, one of which adds to the tissues of the plant, the other remaining in the meristem to repeat the process.

insectivorous plant (see **carnivorous plant**)

inserted Growing out from another part of the plant.

insertion The place where one plant part grows out of another.

in sicco In a dried state (as a herbarium specimen), a phrase used to notify possible differences from descriptions of the plant in its living condition.

in situ In the natural or original position. **[149]**

integument The outer covering of an ovule, which becomes the testa of the seed. **[130, 140, 193]**

intercalary growth The result of meristem activity between the apex and base of a stem or other part of a plant.

intercellular Between the cells.

interfascicular cambium The layer of cambium between the vascular bundles that joins with the fascicular cambium to form a cylinder of meristematic cells in stems and roots. **[82]**

interfoliar Borne among the leaves. **[166]**

intergeneric hybrid A plant produced by crossing species of two different genera, e.g. x *Cupressocyparis leylandii*, a hybrid between *Cupressus macrocarpa* and *Chamaecyparis nootkatensis*.

internodal Between nodes.

internode The part of the stem between two adjacent nodes. **[58, 80, 81, 98]**

interpetiolar Between the petioles.

interspecific hybrid A plant produced by crossing two species within the same genus, e.g. *Mahonia* x *media*, a hybrid between *Mahonia japonica* and *M. lomariifolia*.

intine The inner part of the wall of a pollen grain. **[129]**

intrafloral Within the flower.

intramarginal Within and close to the margin.

intrastaminal Within the stamens.

introduced Brought in from another region.

introrse With the anthers facing and opening inwards, towards the centre of the flower. **[135]**

intrusive-parietal With the placentas projecting

inwards from the walls of the ovary, in some cases almost meeting at the centre.
inverted Reversed, turned upside down.
***in vitro* culture** (= in glass) Studies on living material performed under sterile conditions away from the plant from which it was obtained. (includes micropropagation)
involucel The involucre of a partial umbel in the Umbelliferae; the epicalyx of connate bracteoles at the base of each individual floret in the Dipsacaceae.
involucral Relating to the involucre. **[98]**
involucrate With an involucre.
involucre A ring of bracts surrounding the head of flowers in the Compositae or subtending the umbel in the Umbelliferae **[178]**; in some species of gymnosperms, the whorl of scales subtending the cone.
involute Rolled inwards at the margin, i.e. towards the adaxial surface. **[103]**
irregular Not actinomorphic, sometimes applied to both zygomorphic as well as asymmetric flowers.
isobilateral Divisible into two similar halves; in leaves, having both surfaces similar to each other.
isodiametric Of cells, having equal diameters, i.e. roughly spherical in shape.
isomerous Having an equal number of members in successive series or whorls.
isomorphic Similar in form.
isophyllous Having leaves of only one kind.
isostemonous Having as many stamens as petals.
isthmus The narrow part that connects two broader parts of the same organ. **[91]**

jaculator A hook-like outgrowth from the stalk of the seed which aids its dispersal, as in members of subfamily Acanthoideae in the Acanthaceae. **[68]**
joint (see **node**)
jugate Joined together in pairs.
juvenile foliage The young leaves of e.g. *Eucalyptus* or *Juniperus* which differ in shape and colour from the mature or adult leaves. **[191]**

K (in a floral formula) Calyx, e.g. K5 indicates a calyx composed of 5 sepals.
keel (see **carina**)
keiki In orchids, a plantlet that develops adventitiously on a stem, pseudobulb, or branch of an inflorescence.
kernel The nucellus of an ovule or a seed; the grain of a cereal grass **[163]**; the edible part of a nut within its hard pericarp.
kex The dry, usually hollow stem of a large umbellifer, or the whole umbelliferous plant.
kingdom The taxonomic rank that includes all plants and comprises a number of divisions.
knee root One of the breathing roots that grow upwards from the submerged roots of tropical swamp plants and project above the surface of the water or mud to form an angular structure or 'knee'. In the case of *Taxodium distichum* (Swamp Cypress) they are known as 'cypress knees'. **[53]**
knot The hard tissue formed where a branch grows out of the trunk of a tree, clearly visible in cross-section in timber.
knur, **knurr** A swollen outgrowth from a tree trunk.
labellum A lip, especially the highly modified third petal in the Orchidaceae. **[154, 170]**
labiate Lipped **[128]**; a member of the Labiatae.
lacerate Jagged, irregularly cut as if torn. **[108]**
lachrymiform, **lacrimiform** Tear-shaped, obovate.
laciniate Deeply cut into narrow lobes. **[108]**
lacuna A space or cavity. **[54]**
lacunar Relating to lacunae.
lacunate Having lacunae.
lamina (**blade**) The expanded part of a leaf or frond. **[93, 95, 103]**
laminar Blade-like.
lanate Woolly. **[116]**
lanceolate Lance-shaped. **[105]**
lanose, **lanuginose** Woolly.
lanulose Diminutive of lanose.
lateral At the side. **[54, 56, 57, 61, 144, 164, 175]**
lateral dehiscence The shedding of pollen from a split at the side of the anther.
lateral growth Increase in girth of a stem or root, resulting from the activity of the cambium.
late wood (see **autumn wood**)
latex A juice produced by special cells in many different plants. It is usually milky (as in *Taraxacum*, dandelion) but may be colourless, yellow, orange (as in *Chelidonium majus*, Greater Celandine) or red. In some tropical trees, e.g. *Hevea brasiliensis*, the latex can be collected and processed to form rubber.
laticiferous Bearing latex.
latiseptate Having the partition (septum) across the broadest diameter of the fruit, as in *Lunaria annua* (Honesty). **[185]**
latrorse With the anthers facing and opening sidewards, as in *Begonia cucullata*.
lax Loose or open, not dense.

leaf A lateral outgrowth from the stem, usually consisting of a stalk (petiole) and a flattened blade (lamina). **[101-112]**
leaf blade (see **lamina**)
leaflet A leaf-like segment of a compound leaf. **[94, 107]**
leaf mosaic The overall arrangement of the leaves of a plant, determined by their shape and by phyllotaxy, which allows the maximum amount of light to fall upon each leaf. **[87]**
leaf-opposed Borne on the stem but on the opposite side from a leaf, as the tendrils in some species of vines, e.g. *Parthenocissus*. **[93]**
leaf rosette A circular cluster of leaves at the base of a stem. **[87]**
leaf sheath The lower part of a leaf stalk which more or less encloses the stem. **[102]**
leaf trace The vascular tissue that leads from the vascular system of the stem to the base of a leaf. **[79]**
leg. (**legit**) Collected by, used on a herbarium label and followed by the name of the collector of the specimen concerned.
legume The two-valved fruit formed from a single carpel in most members of the Leguminosae. **[136]**
leguminous Bearing legumes.
lemma The lower of the two bracts enclosing a grass flower. **[162]**
lenticel A pore in the stem that allows gases to pass between the outside atmosphere and the interior of a plant. **[58, 81]**
lenticular (**biconvex**) Lens-shaped, convex on both sides. **[68]**
lepidote Covered with small, fine scales. **[115]**
leptocaul Having a relatively slender woody stem.
leucoplast A colourless plastid that in roots, tubers, and other underground parts of plants can convert sugar into starch and is then termed an amyloplast. It is often capable of developing chlorophyll, as when potato tubers turn green in the presence of light.
liana, liane A woody climber in tropical forests that grows from the ground into the tree canopy.
lid The more or less flat structure that is attached to one side of the mouth of the pitcher-shaped leaves that act as insect traps in species of *Nepenthes*. **[95]**
life cycle The course of development from any given stage in the life of a plant until the same stage is reached again.
ligneous Woody.
lignified Converted into wood. **[82]**
lignin A hard substance found in the thickened cell walls of xylem and sclerenchyma.
ligulate Strap-shaped. **[105, 128]**
ligule A strap-shaped structure, such as the limb of the ray florets in the Compositae **[152]**; the scarious projection from the top of the leaf sheath in grasses. **[102]**
liliaceous Resembling a lily flower. **[128]**
Liliopsida (see **monocotyledons**)
limb The broadened upper part of a separate petal, as distinct from the claw **[145]**; the spreading rim of a gamopetalous flower, as distinct from the tube, as in *Primula veris*. **[147]**
limbate With a distinct edge or rim, especially when of a different colour from the inner part.
limen The rim at the base of the androgynophore in the genus *Passiflora* (passion flower). **[146]**
linear Long and narrow with parallel sides. **[105, 138]**
lineate Marked with lines.
lingulate Tongue-shaped. **[105]**
Linnaean Relating to the Swedish biologist Linnaeus (Carl von Linné, 1707-78), who developed a system of classification for plants and animals involving binomial nomenclature.
lip One of the two divisions of a bilabiate corolla or calyx **[125]**; the labellum in the Orchidaceae. **[170-172]**
lithocyst A cell containing a cystolith.
lithophile (see **lithophyte**)
lithophilous (see **lithophytic**)
lithophyte (**lithophile**) A plant growing amongst or on rocks or on cliff faces.
lithophytic (**lithophilous**) Growing amongst or on rocks or on cliff faces.
littoral Relating to the seashore or lakeside.
lobate (**lobed**) Having one or more lobes. **[138]**
lobe Any division of an organ, especially if the part is rounded. **[109]**
lobed (see **lobate**)
lobulate Having small lobes.
lobule A small lobe.
loc. class. (**locus classicus**) Classical locality, i.e. the place from where the plant concerned was originally collected.
locule (see **loculus**)
loculicidal Splitting at maturity into the loculus, more or less midway between the partitions of the capsule. **[136]**
loculus (**locule**) A compartment of an ovary or an anther. **[130, 136]**
lodicule One of usually two minute scales in a grass flower, generally considered to be a vestigial perianth. **[162]**
loment (see **lomentum**)
lomentose Relating to a lomentum.
lomentum A fruit derived from a single carpel which breaks up into one-seeded portions, as in the genus *Ornithopus* (Leguminosae). **[175]**

long-day plant A plant which needs prolonged periods of light alternating with shorter periods of darkness for the proper development of its flowers and fruit.
longitudinal Lengthwise.
longitudinal dehiscence The shedding of pollen from a lengthwise split in the anther, or seeds from a lengthwise split in a capsule. **[176]**
lorate Strap-shaped, usually more broadly than in ligulate. **[105]**
L.S. Longitudinal section.
lunate Crescent-shaped. **[106]**
lupulin A secretion from the glands present on the fruits (strobiles) of *Humulus lupulus* (Hop). **[182]**
lustrous (glossy) Smooth and shiny.
lyrate Pinnatifid, with the terminal lobe rounded and much larger than the others. **[106]**

m Metre(s).
macrogamete (see **megagamete**)
macronutrients The elements nitrogen, phosphorus, potassium, calcium, magnesium, and sulphur, that are required in relatively large amounts for the formation of plant tissue. (see also **micronutrients**)
macrophyll The term proposed by Raunkiaer for leaves of a large size.
macrophyllous (see **megaphyllous**)
macrophyte A plant, especially an aquatic plant, large enough to be visible to the naked eye.
macrospore (see **megaspore**)
macrothermic (see **megathermic**)
maculate Having spots or markings.
Magnoliophyta (see **Angiospermae**)
Magnoliopsida (see **dicotyledons**)
male flower Having fertile stamens, but only a rudimentary or non-functional gynoecium. **[124, 125, 145, 183]**
male nucleus (sperm cell) One of the two cells into which the generative cell divides, at the time of or sometimes before pollination. **[130]**
malodorous Having an unpleasant smell.
many-n Polyploid.
marcescent Withering, but not falling off. **[166]**
marginal placentation The arrangement in which the placenta extends along one side of the ovary of a free carpel. **[139, 140]**
massula A cluster of pollen grains developed from a single cell. **[205]**
mast The fallen fruit of certain genera in the Fagaceae, e.g. *Fagus* (beech) and *Quercus* (oak), formerly used as food for animals.
matted Tangled into a dense mass.
mealy Covered with a coarse, flour-like powder.
medial, median Relating to the middle. **[170]**
medifixed Attached at the middle.
mediseptate Having the partition (septum) across the middle of a more or less terete fruit, as in *Aubrieta deltoidea*. **[185]**
medulla The central part of an organ; the pith in a young stem. **[82]**
medullary ray (pith ray) One of the sheets of tissue in the stem of a dicotyledon that extend from the medulla or pith to the cortex (primary medullary ray), or outwards and inwards from the vascular cambium (secondary medullary ray). **[82]**
megagamete (macrogamete) In an organism where male and female gametes differ in size, the larger, usually female gamete.
megaphyll The term proposed by Raunkiaer for leaves of the largest size.
megaphyllous (macrophyllous) Large-leaved.
megasporangiate Relating to a megasporangium.
megasporangium The structure in which megaspores are produced. **[208]**
megaspore A spore that develops into a female gametophyte. **[208]**
megasporophyll A specialised leaf in heterosporous plants that bears the megasporangia; in gymnosperms, one of the ovuliferous scales that are arranged round the central axis of a female cone, and correspond to the carpels in a flowering plant. **[189, 194]**
megathermic Requiring much heat for growth and development, as tropical plants.
membranous Thin and semi-transparent like a membrane. **[57]**
mentum In some orchids, e.g. *Dendrobium*, a chin-like projection formed by the base of the column and the lateral sepals. **[172]**
mericarp A portion of a schizocarp which splits away at maturity as a perfect fruit, as in most members of the Geraniaceae and Umbelliferae. **[150, 177]**
meristele One of the strands of vascular tissue, consisting of xylem surrounded by phloem within a sheath of endodermis, that compose a dictyostele. **[204]**
meristem An area of tissue, found especially in the tips of shoots and roots, and in the cambium, that continues to undergo cell-division throughout the life of the plant. **[85]**
meristematic Relating to a meristem.
mesocarp The middle layer of the pericarp. **[179]**
mesochile The middle portion of the labellum in an orchid. **[172]**
mesophyll The parenchymatous, photosynthetic tissue that forms the inner part of the leaf blade between the upper and lower epidermis. In many dicotyledons the upper part consists of the

palisade mesophyll and the lower part the spongy mesophyll. **[111, 190]**
mesophyte A plant that is adapted to grow in a moist habitat, where there is no prolonged drought.
mesophytic Growing in moist habitats.
mesotherm A plant requiring moderate heat for its optimal growth and development.
mesotonic A type of branching in which the shoots nearest the middle of the stem show the greatest development.
metagyny (see **protandry**)
metamorphosis Transformation of one structure into another, as stamens into petals; a change in the type of branching of a tree from generally plagiotropic to generally orthotropic.
metaphloem The phloem that is formed after the protophloem.
metaxylem The xylem that is formed after the protoxylem.
microclimate The climatic conditions existing in a small, localised area.
microgamete In an organism where male and female gametes differ in size, the smaller, usually male gamete.
micronutrients (**trace elements**) The elements that, together with the macronutrients, are necessary for the successful growth of a plant, but which are required only in small amounts. They include iron, manganese, boron, zinc, molybdenum, chlorine, and copper.
micropropagation The development of new plants from very small pieces of plant tissue, e.g. embryos, shoot tips, root tips etc., in an artificial medium and under sterile conditions (included in the concept of *in vitro* culture).
micropylar Relating to the micropyle. **[67]**
micropyle The opening in the integuments of an ovule, through which the pollen tube grows after pollination. **[65, 130, 140]**
microspecies Species founded on minute differences and used mostly for apomictic plants, as in *Taraxacum* (dandelion) and *Hieracium* (hawkweed).
microsporangiate Relating to a microsporangium. **[195]**
microsporangium The spore sac in which microspores are produced. **[189, 193, 194]**
microspore A spore that develops into a male gametophyte, corresponding to a pollen grain in a flowering plant. **[208]**
microsporophyll A specialised leaf in heterosporous plants that bears the microsporangia; in gymnosperms, one of the bracts that are arranged round the axis of a male cone, and correspond to the stamens in a flowering plant. **[189, 194]**
midrib The middle and principal vein of a leaf. **[103, 194]**
mm Millemetre(s).
monad A single pollen grain, not united with others.
monadelphous Having the stamens united in one group by the fusion of their filaments, as in some of the Leguminosae, Polygalaceae and Malvaceae. **[132]**
monandrous Having one stamen, as most orchids.
monarch root A root with a single protoxylem strand in the stele.
moniliform Like a string of beads. **[52, 118]**
monocarpellary (see **monocarpous**)
monocarpic (**hapaxanthic**) Flowering and fruiting once only before dying.
monocarpous (**monocarpellary**) Of a fruit, composed of a single carpel.
monocaulous Having one stem.
monocephalic Bearing one head of flowers, as the scape of plants in the genus *Taraxacum* (dandelion).
monochasial cyme, monochasium A cyme with lateral branching on one side only of the main axis. **[121]**
monochlamydeous (**haplochlamydeous**) Having a perianth of a single whorl, i.e. either the calyx or the corolla is present.
monoclinous (see **bisexual**)
monocolpate Of a pollen grain, having a single colpus.
monocotyledons (**Monocotyledonae**, **Liliopsida**) Flowering plants having one cotyledon. **[79]**
monocotyledonous Having one cotyledon.
monoecious With male and female flowers on the same plant.
monoecism, monoecy The state of being monoecious.
monogeneric Of a family, containing only a single genus.
monograph A systematic account of a particular genus or family.
monomerous Formed of a single unit, as a monocarpous fruit.
monomorphic Occurring in only one form.
monopetalous With only one petal; gamopetalous.
monopodial With a simple main stem or axis, growing by apical extension and bearing lateral branches.
monoporate Of a pollen grain, having one pore.
monospecific Of a genus, containing only a single species.
monosulcate Of a pollen grain, having one groove or furrow.
monotelic With each lateral inflorescence-branch ending in a flower, as a cyme. **[121]**

monothecous With one anther cell.
monotypic Having only one representative, as a genus with one species.
montane Growing in mountainous regions.
morphological Relating to the form of a plant.
morphology The science or study of the form of plants, as distinct from anatomy.
mother bulb A mature bulb that is capable of flowering and producing one or more bulblets (daughter bulbs). **[79]**
mother cell A cell that divides into two daughter cells.
motile Capable of independent movement.
mucilage A sticky substance or solution.
mucilaginous Slimy.
mucro A short straight point. **[110]**
mucronate Ending abruptly in a short, straight point. **[110]**
mucronulate Ending abruptly in a very short, straight point. **[110]**
multiaperturate Of a pollen grain, having many pores.
multicellular Many-celled. **[118]**
multiciliate Having many marginal hairs.
multilocular With many loculi or cells.
multiple fruit (**collective fruit**) A fruit formed from an inflorescence, often including bracts, as in *Ananas* (pineapple) and *Morus* (mulberry). **[182]**
multiseriate In several series, rows or whorls.
muricate Rough with short, hard points. **[115]**
mutant (**sport**) An individual that has arisen as a result of mutation.
mutation A genetic change that may occur spontaneously or may be induced artificially by the use of certain chemicals; a mutant.
mutualism A form of symbiosis in which two different organisms co-exist to their mutual advantage.
mycorrhiza The association of fungi and the roots of plants to their mutual advantage. **[206]**
mycorrhizal Relating to mycorrhiza.
mycotroph A plant that lives in symbiosis with a fungus.
myrmecochorous Having seeds that are dispersed by ants.
myrmecochory Dispersal of seeds by ants.
myrmecophilous Having the stem or root inhabited by ants. **[96]**
myrmecophily Symbiosis between ants and plants; pollination by ants.

n Haploid; a figure preceding the 'n' indicates the number of sets of chromosomes in each cell, e.g. 2n = diploid.
n- (**notho**) A prefix placed before a taxonomic rank to indicate a hybrid.
N (see **nitrogen**)
NaCl (see **sodium chloride**)
naked Lacking a covering, as a flower without a perianth, e.g. *Salix* (willow). **[128]**
napiform Turnip-shaped, e.g. the hypocotyl of *Brassica rapa* (Turnip). **[59]**
narrowly upright With the branches growing up more closely to the trunk. **[72]**
nastic movement A plant movement independent of the direction of the external stimulus, such as the opening or closing of some flowers in response to an alteration in temperature or light intensity. **[86]**
natant Floating in water.
native (see **indigenous**)
naturalised Thoroughly established after introduction from another region.
navicular (**cymbiform**) Boat-shaped, as the united lower petals in flowers of subfamily Papilionoideae in the Leguminosae. **[148]**
nec Nor, nor of.
necrosis Localised death of cells which leaves the surrounding plant tissue unaffected.
nectar A sugar solution, attracting insects or birds to flowers for the purpose of pollination. **[124]**
nectar guide (**honey guide**) Markings on the perianth of a flower, usually consisting of lines or dots, that direct a pollinating insect to the nectary. **[149]**
nectariferous Bearing nectar-secreting glands. **[91]**
nectar pit A depression in which nectar collects.
nectar roll The nectary in carnivorous plants such as *Sarracenia*, in which the edge of the leaf forming the mouth of the pitcher-like trap is rolled over. **[95]**
nectary (**honey gland**) A gland or surface from which nectar is secreted. **[144, 149]**
needle A long, narrow leaf, characteristic of many conifers. **[189-191]**
nervate Nerved or veined. **[104]**
nervation, nervature (see **venation**)
nerve (see **vein**)
nervure A principal vein of a leaf. **[104]**
net-veined Having veins that join together to form a network across the lamina. **[104]**
neuter Having both stamens and gynoecium non-functional.
neutral soils Soils with a pH value of about 7.
nitrogen (**N**) The most abundant gas in the atmosphere and one of the six macronutrients that, together with carbon, oxygen, and hydrogen, comprise the group of elements essential to the life of a plant.

nitrogen fixation A process carried out by bacteria which convert gaseous nitrogen into compounds that can be assimilated by the plant.
nocturnal Occurring at night.
nodal Relating to a node.
nodding (see **nutant**)
node (joint) The point on a stem where one or more leaves are borne. **[58, 80]**
nodose Knobby.
nodulation The formation of nodules on the roots of certain plants. **[52]**
nodule A small, rounded structure on the roots of plants, especially those of the Leguminosae that contain nitrogen-fixing bacteria. **[52]**
nodulose Bearing nodules. **[52]**
nom. cons. (**nomen conservandum**) Conserved name, a plant name invalid under the rules of the I.C.B.N. but which has been retained in order to avoid further (possibly wide-ranging) changes in the nomenclature.
nomenclature The naming of plants, especially the precise usage formulated in the I.C.B.N.
nom. illeg. (**nomen illegitimum**) Illegitimate name, a plant name invalid under the rules of the I.C.B.N.
nom. nud. (**nomen nudum**) Bare name, a new name published without any description of the plant in question, and therefore invalid under the rules of the I.C.B.N.
non Not, not of.
nonaploid (9n) Having nine sets of chromosomes in each cell.
non-endospermic seed A seed in which the nutritive tissue is absorbed more rapidly into the developing embryo so that the process is completed by the time of germination, e.g. *Vicia faba* (Broad Bean) and *Phaseolus vulgaris* (French Bean). **[65]**
non-vascular Not containing vessels.
nose The pointed end of a bulb. **[57]**
nothogenus The rank given to a hybrid genus produced by the crossing of two or more different genera, and usually indicated by a multiplication sign placed before the name, e.g. x *Cupressocyparis* (*Cupressus* x *Chamaecyparis*).
nothotaxon A unit of classification for hybrid plants. The highest rank permitted by the I.C.B.N. is nothogenus. Names of all hybrid ranks consist of the normal taxonomic rank prefixed by 'notho'.
nothospecies The rank given to a hybrid species produced by the crossing of two or more species within the same genus, and usually indicated by a multiplication sign placed before the specific epithet, e.g. *Viburnum* x *bodnantense* (*V. farreri* x *V. grandiflorum*).
nucellus The mass of cells within an ovule that contains the embryo sac. After fertilisation, it may either be absorbed or may persist to form a perisperm. **[140]**
nuciferous Bearing nuts.
nuciform Nut-like in form.
nucleus The usually ovoid or spherical structure in a plant cell that contains the chromosomes. **[129, 130]**
nucule A nutlet. **[178]**
numerous Many, often indefinite in number.
nuptial nectary A nectary situated near to the reproductive organs of the flower.
nut A dry, one-seeded, indehiscent fruit with a woody pericarp. **[98, 178]**
nutant (**drooping, nodding**) Bending over and pointing downwards. **[72, 73]**
nutlet A small nut, sometimes applied to an achene or part of a schizocarp. **[178]**
nutrient One of the substances necessary for plant growth and development. (see **macronutrients** and **micronutrients**)
nyctinasty (**sleep movement**) The response of plant parts, especially flowers and leaves, to night-time darkness. **[86]**

obconical (see **turbinate**)
obcordate Inversely heart-shaped, broadest towards the emarginate apex and tapering to the stalk. **[106]**
obdeltoid Triangular, with the apex truncate and tapering to the stalk. **[105]**
obdiplostemonous Having the stamens in two whorls, the inner whorl opposite the sepals and the outer whorl opposite the petals. **[131]**
oblanceolate Inversely lanceolate, broadest towards the apex and tapering to the stalk. **[105]**
oblate (see **subglobose**)
oblique Unequal, as a leaf with one side extending below the other; ascending, as tree branches that slope upwards. **[73]**
oblong Longer than broad with more or less parallel sides. **[105]**
obovate Inversely ovate, broadest towards the apex and tapering to the stalk. **[105]**
obsolescent Having dwindled to a rudimentary state or vanished altogether.
obtrullate Inversely trullate, i.e. with the two longer sides meeting at the base.
obturator An outgrowth from the placenta over the micropyle that nourishes the pollen tube and guides it towards the ovule. **[68]**
obtuse Blunt. **[109, 110]**

obvolute With half of one leaf wrapped round half of another leaf in the bud.
occlusion The process by which wounds in trees are healed by the formation of callus.
ochrea A tubular sheath, formed by the fusion of two stipules at the nodes of many plants in the Polygonaceae. **[102]**
ochreate Having ochreae. **[102]**
octamerous Having the parts of the flower in eights.
octandrous Having eight stamens.
octoploid (**8n**) Having eight sets of chromosomes in each cell.
odd-pinnate (see **imparipinnate**)
odoriferous Having a smell, especially a fragrant one.
odorous Having a distinct smell.
offset, offshoot A short runner producing a new plant at its tip, as in the genus *Sempervivum* (Crassulaceae). **[60]**
oleaceous Relating to the Oleaceae.
oleiferous Producing oil.
oligostemonous (**paucistemonous**) Having few stamens.
ombrophile (see **ombrophyte**)
ombrophobe A plant intolerant of prolonged rainfall.
ombrophyte (**ombrophile**) A plant adapted to grow in places with prolonged rainfall.
ombrophytic Growing in rainy habitats.
oogamous Involving the union of gametes of dissimilar size.
oogamy The union of gametes of dissimilar size, usually a small motile male gamete and a large non-motile female gamete.
oosphere A female gamete.
oospore The zygote formed from a fertilised oosphere. **[208]**
op. cit. (**opere citato**) In the work cited, a reference to a publication mentioned earlier in the same account.
open aestivation With the flower parts in the bud not touching.
operculate Having an operculum. **[176]**
operculum The lid of a circumscissile fruit; the membranous cover of the nectar-secreting ring in the genus *Passiflora* (passion flower). **[146]**
opposite leaves With two leaves at a node, one on each side of the stem or axis. **[101]**
opposite vernation With the leaves in the bud facing each other but not appressed.
oppositipetalous Situated before a petal.
oppositisepalous Situated before a sepal.
orbicular Circular. **[105]**
orchid A member of the Orchidaceae. **[169-172]**
orchidaceous Resembling an orchid flower. **[128]**
order A taxonomic rank comprising a group of families. Names of orders end in '-ales'.
ornamentation The layer of spines on the tectum of a pollen grain. **[129]**
ornithophilous Depending on birds to convey pollen for fertilisation.
ornithophily Pollination by birds.
orthotropic Growing directly towards the source of the stimulus (positively orthotropic) or directly away from the source of the stimulus (negatively orthotropic).
orthotropism The growth movement of a plant or part of a plant directly towards or directly away from the source of the stimulus.
orthotropous (**atropous**) Having the ovule borne on a straight funicle, and with the micropyle in a line with it. **[140]**
osmosis The movement of molecules from a solution of low concentration to one of higher concentration through a semi-permeable membrane until both solutions are of the same concentration.
ostiole, ostiolum A pore, especially one which acts as an outlet for spores or gametes. **[183]**
outgrowth A structure growing out from the main body.
oval Broadly elliptical, but ellipsoid when applied to habit. **[72, 85, 105]**
ovary The lower part of a carpel (or carpels) which contains the ovules. **[130, 135, 136, 139]**
ovate With the outline egg-shaped. **[105]**
oviform, ovoid, ovoidal Egg-shaped. **[192]**
ovulate cone A female cone.
ovule A structure which, after fertilisation, develops into a seed. **[130, 136, 139, 140]**
ovuliferous Bearing ovules. **[189]**
ovum (**egg**) A non-motile female gamete. **[130, 199]**

P (in a floral formula) Perianth, e.g. P6 indicates a perianth composed of 6 perianth segments.
pachycaul Thick-stemmed, as palms and cycads, species of *Acropogon* native in New Caledonia, and tree-like herbs with grossly stout stems such as some species of *Lobelia* (giant lobelia) and *Dendrosenecio* (giant groundsel) found in the montane forests of Africa.
pad In the fern genus *Platycerium*, the shield or saddle-shaped sterile fronds at the base of the plant. **[201]**
paired (see **geminate**).
palate The swollen part of the lower lip of a gamopetalous flower which almost or entirely closes the throat, as in the personate flower of *Antirrhinum*. **[128]**

palea The upper of the two bracts enclosing a grass flower. **[162]**
paleaceous Chaffy. **[116]**
palaeobotany The study of fossil plants.
palisade mesophyll The layer(s) of elongated cells that are arranged fence-like immediately beneath and at right-angles to the epidermis in mesophytic dicotyledon leaves. **[111]** (see also **spongy mesophyll**)
palman Of palms, the undivided central part of a fan palm leaf, e.g. *Chamaerops*. **[168]**
palmate Divided to the base into separate leaflets, all the leaflets arising from the end of the leaf stalk **[107]**; having the veins radiating from the end of the leaf stalk to the tips of the lobes. **[104]**
palmatifid More or less hand-shaped, with the lobes of the leaf extending about half-way to the base. **[106]**
palmatipartite More or less hand-shaped, with the lobes of the leaf extending from about half to two-thirds of the way towards the base.
palmatisect More or less hand-shaped, with the lobes of the leaf extending almost to the base. **[106]**
palus One of the shorter filaments in the corona of plants in the Passifloraceae. **[146]**
palynology The study of pollen grains and other spores to provide information on the distribution of species in earlier times and to assist in dating geological formations and archaeological remains.
pandurate Fiddle-shaped. **[106]**
panicle A much-branched inflorescence. **[122]**
paniculate In the form of a panicle.
pannose Felt-like, composed of densely matted woolly hairs. **[117]**
papilionaceous Butterfly-like, as the flowers of subfamily Papilionoideae in the Leguminosae, in which the corolla consists of a standard petal that encloses two wing petals in bud, and two lower petals more or less united to form a keel, e.g. *Lathyrus*. **[148]**
papilla (papula) A small, nipple-shaped projection. **[138]**
papillate Bearing papillae. **[115]**
papilliform Nipple-shaped.
papillose (papulose) Covered in papillae.
pappus The specialised calyx of hairs or scales occurring mainly in the Compositae. **[152]**
papula (see **papilla**)
papulose (see **papillose**)
parachute mechanism A form of seed dispersal (in e.g. *Asclepias*) or fruit dispersal (in e.g. *Taraxacum* and other members of the Compositae) in which a feathery appendage enables the seed or fruit to be carried away by the wind, sometimes for great distances.
paracladium A portion of an inflorescence, which has the same structure as the whole inflorescence.
parallel With the veins remaining more or less the same distance apart along much of the leaf, a characteristic of the veins in the leaves of the majority of monocotyledons. **[79, 104]**
paraphysis (plural **paraphyses**) A sterile filament growing amongst the sporangia of cryptogams. **[204]**
parasite A plant that lives on another plant and derives its nourishment from it, e.g. *Cuscuta* (dodder). **[53]**
parasitic Growing on another plant and deriving nourishment from it.
parenchyma Succulent tissue, consisting of more cr less isodiametric, thin-walled cells, often with intercellular spaces, that is found in e.g. the softer parts of leaves, the pulp of fruits and the pith of stems. **[54, 79]**
parenchymatous Relating to or consisting of parenchyma.
parietal placentation The arrangement in which the placentas develop along the fused margins of the carpels of a unilocular ovary. **[139]**
paripinnate (even-pinnate) Having an equal number of leaflets and lacking the terminal leaflet. **[107]**
parted Cut, but not quite to the base.
parthenocarpic Bearing fruits produced without preliminary fertilisation.
parthenocarpy The formation of fruit without preliminary fertilisation and usually without development of seeds.
partial parasite (see **hemi-parasite**)
patent Spreading.
pathogen An organism that causes disease.
paucistemonous (see **oligostemonous**).
pectinate Pinnatifid, with narrow segments set close like the teeth of a comb. **[108, 159]**
pedate Palmately lobed or divided, but with the basal lobes again divided. **[107]**
pedicel The stalk of a single flower. **[98, 147]**
pedicellate Of a flower, stalked.
peduncle The stalk of an inflorescence. **[98, 147]**
pedunculate With a peduncle.
peg root One of the breathing roots that grow upwards from the submerged roots of tropical swamp plants and project upright above the surface of the water or mud. **[53]**
pellucid Translucent, allowing the passage of light.
peloria The abnormal development of a typically zygomorphic flower so that it becomes actinomorphic. **[97]**

peltate Shaped like a disc and attached at the centre of its lower surface to the stalk. **[106, 117]**
pendent, pendulous (pensile) Hanging down. **[73, 170]**
pendulous placentation (see **apical placentation**)
penicillate Streaked, as if with a pencil or brush.
penninerved Having veins that branch pinnately. **[104]**
pensile (see **pendent**)
pentadelphous With five groups of stamens, as some species of *Hypericum*.
pentamerous Having the parts of the flower in fives.
pentandrous Having five stamens.
pentaploid (5n) Having five sets of chromosomes in each cell.
pentarch root A root with five protoxylem strands in the stele. **[55]**
pepo A unilocular, many-seeded, hard-walled berry that forms the fruit of *Cucurbita pepo* (Marrow), *Cucumis melo* (Melon), and some other members of the Cucurbitaceae. **[180]**
perennate To continue to live for a number of years, as a perennial plant.
perennating Surviving from one growing season to another, usually with a dormant (resting) period in between.
perennation The act of living for a number of years.
perennial Living for a number of years; a plant that lives for a number of years.
perfect flower A flower in which both the androecium and the gynoecium are functional.
perfoliate With the bases of two opposite, sessile leaves connate round the stem so that the stem appears to pass through a leaf blade. **[102]**
pergamentaceous Parchment-like.
perianth (perigone) A collective term for the outer, non-reproductive parts of a flower, often differentiated into calyx and corolla.
perianth lobe One of the free parts of the perianth when the lower portions are united into a tube. **[143]**
perianth segment One of the parts of the perianth, often used to describe segments which closely resemble each other, as in the genus *Tulipa* (tulip). **[79, 144]** (see **tepal**)
perianth tube The tube formed when the lower portions of the perianth segments are united. **[143]**
periblem The intermediate layer of the apical meristem which develops into the cortex.
pericarp The fruit wall that has developed from the ovary wall. **[66]**
pericycle The outermost layer of cells of the stele in a stem or root. **[55, 80, 190]**
periderm Secondary protective tissue, consisting of phellem (cork), phellogen (cork cambium), and phelloderm, that often replaces the epidermis in older stems and roots. **[82, 85]**
perigone (see **perianth**)
perigynium The utricle which encloses the female flower in some members of the Cyperaceae. **[165]**
perigynous With the sepals, petals and stamens inserted around the ovary on the hypanthium, a concave structure developed from the receptacle. **[135]**
perisperm The albumen of a seed formed outside the embryo sac.
perpetual Flowering several times in a season, as some rose cultivars.
persistent Remaining attached.
personate Bilabiate with a prominent palate. **[128]**
perular scale One of the basal scales of leaf buds which may persist for some time after the development of the shoot.
petal A single segment of the corolla. **[150]**
petaloid Petal-like, as the style in the genus *Iris* **[98, 128, 143]**; a petal-like structure bearing distorted anthers, situated between the normal petals and the stamens in semi-double and double flowers.
petiolar Relating to a petiole.
petiolate Having a leaf stalk. **[102]**
petiole A leaf stalk. **[57, 91-94]**
petiolule The stalk of a leaflet in a compound leaf.
petrophilous Adapted to growing in a rocky environment.
pH A value of hydrogen-ion concentration which allows the acidity or alkalinity of a solution to be measured on a scale from 1 (extremely acid) to 15 (extremely alkaline). (see **acid soils**, **neutral soils**, and **alkaline soils**)
phanerogam A plant belonging to the Phanerogamia, a division in former classifications that included all the seed-bearing plants, now called spermatophytes.
phanerophyte A tall, woody or herbaceous perennial, with resting buds more than 25 cm above soil level. **[75]**
phellem (cork) A spongy, protective layer of thin-walled cells impregnated with suberin that is formed from cork cambium, and often replaces the epidermis in stems and roots as they grow older. **[82]**
phelloderm One or more layers of thin-walled cells formed from the inner side of the phellogen. **[82]**
phellogen (cork cambium) The layer of meristematic cells lying just beneath the surface of a stem or root, that forms phellem (cork) on its outer side and phelloderm on its inner side. **[82]**
phenetic classification A type of classification which expresses relationships between plants in

terms of their visible or otherwise measurable physical and biochemical characteristics.
phloem (**bast**) The vascular tissue that conducts sap containing nutrients produced by photosynthesis from the leaves to other parts of the plant. **[82]**
photonasty The response of a plant to a change in light intensity, e.g. the opening and closing of the flowers.
photoperiodism The response of a plant to the relative duration of day and night, especially in regard to flowering.
photosynthesis The process by which green plants convert carbon dioxide and water into carbohydrates in the presence of sunlight.
photosynthetic Relating to photosynthesis. **[53]**
phototropic Turning towards the light source (positively phototropic) or away from it (negatively phototropic).
phototropism The growth movement of plants in response to the stimulus of light. **[85]**
phyllary One of the involucral bracts that surround the head of flowers in the Compositae. **[152]**
phylloclade (see **cladode**)
phyllode A petiole taking on the form and functions of a leaf, as in the genus *Acacia*. **[91]**
phyllopodium An outgrowth on the rhizome of some ferns on which a frond arises. **[202]**
phyllotaxis, phyllotaxy The arrangement of leaves on an axis or stem. **[87]**
phylogenetic classification A type of classification which expresses supposed relationships between plants in terms of their evolutionary history.
phylogeny The relationships between plants as determined by their evolutionary history.
phylum (see **division**)
phytochrome A light-sensitive pigment in plants that is involved with certain developmental processes such as photoperiodism, reversal of etiolation, and the germination of some seeds and spores.
phytomer A bud-bearing node, the smallest structural unit of a plant which is capable of reproducing vegetatively.
pigment The natural colouring matter in plant tissues. (see **chloroplast** and **chromoplast**)
piliferous layer (**epiblem, epiblema, rhizodermis**) The epidermis of a young root that bears the root hairs. **[54, 55]**
pilose Softly hairy. **[116]**
pinetum A collection of conifers, especially *Pinus* (pine) species and varieties, for scientific or ornamental purposes.
pin-eyed One of the two forms of a dimorphic flower, e.g. *Primula vulgaris* (Primrose), where the style is long and the stamens are below the stigma. **[147]** (see also **thrum-eyed**)
pinna A primary division or leaflet of a compound leaf. **[107]**
pinnate Having separate leaflets along each side of a common stalk **[107]**; having separate veins along each side of the midrib of a leaf. **[104]**
pinnatifid Pinnately lobed, the lobes extending from about a quarter to half-way towards the rachis. **[106]**
pinnatipartite Pinnately divided, the divisions extending from about half to two-thirds of the way towards the rachis. **[106]**
pinnatisect Pinnately divided, the divisions extending almost to the rachis. **[106]**
pinnule A secondary division of a pinnate leaf. **[199]**
pinnulet A segment of a pinnule. **[200]**
Pinopsida (**Coniferae**) The class that comprises the conifers. (see **Gymnospermae**)
pistil A single carpel in an apocarpous flower, or the gynoecium in a syncarpous flower. **[136]**
pistillate (**carpellate**) Having only female organs.
pistillode A sterile pistil.
pitcher plant A carnivorous plant that obtains its nutrients from insects lured into traps formed by its pitcher-shaped leaves, e.g. *Sarracenia*, *Darlingtonia*, *Nepenthes*, etc. **[95]**
pith The central column of spongy, parenchymatous tissue in the stems of dicotyledons and certain monocotyledons such as *Juncus* (rush); the medulla. **[55, 80]**
pith ray (see **medullary ray**)
pitted With small depressions on the surface.
placenta The part of the ovary to which the ovules are attached. **[130, 139]**
placental Relating to a placenta.
placentation The arrangement of placentas in an ovary. **[139, 140]**
plagiotropic Growing at an angle towards or away from the source of the stimulus.
plagiotropism The growth movement of a plant or part of a plant at an angle to the source of the stimulus.
plantlet A small plant, as those formed on the leaves of some species of flowering plants, e.g. *Kalanchoe*, or on the fronds of some ferns. **[61]**
plastid A specialised structure in the cytoplasm of a plant cell. (see **chloroplast, chromoplast** and **leucoplast**)
plectostele A type of protostele, in which the xylem and phloem form more or less parallel bands.
pleiochasium A cymose inflorescence in which the main axis has more than two lateral branches. **[123]**
plerome The innermost layer of an apical meristem which develops into the central vascular tissue.
plica A fold.

plicate Folded more than once lengthwise. **[103]**
plumose Feathery. **[53, 138]**
plumule The young shoot as it emerges from the seed on germination, usually after the appearance of the radicle. **[65, 67, 86]**
pluricarpellate Having several carpels.
plurilocular Composed of several or many loculi or compartments.
pneumatophore One of the breathing roots that grow upwards from the submerged roots of mangroves and other tropical swamp plants. They project above the surface of the water or mud as peg roots or knee roots, and have lenticels in their bark that allow air to pass through into the root system. **[53]**
pocket nectary A type of nectary found in e.g. *Ranunculus repens*, in which nectar is secreted beneath a small scale at the base of a petal. **[149]**
pod A dry, many-seeded, dehiscent fruit, particularly the legume in plants of the Leguminosae. **[175]**
podium A base or supporting structure, e.g. the stalk bearing the strobili in *Lycopodium*. **[206]**
poikilohydrous Of a plant, having its water content determined by the wetness or dryness of its surroundings.
polar nuclei The two nuclei that move from the poles of the embryo sac to its centre. **[130]**
pole One of two opposite areas of a more or less spherical structure, e.g. one of the two opposite areas of a pollen grain that are free from apertures. **[129]**
pollen The small grains which contain the male reproductive cells of the flower. **[129]**
pollen flower A flower without nectar that attracts insects by its pollen.
pollen sac One of the two portions into which a theca or anther cell is divided.
pollen tube An outgrowth from a germinating pollen grain which carries the male gamete(s) down the style to an ovule in the ovary. **[130]**
pollinarium A collective term for the pollinia, stipe, and viscidium in flowers of the Orchidaceae. **[171]**
pollination The placing of pollen on the stigma or stigmatic surface. **[130]**
polliniferous Bearing pollen.
pollinium A pollen mass, as in Orchidaceae and Asclepiadaceae. **[151, 171]**
polyad A group of more than four pollen grains.
polyadelphous With more than two groups of stamens, as in some species of the genus *Hypericum*. **[134]**
polyandrous (**polystemonous**) Having numerous stamens.
polyarch root A root with many protoxylem strands in the stele. **[55]**
polygamo-dioecious Functionally dioecious, but bearing a few flowers of the opposite sex or a few perfect flowers on the same plant.
polygamo-monoecious Polygamous, but in the main monoecious.
polygamous Bearing both unisexual and bisexual flowers on the same plant.
polygynous Having many styles.
polymerous With numerous members in each series or whorl.
polymorphic Having several or many forms.
polymorphism The occurrence of different forms of a plant species within a particular population.
polypetalous (see **apopetalous**)
polyphyllous Having separate petals and sepals.
polyploid (**many-n**) Having more than the usual two sets of chromosomes in each cell.
polysepalous (see **aposepalous**)
polystemonous (see **polyandrous**)
polystichous Arranged in several or many rows.
polytelic An inflorescence in which the branches do not end in a flower.
pome A fruit consisting of a core, formed by several united carpels, enclosed within a firm, fleshy receptacle, as in the genus *Malus* (apple), and other members of subfamily Maloideae in the Rosaceae. **[181]**
pomology The science and practice of fruit culture.
pore A small, usually round aperture. **[129, 176]**
poricidal, porose Opening by pores, as the anthers in many members of the Ericaceae and the Polygalaceae, or the capsules in some genera, e.g. *Papaver* (poppy). **[133, 176]**
porogamy The entry of the pollen tube into the ovule through the micropyle.
posterior Back, towards the axis. **[144]**
postichous On the posterior side, next the axis.
pouch A bag-shaped structure.
p.p. (**pro parte**) In part, added to a plant name to indicate that the name now only partially covers the former concept of the taxon.
praemorse Irregularly truncate, appearing as if bitten off at the apex. **[110]**
precocious flower A flower that opens early in the season, before the leaves appear.
prickle A sharp-pointed outgrowth from the superficial tissues of the stem, as in the genus *Rosa*. **[94]**
primary meristem Tissue derived from the apical meristem, and appropriate to the part concerned.
primary phloem Phloem tissue derived from the procambium during the growth of a vascular plant.
primary xylem Xylem tissue derived from the procambium during the growth of a vascular plant.

procambium The cells of primary meristem which differentiate into primary xylem, primary phloem, and cambium.
procumbent Lying along the ground. **[71]**
progeny Offspring, immediate descendents.
prolepsis Growth of a bud into a lateral shoot after a period of dormancy.
proleptic Growing into a lateral shoot from a dormant bud.
proliferous Bearing gemmae or plantlets. **[61]**
prolification (see **proliferation**)
proliferating Producing buds in the axils of perianth segments.
proliferation In some angiosperms the development of buds in the axils of perianth segments, or in some gymnosperms the development of a leafy shoot from a female cone.
prone (see **prostrate**)
propagation The multiplication of plants by seeds or various kinds of vegetative material, either under natural conditions or under the sterile conditions of *in vitro* culture.
propagule Any structure capable of giving rise to a new plant by sexual or asexual means. **[61]**
prophyll The lowest of the papery structures attached to the scape in the genus *Crocus* (Iridaceae). **[56]**
prop root (stilt root) An adventitious root that grows out from the lower part of a stem into the soil to support that stem, or grows down from a lower branch into the soil to support that branch. **[52, 55]**
prostrate (prone) Lying flat. **[73]**
pro syn. (pro synonymo) A phrase used after a plant name to indicate that it was first published as a synonym.
protandrous Having stamens which mature and shed their pollen before the stigmas of the same flower become receptive.
protandry The state of being protandrous.
prothallial Relating to the prothallus in ferns and fern allies; the kind of cells that occur within the wall of the pollen grain of e.g. *Pinus* (pine). **[189]**
prothallus The plantlet that develops from the spore of a fern or other vascular cryptogam. It lacks true roots, stems, and leaves, but bears male and/or female sex organs, and is attached to the soil by rhizoids **[199, 206-208]**; the group of prothallial cells in the ovule of a gymnosperm.
protogynous Having stigmas which become receptive before the stamens of the same flower mature and shed their pollen.
protogyny The state of being protogynous.
protophloem The first phloem that is formed from the procambium.
protoplasm The living material within a plant cell, consisting of the nucleus and plastids embedded in cytoplasm.
protostele A type of stele that lacks pith, and consists only of xylem and phloem. **[204]** (see **haplostele**, **actinostele**, and **plectostele**)
protoxylem The first xylem that is formed from the procambium.
provernal Occurring in early spring.
proximal Situated near to the point of attachment.
pruinose Having a bloom.
pseudanthium A group of small flowers that collectively simulate a single flower, as *Cornus florida*, where the flowers are surrounded by large petaloid bracts. **[125]**
pseudobulb A bulb-like enlargement of the stem in orchids. **[169]**
pseudobulbous Having a pseudobulb.
pseudocarp (false fruit) A structure comprising the mature ovary combined with some other part of the plant, as the 'hip' in the genus *Rosa*. **[181]**
pseudocephalium The dense mass of hair at the top of the stem in certain cacti.
pseudostipule One of the lowermost leaflets in the compound leaves of some dicotyledons and therefore very close to the point of insertion of the leaf and the true stipules if these are present.
pseudoterminal Of a bud or flower, giving the appearance of being at the apex of the stem, but in fact axillary.
Pteridophyta, pteridophytes Ferns and fern allies. **[199-208]**
ptyxis The way in which an individual leaf is folded within a vegetative bud. **[103]** (see also **vernation**)
puberulent Minutely pubescent.
puberulous Slightly hairy. **[116]**
pubescence Hairiness.
pubescent (downy) Covered in soft hairs. **[117]**
pullulate To germinate, bud, or sprout.
pulse The edible seeds of plants in the Leguminosae cultivated as food crops, such as peas, beans, lentils, etc.
pulvinate Cushion-like.
pulvinus An enlarged portion of the petiole, at its base in *Mimosa pudica* (Sensitive Plant), a member of the Leguminosae **[94]**, or at its junction with the lamina in members of the Marantaceae. **[92]**
punctate Marked with dots, depressions, or translucent glands. **[115]**
punctiform In the form of a dot or point.
punctum A dot-like marking on any plant organ.
pungent Ending in a stiff, sharp point. **[110]**
pusticulate Bearing minute, pimple-like protuberances.

pustulate Covered with pustules. **[115]**
pustule A pimple-like projection from the surface.
pyramidal Conical in habit. **[72, 73]**
pyrene The stone of a drupe, consisting of the seed surrounded by the hard endocarp. **[179]**
pyriform Pear-shaped.
pyxidium (see **pyxis**)
pyxis (plural **pyxides**) A capsule with circumscissile dehiscence, as in the genus *Anagallis* (Primulaceae). **[176]**

quadrifoliate Four-leaved.
quadrifoliolate With four leaflets. **[205]**
quadrigeneric Composed of four different genera.
quadrilocular Having four loculi or compartments. **[136]**
quadripartite Divided almost to the base into four parts.
quincuncial Partially imbricated of five parts, two being exterior, two interior, and the fifth having one margin exterior and the other interior, as the sepals in the genus *Rosa*.
quinquelocular Having five loculi or compartments.

race A group of individuals within a species that show some similarities but are not sufficiently distinct to constitute a separate species.
raceme An indeterminate inflorescence with pedicillate flowers. **[121]**
racemose In the form of a raceme; bearing racemes. **[98]**
rachilla A secondary axis in the inflorescence of grasses. **[162]**
rachis (plural **rachides** or **rachises**) The axis of a compound leaf or an inflorescence. **[194, 199]**
radial Extending from a common centre. **[159]**
radially symmetric (see **actinomorphic**)
radiate Spreading from a common centre **[138]**; having a capitulum with ray florets, as in some members of the Compositae. **[153]**
radical Of leaves, arising directly from the rootstock. **[164]**
radicant Rooting.
radicel A small root, a rootlet.
radicle The young root as it emerges from the seed, normally the first organ to appear on germination. **[65, 66, 86]**
radius One of the longer filaments in the corona of plants in the Passifloraceae. **[146]**
rambler Any of the cultivated varieties of *Rosa* (rose) which straggle over adjacent vegetation, fences, walls, etc.
ramentum One of the thin, dry scales that often occur on the stems and leaves of ferns. **[201]**
ramet An individual member of a clone.
ramiflory The production of flowers on the branches of trees (included in the term cauliflory).
ramose Branching, having many branches.
rank A vertical row of lateral organs; the name of a taxonomic unit, e.g. genus, species, etc.
raphe The longitudinal ridge that represents the part of the funicle that is fused to the ovule or seed.
raphide One of the needle-shaped crystals, usually of calcium oxalate, found in the cells of some plants.
Raunkiaer A Danish ecologist (Christen Raunkiaer, 1860-1938) who devised a classification of plants based on the position of their perennating buds in relation to the soil surface, and a classification of leaves according to their surface area. **[75]**
ray A primary branch of an umbel **[122]**; the extended, strap-like portion of a ray floret in the Compositae. **[152]**
ray floret One of the outer, irregular flowers in the flower heads of some plants in the Compositae. **[152]**
raylet A smaller, secondary branch of an umbel. **[122]**
reaction wood Wood of distinctive structure formed by trees in branches and inclined trunks in order to maintain them in the appropriate position. (see **compression wood** and **tension wood**)
receptacle (**thalamus, torus**) The end of the stem which bears the flower parts. **[148, 181]**
receptacular Relating to the receptacle. **[152]**
recurved Curved backwards.
reduplicate Folded outwards or downwards. **[167]**
reflexed Bent abruptly backwards. **[147, 192]**
regma (plural **regmata**) A type of schizocarp in which the mericarps dehisce by elastic movement to release their seeds, as in *Geranium*.
regular (see **actinomorphic**)
relict A plant that has survived from a previous age or, because of changed circumstances, is growing in a place geographically remote from its most closely related species.
reniform Kidney-shaped. **[106, 182]**
repand With a slightly sinuate margin. **[108]**
repent (**reptant**) Creeping along the ground. **[71]**
replicate Explicative, with the leaf margins in the bud folded back, as in some species of the genus *Galanthus* (snowdrop); of a cell, to reproduce itself.
replum (**false septum**) A partition formed by outgrowths of the placentas, as in fruits of the Cruciferae. **[184]**
reproduction The production of new individuals by existing ones.

reptant (see **repent**)
resin A sticky, sometimes fragrant substance, insoluble in water, which is secreted by special cells in conifers for protective purposes.
resin canal, resin duct One of the minute tubes found in the leaves and sometimes wood of conifers that contain resin-producing cells. **[190]**
resiniferous Bearing resin.
resinous Containing or bearing resin; resembling resin.
resting (see **perennating**)
resupinate Inverted, turned upside down. In the Orchidaceae, many of the flowers are twisted through 180 degrees, so that the position of the upper and lower petals is reversed. **[170]**
resupination Inversion.
reticular Net-like in appearance or construction.
reticulate Marked with a network pattern. **[104, 175]**
retinaculum The band connecting a pollinium to the corpusculum in the Asclepiadaceae. **[151]**
retrorse Curved or bent downwards or backwards, away from the apex.
retuse Slightly notched at the apex. **[110]**
revolute Rolled downwards at the margin, i.e. towards the abaxial surface. **[103]**
rhachis (see **rachis**)
rheophyte A plant confined to flowing water.
rhipidium A fan-shaped cyme, with axes only on one plane, branching alternately to one side and the other. **[123]**
rhizodermis (see **piliferous layer**)
rhizoid One of the hair-like roots on a prothallus. **[199]**
rhizomatous Resembling or possessing a rhizome.
rhizome A root-like stem, lying horizontally on or situated under the ground, bearing buds or shoots and adventitious roots. **[58, 59]**
rhizome tuber (see **stem tuber**)
rhizophore In the genus *Selaginella*, a leaf-less branch, which arises from a fork in the stem and grows downwards, putting out roots when it reaches the ground. **[208]**
rhomboid More or less diamond-shaped, with four equal sides. **[106]**
rigid Stiff.
riparian Growing on the banks of rivers and streams.
rod (**column**) One of the elongated structures that form part of the exine of a pollen grain, and radiate from its base. **[129]**
root The lower portion of the axis of a plant, usually branched, that anchors it in the soil and enables it to absorb nutrients. **[51-59]**
root cap A hollow cone of cells covering the apical meristem of a root tip and protecting it from damage as it is pushed through the soil. **[54, 66]**
rootstock A frequently subterranean stem or rhizome. **[199]**
root sucker A shoot that arises from an adventitious bud on the root, either naturally or as a consequence of wounding; a haustorium. **[54]**
root tuber The swollen part of a root. **[51]**
rosette (see **leaf rosette**)
rostellum The beak-like extension of the stigma in some members of the Orchidaceae. The pouch-like base (bursicle) breaks, under pressure from a visiting insect, to expose the viscidia. **[170]**
rostrate With a beak.
rostrum A pointed, beak-like projection, e.g. the thickened base of the style in the genus *Geranium* **[150]**, or the structure in the fruit of some members of the Compositae by which the pappus is attached to the cypsela. **[153]**
rosulate Arranged in a rosette. **[101]**
rotate Gamopetalous, with a short tube and spreading lobes. **[128]**
round Circular. **[85]**
rounded Having a more or less semicircular outline at the apex or base. **[72, 109, 110]**
rubiginous Rust-coloured.
ruderal Growing in waste places; a plant that grows in waste places.
rudimentary Not fully developed, vestigial. **[145]**
rufous Reddish brown.
rugose Wrinkled. **[115]**
rugulose Finely wrinkled.
ruminate Looking as though chewed.
runcinate With sharply cut divisions directed backward towards the base. **[106]**
runner A creeping stem, rooting and giving rise to plantlets at the nodes. **[60]**
rupicolous (**saxatile**, **saxicolous**) Growing amongst rocks.
russet A variety of apple or potato which has a reddish brown, usually rough skin.

sac A small pouch or bag-shaped structure.
saccate Pouched or bag-like. **[128]**
sagittate Arrowhead-shaped. **[106, 109]**
salicetum A collection of *Salix* (willow) species and varieties for scientific or ornamental purposes.
saline Containing chemical salts, especially sodium chloride.
salt gland One of the epidermal glands in the leaves of many salt marsh plants and mangroves that secrete sodium chloride. **[112]**
salverform, salver-shaped (see **hypocrateriform**)

samara A dry, indehiscent, winged fruit as in the genus *Fraxinus* (ash). **[178]**
samaroid Resembling a samara.
sap The juice of a plant.
saprophyte A plant that lives on decaying organic matter.
saprophytic Living on decaying organic matter.
sapwood The outer, younger layers of wood in the trunk or branch of a tree or in the stems of a shrub, usually paler in colour than the heartwood, and with living cells that are still capable of conducting sap. **[82]**
sarcotesta A fleshy seed coat.
sarcotestal Having a sarcotesta. **[67]**
sarmentose Bearing long, flexuous runners or stolons.
saxatile, saxicolous (see **rupicolous**).
scaberulous Slightly rough.
scabrid Somewhat scabrous.
scabrous With the surface rough to the touch, due to minute projections. **[116]**
scalariform Ladder-like, or having ladder-like markings.
scale, scale leaf A reduced leaf, usually membranous, and often found covering buds, bulbs and corms. **[56, 58, 59]**
scandent Climbing.
scape A leafless stalk, arising from the ground, which bears one or more single flowers, e.g. *Hyacinthoides* (bluebell), or a head of flowers, e.g. *Taraxacum* (dandelion). **[155]**
scapose Having a scape. **[74]**
scar The mark left when a part has fallen from its place of attachment. **[81, 162, 181]**
scarious Thin, dry and membranous.
schizocarp A fruit derived from a syncarpous ovary which breaks up at maturity into one-seeded portions (mericarps), as in most members of the Geraniaceae (5 mericarps) and Umbelliferae (2 mericarps). **[177]**
schizocarpic Bearing schizocarps.
scion A young shoot which is inserted into a rooted stock in grafting.
sciophilous Shade-loving.
sclereid Short, blunt-ended cells that, together with the fibres, compose the sclerenchyma.
sclerenchyma Strengthening tissue consisting of two kinds of cells (fibres and sclereids) with thick, often lignified walls, that supports the softer tissues of the plant. **[79, 112]**
sclerenchymatous Relating to or consisting of sclerenchyma.
sclerophyll A leaf that is tough and leathery due to well developed sclerenchyma, usually evergreen; a plant with such leaves, typical of trees and shrubs in warm, dry climates.
sclerophyllous Having tough, leathery, usually evergreen leaves.
sclerotic Hardened, stony in texture.
scorpioid cyme (see **cincinnus**)
scrambler A rather weak-stemmed climbing plant that sprawls over other plants, fences, and walls.
scrobiculate Having the surface pitted with small rounded depressions.
scurfy Covered with flake-like scales.
scutate Shaped like an oblong shield.
scutellum A more or less shield-shaped structure, situated between the endosperm and the embryo of a grass, that at germination secretes enzymes into and absorbs sugars from the endosperm. **[66]**
secondary thickening The additional xylem and phloem produced by the vascular cambium. **[82]**
secretion Fluid produced by and released from a gland. **[124]**
sect. Section.
section A subdivision of a genus, e.g. *Primula* (sect. Bullatae).
secund Directed towards one side. **[126]**
seed A unit of sexual reproduction developed from the fertilised ovule. **[65-68]**
seed apomixis (see **agamospermy**)
seed leaf (see **cotyledon**)
seedling A young plant that has grown from a seed. **[65-67]**
seed plant A member of the Spermatophyta.
segment One of the divisions of an organ. **[128, 168]**
segregate species A species, distinguished often on the basis of minute characters, within an aggregate species.
seismonasty The response of a plant to being touched, e.g. *Mimosa pudica* (Sensitive Plant). **[94]**
self-fertilisation (see **autogamy**)
self-incompatible Of a cultivar, producing viable pollen and a functional ovary, but unable to produce fruit when self-pollinated, although it may be effective in the pollination of another cultivar.
self-sterile Unable to produce viable progeny by self-fertilisation.
semi-amplexicaul Clasping the stem, but only to a small extent. **[102]**
semi-double flower A flower having petals or petaloids intermediate in number between the typical and the double forms. **[127]**
semi-inferior (**half-epigynous, half-inferior**) Of an ovary, the lower part of which is embedded in the pedicel but the upper part is free.
seminal Relating to the seed.
semi-parasite (see **hemi-parasite**)
semi-parasitic (see **hemi-parasitic**)

senescent Becoming old, and thereby losing the power of cell division and growth.
sens. (sensu) In the sense of.
sens. lat. (sensu lato) In a broad sense.
sens. strict. (sensu stricto) In a narrow sense.
sepal A single segment of the calyx. **[150]**
sepaloid Sepal-like.
septate Divided into compartments by walls or partitions.
septicidal Splitting at maturity along or into the partitions (septa) of the capsule. **[136]**
septifragal With the valves at maturity breaking away from the partitions (septa) of the capsule. **[136]**
septum A partition, as that separating the loculi in an ovary. **[136, 185]**
ser. Series.
sericeous Silky. **[116]**
series A subdivision of a genus, e.g. *Iris* (ser. Laevigatae).
serotinal Occurring in late summer.
serotinous Of cones, remaining closed long after reaching maturity; of plants, releasing seeds only after burning.
serrate With a saw-toothed margin. **[108]**
serrulate Minutely serrate. **[108]**
sessile Not stalked. **[102, 138]**
seta A bristle.
setiferous Bearing bristles.
setose Bristly. **[116]**
sexaperturate Having six pores.
shade plant A plant adapted to living in low light intensities.
sheath A tubular covering. **[66]**
sheathing With at least the base of the leaf or stipule forming a tube that more or less encloses the stem or stalk. **[91, 92, 102]**
shoot A developed bud or young, green stem. **[58]**
shrub A woody, perennial plant, generally smaller than a tree, and with several stems arising from ground level. **[71]**
silicle, silicula The capsular fruit of the Cruciferae when less than three times as long as broad. **[184]**
siliqua, silique The capsular fruit of the Cruciferae when at least three times as long as broad. **[184]**
siliquiform In the form of a siliqua.
silks The long, silk-like styles in the female inflorescence of *Zea mays* (Maize, Sweet Corn). **[163]**
simple Of one piece. **[105, 106, 118, 121, 122]**
single flower A flower having the typical number of petals or petaloids. **[127]**
sinker A shoot that grows downwards from a bulb or a corm and produces a bulb or a corm at its apex **[60]**; in parasitic plants, an outgrowth of the haustorium that extends into the tissues of the host plant. **[53]**
sinuate Having the blade of the leaf flat but with the margin winding strongly inward and outward. **[108]**
sinus The recess between two teeth or lobes, e.g. on a leaf margin. **[108]**
siphonostele A type of stele in which the xylem and phloem surround a central core of pith. **[204]** (see **amphiphloic** and **ectophloic**)
s.l. (sensu lato) In a broad sense.
sleep movement (see **nyctinasty**)
soboliferous Forming clumps.
sodium chloride (NaCl) Common salt.
softwood Wood obtained from conifers.
solenostele (see **amphiphloic siphonostele**)
solitary Borne singly.
solute A substance that is dissolved in water or another liquid.
soriferous Bearing sori.
sorosis (plural **soroses**) A fleshy, multiple fruit, e.g. *Ananas* (pineapple) or *Morus* (mulberry). **[182]**
sorus A group of sporangia. **[199]**
sp. Species (singular).
spadix (plural **spadices**) A spike with a fleshy axis, as in Araceae. **[124]**
spathe A large bract subtending and often enclosing a flower or an inflorescence. **[61, 124]**
spathulate, spatulate Spatula-shaped. **[106, 151]**
species (plural **species**) A group of closely related, mutually fertile individuals, showing constant differences from allied groups, the basic unit of classification. The names of species are written nowadays with a small initial letter, but formerly the names of some species, e.g. those derived from personal names, were written with a capital initial letter.
specific epithet The second element, usually an adjective, in the binomial name of a plant, which follows the generic name and serves to distinguish one species from others in the same genus.
Spermatophyta, spermatophytes Seed-bearing plants, comprising the Angiospermae and the Gymnospermae.
spermatozoid (see **antherozoid**)
sperm cell (see **male nucleus**)
spicate Spike-like. **[121]**
spicule A small or secondary spike.
spike An indeterminate inflorescence with sessile flowers. **[121]**
spikelet A unit of the inflorescence in grasses, consisting of one or more flowers subtended by a common pair of glumes. **[162]**

spine A sharp woody or hardened outgrowth from a leaf, sometimes representing the entire leaf, or from a fruit. **[94, 96, 159, 176]**
spinescent Ending in a spine or sharp point.
spinose Spiny. **[108]**
spinule A small spine.
spinulose Bearing small spines.
spiralled cyme (see **cincinnus**)
spirally arranged With the leaves arranged alternately along a stem so that a line joining their points of attachment would form a spiral. **[87]**
spongy mesophyll The tissue with a conspicuously open texture lying beneath the palisade layer in the leaves of mesophytic dicotyledons, and consisting of cells with large intercellular spaces. **[111]**
sporangiophore The stalk of a sporangium, as in the genus *Equisetum* (horsetail). **[207]**
sporangium A spore-case, the structure in which the spores are produced. **[199]**
spore The reproductive unit in ferns, produced in the sporangium and developing into a prothallus. **[199]**
sporeling A very young fern plant that has developed from the prothallus. **[199]**
sporiferous Bearing spores.
sporocarp The body enclosing the sporangia, especially the hard, rounded structure near the base of the leaves in some aquatic ferns, e.g. *Pilularia*. **[205]**
sporophyll A specialised leaf that bears sporangia. **[206]**
sporophyte The mature stage in the life-cycle of ferns when the cells are diploid or, more rarely, polyploid. **[199]**
sport (see **mutant**)
spp. Species (plural).
spring wood (**early wood**) Pale-coloured wood, with large xylem cells, that is produced early in the growing season. **[82]**
sprout A bud or newly grown shoot.
spur A slender projection from a plant part, especially the tubular extension, often nectariferous, of the calyx or corolla that occurs in *Aconitum*, *Aquilegia*, *Linaria*, *Viola* etc. **[144]**; a short, leafy branch of a tree, often with flowers and fruits in clusters at closely-spaced nodes. **[181]**
squamiform Scale-like, as the adult leaves of *Juniperus*. **[191]**
squamose Covered with large, coarse scales. **[115]**
squamulose Covered with small scales.
square Having four sides. **[85]**
squarrose With a rough surface due to projecting hairs or scales **[115]**; in trees and shrubs, with branches projecting more or less at right-angles to the main stem.
squarrulose Minutely squarrose.
ssp. (plural **sspp.**) Subspecies.
s. str. (**sensu stricto**) In a narrow sense.
stamen One of the male sex organs, usually consisting of anther, connective, and filament. **[124, 131-135]**
staminal Relating to stamens. **[149]**
staminate Having only male organs.
staminodal (**staminodial**) Relating to staminodes.
staminode A sterile stamen. **[134, 154, 170]**
staminodial (see **staminodal**)
standard The large upper petal (vexillum) in the flowers of plants of subfamily Papilionoideae in the Leguminosae **[148]**; one of the three more or less erect inner perianth segments in flowers of the genus *Iris* **[156]**; a tree, especially a fruit tree, trained so that it has an upright stem free of branches.
starch grain A rounded or irregular mass of starch within a chloroplast or other plastid.
stele (**vascular cylinder**) The central core of the stem and root in a vascular plant, consisting of vascular tissue and often associated tissue such as pith, pericycle etc. **[54, 79]**
stellate Star-shaped. **[117]**
stellular, stellulate Shaped like a small star.
stem The main supporting axis of a plant. **[57, 59, 79]**
stem leaf A leaf borne on the stem as opposed to being at its base.
stem tuber (**rhizome tuber**) The swollen end of an underground stem, e.g. a potato tuber. **[58]**
sterile Unable to reproduce sexually. **[124]**
stigma The apex of the style, usually enlarged, on which the pollen grains alight and germinate. **[130, 136-138]**
stigma flap In the genus *Iris*, a small flap on the underside of the style branch. **[156]**
stigmatic Relating to the stigma. **[143]**
stilt root (see **prop root**)
stimulus A substance or action that causes a response in the plant. **[85]**
stinging hairs Stiff, tubular hairs, filled with irritant substances, as in *Urtica dioica* (Stinging Nettle). When the tip of the hair is broken off, e.g. by an animal, pressure on the saccate base of the hair forces the contents out into the animal's skin. **[118]**
stipe A stalk, especially the caudicle in an orchid flower, or the petiole of a fern or palm frond. **[168, 171, 185, 199]**
stipel The stipule of a leaflet.
stipitate Having a stipe, or borne upon one.
stipular Relating to a stipule. **[93, 94]**

stipulate Having stipules. **[102]**
stipule A leafy outgrowth, often one of a pair, arising at the base of the petiole. **[80, 92-94, 96, 102]**
stock The rooted stem into which the scion is inserted when grafted.
stolon A lateral stem growing horizontally at ground level, rooting at the nodes and producing new plants from its buds, as in the genus *Fragaria* (strawberry).
stoloniferous Bearing stolons. **[71]**
stoma (plural **stomata**) One of the small pores, found most often in the epidermis on the lower surface of leaves but also on young stems, which allow gases to pass in and out of the plant. **[111, 190]**
stomatal Relating to stomata.
stomium The opening in the annulus of the sporangium through which the spores are released. **[199]**
stone (see **pyrene**)
stone cell A strongly lignified type of sclereid found in the flesh of fruits in the genus *Pyrus* (pear).
storage organ The swollen part of a plant where food is stored in the form of starch or sugar, e.g. a tuber, bulb, or corm. **[56-59]**
stramineous Straw-coloured.
striate Having fine, longitudinal lines, grooves, or ridges.
strigose Bearing stiff hairs or bristles. **[116]**
strobile, strobilus (**cone**) The reproductive structure in gymnosperms **[207]**. In conifers, this consists of an ovoid, cylindrical, or spherical cluster of sporophylls (cone scales) arranged round a central axis **[189]**. The term is sometimes applied to e.g. the papery, cone-shaped fruits of *Humulus lupulus* (Hop). **[182]**
strobiloid Resembling a strobilus.
stroma The colourless material inside a chloroplast in which the chlorophyll necessary for photosynthesis is embedded.
stylar Relating to the style. **[156]**
style The often elongated apical part of a carpel or gynoecium that bears the stigma at its tip. **[66, 124, 130, 136-138]**
stylopodic Arising from an enlarged base. **[137]**
stylopodium The enlargement at the base of the styles in some members of the Umbelliferae. **[137]**
subapical Almost at the apex.
subbasal Almost at the base.
subclass A subdivision of a class. Names of subclasses end in '-idae'.
subcompound More or less compound.
subcordate More or less heart-shaped. **[109]**
subcuneate More or less wedge-shaped.
subdivision The rank immediately below division. Names of subdivisions end in '-phytina'.
subequal Almost equal.
suberin A mixture of fatty substances present in the cell walls of cork that renders them waterproof and resistant to decay.
suberised Converted into cork.
suberose Corky.
subfamily A subdivision of a family. Names of subfamilies end in '-oideae'.
subform, subforma The lowest taxonomic rank, a subdivision of a form or forma.
subg. Subgenus.
subgenus (plural **subgenera**) A subdivision of a genus, e.g. *Prunus* (subg. Cerasus).
subglabrous Almost without hairs.
subglobose (**depressed-globose, oblate**) Almost globular, but flattened at the ends of the axis.
subkingdom A subdivision of kingdom, a taxonomic rank.
suborbicular Almost circular.
suborder A subdivision of an order. Names of suborders end in '-ineae'.
subpetiolar, subpetiolate Under the petiole, e.g. in the genus *Platanus* (plane), the petiole is expanded at the base to form a hood over the axillary bud. **[91]**
subsessile Almost devoid of a stalk.
subshrub (**suffrutex, undershrub**) A low shrub, sometimes with partially herbaceous stems. **[71]**
subsp. (plural **subspp.**) Subspecies.
subspecies (plural **subspecies**) A subdivision of a species, often used for a geographically or ecologically distinct group of plants.
subtend To stand below and close to, to extend under. **[98]**
subtribe A subdivision of a tribe. Names of subtribes end in '-inae'.
subulate Awl-shaped, tapering from the base to the apex. **[106]**
subvalvate Almost valvate.
subvar. Subvariety.
subvariety A subdivision of a variety.
subvars. Subvarieties.
succulent Fleshy and juicy.
sucker A shoot of subterranean origin. **[60]** (see also **root sucker**)
sucker disc The expanded tip of some tendrils that enables them to adhere to fences, walls etc., e.g. *Parthenocissus* (Vitaceae). **[93]**
suffrutescent Woody only at the base of the stem.
suffrutex (see **subshrub**)
suffruticose Woody in the lower part of the stem.
sulcate Grooved or furrowed.
sulcus A groove or furrow.
summer wood (see **autumn wood**)

superior Above, as when the ovary is situated above the other floral parts on the receptacle. **[135]**
superposed Placed above or on another structure.
supervolute In ptyxis, having one margin rolled within the other. **[103]**
supra-axillary Growing above an axil.
suprafoliar Borne above the leaves. **[166]**
surcurrent Running up, as when the base of the leaf is prolonged up the stem as a wing. **[202]**
surmounted Capped or crowned. **[136]**
suspensor In spermatophytes and certain pteridophytes, the group of cells within the embryo sac which pushes the embryo down into the developing endosperm. **[208]**
suture A seam or line of joining. **[136, 176]**
switch plant A plant of dry places with long, thin stems, which at first bear a few leaves but subsequently, being green, take over the process of photosynthesis.
syconium A multiple hollow fruit, as in the genus *Ficus* (fig). **[183]**
syllepsis Growth of a bud into a lateral shoot without any period of dormancy.
sylleptic Growing from a bud into a lateral shoot without any delay.
symbiont Either one of a pair of organisms involved in symbiosis.
symbiosis The arrangement whereby two different organisms (symbionts) co-exist, not necessarily to their mutual advantage, though often used in this respect. (see also **commensalism** and **mutualism**)
sympatric Of plant species or populations, having a common or an overlapping distribution.
sympetalous (see **gamopetalous**)
sympodial With the main stem or axis ceasing to elongate but growth being continued by the lateral branches.
syn. Synonym.
synandrium An androecium coherent by the anthers, as in some members of the Araceae.
synandrous With coherent anthers.
synangium A compound structure formed by the fusion of two or more sporangia in certain tropical ferns, e.g. *Platycerium*, and fern allies, e.g. *Psilotum*, or by the fusion of groups of pollen sacs in certain cycads. **[205]**
syncarp, syncarpium A fleshy, multiple fruit with united carpels. **[182]**
syncarpous Having united carpels.
synergid One of the two haploid nuclei that, together with the ovum, lie at the micropylar end of the embryo sac. **[130]**
synflorescence A compound inflorescence, composed of a terminal inflorescence (florescence) and one or more lateral inflorescences (co-florescences).
syngamy Sexual reproduction involving the fusion of a male and a female gamete.
syngenesious With anthers united into a tube, but filaments free, as in flowers of the Compositae. **[132]**
synonym Another name for the same taxon, either an alternative name that is valid under a different classification, or a name now invalid according to the I.C.B.N. that has been superceded by a later name.
synsepal (see **synsepalum**).
synsepalous (see **gamosepalous**)
synsepalum (**synsepal**) A structure formed by the joining of two or more sepals, especially in certain genera of the Orchidaceae, e.g. in *Masdevallia* all three sepals are connate to some extent, and in *Paphiopedilum* and most species of *Cypripedium* the two lateral sepals are united. **[170]**
syntepalous Having the tepals united.
systematic botany (see **taxonomy**).

t., tab. (**tabula**) A plate, a full-page illustration.
tannin An acidic substance, soluble in water, with a bitter taste and astringent properties, that is present in a number of plants, especially in the bark of species of *Quercus* (oak).
tanniniferous Bearing tannin.
tapetum The layer of cells that forms a nutrient tissue round the pollen mother-cells in the anthers of flowering plants or the spore mother-cells in pteridophytes. **[129]**
taproot A strongly developed main root which grows downwards bearing lateral roots much smaller than itself. **[51, 57, 59, 80]**
tassel The panicle comprising the male inflorescence in *Zea mays* (Maize, Sweet Corn). **[163]**
taxon (plural **taxa**) A unit of classification of any rank, e.g. *Bellis perennis* (species); *Bellis* (genus); Compositae (family).
taxonomy (**systematic botany**) The aspect of botany that deals with the identification, classification and nomenclature of plants.
tectum The structure forming a roof over the rods in the exine of a pollen grain. **[129]**
tendril A twining, thread-like structure produced from a stem or leaf that enables a plant to hold its position securely. **[91, 93]**
tension wood A kind of reaction wood found on the upper sides of the branches and inclined trunks of hardwood trees, and characterised by a greatly thickened inner layer of the cell walls, which

may separate from the rest of the cell wall or may enlarge to fill the lumen.

tenuinucellate Having a nucellus composed of little more than an epidermis and an embryo sac.

tepal One of the petals or sepals of a flower in which all the perianth segments closely resemble each other. **[79]**

terete Like a slender, tapering cylinder, and more or less circular in any cross-section. **[137]**

terminal At the apex or end. **[56, 81, 98, 192]**

ternate In threes.

terrestrial Growing on the ground.

tessellate, tessellated Chequered, having markings or colours arranged in squares. **[104]**

testa The seed coat, a hard covering that has developed from the integument of an ovule after fertilisation. **[65]**

testiculate Shaped like the tubers of certain terrestrial orchids, e.g. *Ophrys*. **[51]**

tetrad A group of four pollen grains. **[171, 208]**

tetradynamous Having four long stamens and two short ones, as in the Cruciferae. **[132]**

tetrahedral Four-sided, as a pyramid.

tetramerous Having the parts of the flower in fours.

tetrandrous Having four stamens.

tetraploid (**4n**) Having four sets of chromosomes in each cell.

tetrarch root A root with four protoxylem strands in the stele. **[55]**

thalamus (see **receptacle**)

thallus A plant body not differentiated into leaves and stem. **[199]**

theca (**anther cell**) One of the usually two lobes of an anther in which pollen is produced. At first, each anther cell is divided into two portions (pollen sacs), making four in all, but before anthesis the tissue separating each pair disintegrates, and the anther then becomes two-celled. In e.g. the family Malvaceae, the anther has only one lobe. **[133, 170]**

thermonasty The response of a plant to a change in temperature. **[86]**

thermotropic Turning towards the heat source (positively thermotropic) or away from it (negatively thermotropic).

thermotropism The movement of a plant in response to heat from a particular direction.

therophyte An annual, a plant which survives the unfavourable season of the year in the form of seeds.

thigmotropic (**haptotropic**) Of a plant, turning towards the object with which it has come into contact (positively thigmotropic), as many climbing stems and tendrils, or away from it (negatively thigmotropic).

thigmotropism The movement of a plant in response to physical contact with another object.

thorn A short, pointed branch. **[94]**

throat The region of a corolla or calyx with united segments where the lower, tubular portion expands into the upper, spreading part. **[125]**

thrum-eyed One of the two forms of a dimorphic flower, e.g. *Primula vulgaris* (Primrose), where the style is short and the stamens are above the stigma. **[147]** (see also **pin-eyed**)

thyrse A mixed inflorescence in which the main axis is indeterminate and the secondary and ultimate axes are determinate or cymose. **[123]**

thyrsoid Resembling a thyrse.

tiller A lateral shoot arising at ground level from the stems of grasses. **[160]**

tissue An aggregation of cells with a similar form and function that form the material of which a particular part of a plant is composed.

tomentose Densely covered in soft hairs. **[116]**

tomentum The dense covering of hairs on a plant or a particular organ.

tooth One of the small, pointed projections that form the margin of many leaves, or sometimes the apex of a mature capsule. **[176]**

topocline A type of cline relating to variations in a particular taxon throughout its geographical range.

topodeme A deme occupying a particular geographical area.

topogamodeme A gamodeme occupying a precise locality.

torose Cylindrical, with swellings or contractions at intervals.

torulose Diminutive of torose. **[185]**

torus (see **receptacle**).

trabecula A rod-like structure, e.g. one of the strands of sterile tissue dividing the cavity in the sporangia of *Isoetes* (quillwort). **[205, 208]**

trace elements (see **micronutrients**)

tracheid An elongated structure derived from a single cell that develops lignified walls and is able to conduct water and solutes in gymnosperms and other woody plants.

Tracheophyta, tracheophytes (see **vascular plants**)

translator The clip (corpusculum) and bands (retinacula) which connect the pair of pollinia in the Asclepiadaceae. **[151]**

translocation Transport of dissolved substances within the xylem and phloem of a plant.

translucent Allowing light to pass through, but not transparent.

transpiration The loss of water vapour into the atmosphere through the stomata.

transverse Crosswise.

trapeziform, trapezoid Shaped like a trapezium, i.e. with only two of its four sides parallel.
traumatonasty The movement of a plant in response to wounding.
tree A woody, perennial plant, usually tall, with a single bole or trunk that bears a crown of branches. **[71]**
tree canopy The cover formed by the crowns of the tallest trees in a wood or forest.
tree ring (see **annual ring**)
triadelphous With three groups of stamens, as the flowers in some species of *Hypericum*.
triandrous Having three stamens, as in the Iridaceae.
triangular Having three angles and sides. **[85, 178]**
triaperturate Having three pores.
triarch root A root with three protoxylem strands in the stele. **[55]**
tribe A subdivision, usually of a large subfamily, but sometimes a family may be divided directly into tribes. Names of tribes end in '-eae'.
trichome Any hair-like growth, glandular or eglandular, from the epidermis.
trichotomous Branching into three.
tricolpate Of a pollen-grain, having three colpi.
tricolporate Of a pollen grain, having three composite apertures, each consisting of a colpus and a pore.
tridentate With three teeth. **[110]**
trifid Divided to about half-way into three parts.
trifoliate Three-leaved.
trifoliolate With three leaflets. **[107]**
trigeneric Composed of three different genera, as the orchid x *Brassolaeliocattleya*, a hybrid genus produced by crossing a species of *Brassia* with one of *Laelia* and one of *Cattleya*.
trigger hairs Sensitive hairs, which when touched, cause a particular reaction. In e.g. *Dionaea muscipula* (Venus's Fly-trap), an insect touching one of the trigger hairs on the inner surface of the leaf will cause the two lobes to close rapidly, trapping the prey inside. Special glands then secrete digestive enzymes which act on the body of the insect, producing products from which the plant is able to obtain nutrients. **[118]**
trigonous Three-angled.
trijugate Of a compound leaf, having three pairs of leaflets. **[167]**
trilobate, trilobed With three lobes.
trilocular Having three loculi or compartments. **[171]**
trimerous Having the parts of the flower in threes, as in the Iridaceae.
trimorphic Occurring in three forms.
trioecious Having male, female, and bisexual flowers on different plants of the same species.
tripartite Divided almost to the base into three parts.
tripinnate Bipinnate, with the secondary leaflets again pinnate. **[107]**
tripinnatifid Literally bipinnatifid, with the secondary leaflets again pinnatifid, but frequently the primary division is pinnate. **[200]**
triplinerved Having three main veins.
triploid (3n) Having three sets of chromosomes in each cell.
triporate Of a pollen grain, having three pores.
triquetrous Triangular in cross-section, with sharp angles sometimes produced by concave sides. **[165]**
triradiate Having three rays. **[206]**
tristyly The occurrence of three different lengths of style in flowers of the same plant species, as *Lythrum salicaria* (Purple Loosestrife).
trisulcate Of a pollen grain, having three grooves or furrows.
tritegmic Of an ovule, having three integuments.
tropic, tropical Occurring in thc tropics, the region extending to about 23 degrees on either side of the equator.
tropism A directional response by a plant to a stimulus, positive if towards the source of the stimulus or negative if away from it.
tropophyte A plant adapted to a seasonal climate, and surviving periods unfavourable for growth by forming resting buds.
trullate Trowel-shaped. **[105]**
trumpet-shaped Narrowly tubular, ending in a flared limb. **[128]**
truncate Appearing as if cut off at the base or apex. **[109, 110]**
trunk The bole, or upright main stem of a tree which bears a crown of branches. **[71, 194]**
truss A cluster of flowers or fruit growing on a single stalk.
T.S. Transverse section.
tube nucleus The nucleus of the vegetative cell, the larger of the two cells into which the nucleus of the pollen grain divides while still in the pollen sac. **[130]** (see also **generative cell**)
tuber An underground stem or root, swollen with reserves of food. **[58]**
tubercle A small tuber; a small, rounded projection. **[137, 159]**
tubercled, tuberculate (see **verrucose**)
tuberous Resembling or producing tubers.
tubular Cylindrical and hollow. **[128]**
tubular floret A regular floret, usually a disc floret in the Compositae. **[152]**
tunic The outer, dry, papery covering of a bulb or corm, often fibrous or reticulate. **[57]**

tunica The outer layer or layers of cells in an apical meristem, which divide to produce the epidermis of the shoot.
tunicate Consisting of a number of concentric layers, the outermost usually dry and membranous, as the bulb of *Allium cepa* (Onion). **[57]**
turbinate (**obconical**) Top-shaped, inversely conical. **[68]**
turgid Of cells, swollen and rigid as a result of the uptake of water.
turgor Rigidity of cells resulting from their uptake of water.
turion An underground bud or shoot which develops into an aerial stem; a winter bud, characteristic of many aquatic plants, that contains food material and becomes detached from the parent plant, either floating or resting at the bottom of the water until favourable conditions stimulate its development and growth into a new plant. **[60, 75]**
tussock A clump or tuft, especially of a grass.
twig One of the smallest divisions of the branch of a tree. **[81]**
tylose, tylosis In woody plants, an outgrowth of the wall of a parenchyma cell which protrudes into an adjacent duct.

uliginous Growing in swamps.
umbel An inflorescence in which the pedicels arise from the same point on the peduncle. An umbel can range from being flat-topped, as in *Daucus carota* (Carrot) to almost spherical, as in *Allium cepa* (Onion). **[122]**
umbellate In the form of an umbel; bearing umbels.
umbellifer A member of the Umbelliferae.
umbelliferous Bearing an umbel or umbels, typically a member of the Umbelliferae.
umbelliform In the form of an umbel.
umbo A boss or protuberance, especially that which occurs on a cone scale. **[192]**
umbraculate Parasol-shaped. **[143]**
unarmed Without spines, prickles or thorns.
uncinate Hooked.
undershrub (see **subshrub**)
understorey A layer of vegetation composed of shrubs and small trees lying between the tree canopy and the ground cover in a wood or forest.
undulate With a wavy margin curving up and down. **[108]**
unguiculate Narrowed at the base into a claw. **[145]**
uniaperturate Having one pore.
unicarpellate Having one carpel.
unicellular One-celled. **[118]**
unijugate Of a compound leaf, having one pair of leaflets as in *Lathyrus odoratus* (Sweet Pea). **[91]**
unilateral One-sided.
unilocular Having a single loculus or compartment. **[136]**
uniovulate Having a single ovule.
uniseriate Arranged in a single row or series. **[185]**
unisexual (**diclinous**) Having only male or female organs in the flower. The male and female flowers may be on separate plants (dioicism) or on the same individual (monoecism). **[145]**
united Joined. **[124, 143]**
unitegmic Of an ovule, having one integument.
upright Erect; with the branches growing upwards. **[72]**
urceolate Urn-shaped, pitcher-shaped. **[128]**
utricle A bladder-like structure, especially the membranous sac in the fruit of the genus *Carex* (sedge). **[143, 165]**

vaginate Sheathed. **[102]**
vallecular canal Any of the canals in the stem of species of the genus *Equisetum* (horsetail), each one opposite a groove on the surface. **[207]**
valvate When similar parts of the plant meet exactly without overlapping **[81]**; in anthers or fruits, opening by valves. **[133, 176]**
valve One of the pieces into which an anther splits at maturity to release the pollen; one of the pieces into which a fruit splits at maturity to release the seeds. **[176]**
valvular dehiscence Splitting open by means of valves. **[175, 176]**
var. Variety.
variation Divergence from the normal state, usually due to genetic differences or influence of the environment.
variety A rank used to designate a group of plants varying in flower colour, habit, or some other way.
vars. Varieties.
variegated Having two or more colours in the leaves.
vascular Relating to the vessels that convey water and nutrients within the plant, i.e. the xylem and phloem.
vascular bundle One of the strands of tissue that conduct water and nutrients within the plant, consisting of xylem on the inside and phloem on the outside, separated by a layer of cambium. **[55, 79-82, 85, 111, 129, 136]**
vascular cambium The layer of meristematic cells lying between xylem and phloem that produces

additional xylem and phloem, resulting in the lateral growth of a stem or root. **[80, 82]**
vascular cryptogam A cryptogam that resembles flowering plants in having a vascular system, as a fern or a fern ally. **[199-208]**
vascular cylinder (see **stele**)
vascular plants (**Tracheophyta**, **tracheophytes**) Plants that have a vascular system of xylem and phloem to conduct water and nutrients, i.e. members of the Pteridophyta and Spermatophyta.
vascular ray One of the secondary medullary rays, formed by the vascular cambium, that supplement the primary medullary rays.
vascular strand (see **vascular bundle**)
vascular system (**vascular tissue**) The conducting tissue in a vascular plant, consisting of xylem and phloem, separated by a layer of cambium. **[54]**
vegetative Relating to the non-flowering parts of a plant.
vegetative apomixis A form of apomixis in which plants reproduce vegetatively, by rhizomes, stolons, runners, stem tubers, or bulbils.
vegetative propagation Asexual reproduction, i.e. by means of bulbs, rhizomes, runners etc. rather than by seed. **[60, 61]**
vein (**nerve**) A strand of vascular tissue in a leaf or other flat organ. **[79, 111]**
veinlet A small vein. **[202]**
velamen The outer layer of the aerial roots of epiphytic orchids and aroids, consisting of thick-walled, non-living cells. **[54]**
velum The membranous indusium in the genus *Isoetes* (quillwort). **[205]**
velutinous Velvety. **[116]**
venation The arrangement of veins, as in a leaf. **[104]**
ventral The side of an organ facing towards the axis, adaxial. **[133, 176]**
ventricose Swollen or inflated on one side, as the corolla in some members of the Labiatae and Scrophulariaceae.
ventriculose Slightly ventricose.
vernacular name The common name for a plant in any language, as opposed to the scientific name.
vernal Occurring in spring.
vernation The arrangement of leaves in a vegetative bud. **[104]** (see also **ptyxis**)
verrucose (**tubercled**, **tuberculate**) Bearing small, wart-like projections. **[115, 137]**
versatile With the anther attached at the middle and turning freely on its filament. **[133]**
verticil (**whorl**) The arrangement of three or more organs in a circle round the axis. **[101]**
verticillaster (**false whorl**) A deceptive kind of inflorescence found in the Labiatae, which gives the appearance of a whorl but which in reality consists of two dichasial cymes on opposite sides of the stem. **[122]**
verticillate Arranged in a whorl. **[101]**
vesicle A small bladder or sac.
vesiculose Covered with small bladders. **[115]**
vespertine Of flowers, opening in the evening.
vessel In angiosperms, one of the tubular structures forming the xylem and phloem that conduct water and nutrients throughout the plant.
vestigial Imperfect development of an organ which was fully developed in some ancestral form.
vestiture A covering of hairs or scales. **[115-118]**
vexillary (see **imbricate-descending**)
vexillum The standard, the large upper petal in the flowers of plants of subfamily Papilionoideae in the Leguminosae. **[148, 175]**
viable Of seeds, able to germinate.
vicariad, **vicariant** One of two similar taxa occupying separate geographical areas, e.g. *Jasminum nudiflorum* or *J. mesnyi*.
villose, **villous** Covered with long, shaggy hairs. **[117]**
vine A woody or herbaceous plant with a long, climbing, scrambling, or trailing stem, especially one belonging to the grape family (Vitaceae).
virgate Long, straight, and slender.
virgulate Diminutive of virgate.
viscid (**viscous**) Sticky.
viscidium (**viscid disc**) The sticky disc at the base of the caudicle in the Orchidaceae which adheres to the head of a visiting insect. **[171]**
viscin A sticky substance surrounding the seeds in members of the Loranthaceae.
viscous (see **viscid**)
vitta An oil-canal in the fruits of the Umbelliferae. **[177]**
viviparous Reproducing by vivipary. **[61]**
vivipary The production of buds that form plantlets while still attached to the parent plant; the production of seeds that germinate within the fruit. **[61]**
volute Rolled up.

weeping With the branches bending over and hanging down. **[72]**
whorl (see **verticil**)
whorled (see **verticillate**)
window, **window pane** (see **fenestration**)
wing A flat, often dry and membraneous extension to an organ **[175, 184]**; one of the lateral petals (alae) in the flowers of plants of subfamily Papilionoideae in the Leguminosae. **[148]**
winged (see **alate**).

winter annual An annual plant that germinates in the autumn, and survives the winter as a rosette of leaves.

witches' broom An abnormally dense tuft of twigs in a tree, resulting from an attack by fungi, mites, or viruses. **[97]**

wood The secondary xylem of trees; the constructional material obtained from felled trees. **[81]** (see **softwood** and **hardwood**)

x Before a botanical plant name, indicates a sexual hybrid; before a number, indicates the degree of enlargement.

xanthophylls A group of yellow pigments, belonging to the carotenoids, that occur in the chromoplasts of plant cells.

xenogamy A form of allogamy, in which the ovules of a flower are fertilised by pollen from a flower on a different plant.

xerad (see **xerophyte**)

xeromorphic Having the characteristics of plants growing in dry places, i.e. reduced or succulent leaves and stems, dense hairiness or a thick cuticle.

xerophyte (xerad) A plant that is adapted to grow in a dry habitat. **[160]**

xerophytic Growing in dry habitats. **[160]**

xylem The vascular tissue that conducts water and minerals from the roots to other parts of the plant. **[54, 79]**

xylocarp A hard, woody fruit.

zygomorphic (bilaterally symmetric) Divisible through the centre of the flower in only one longitudinal plane for the halves of the flower to be mirror images, as many members of the Labiatae and Scrophulariaceae. **[126, 144]**

zygote The cell (usually diploid) that results from the fusion of a male and a female gamete.

zygotic Relating to a zygote.

Illustrations

1. Roots, Storage Organs and Vegetative Reproduction

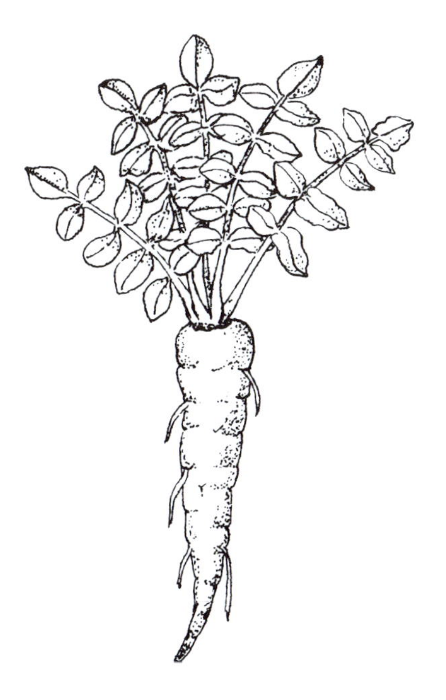

Roots 1

Fibrous roots of *Cerastium* sp.

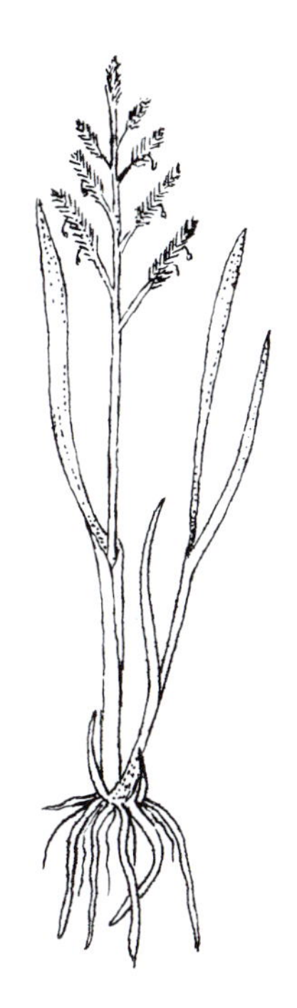

Adventitious roots of a grass

Taproot of *Pastinaca sativa*

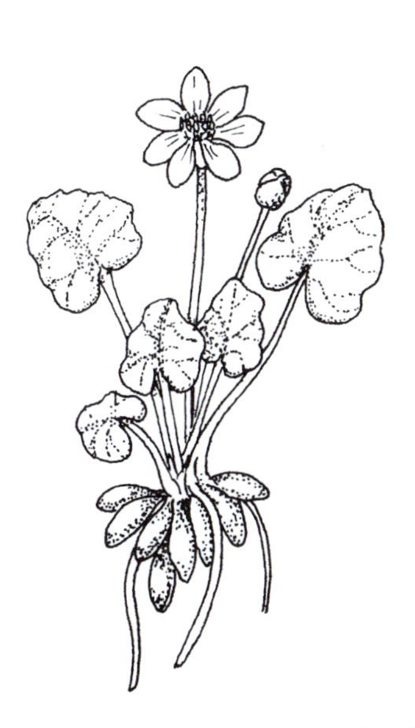

Root tubers of *Ranunculus ficaria*

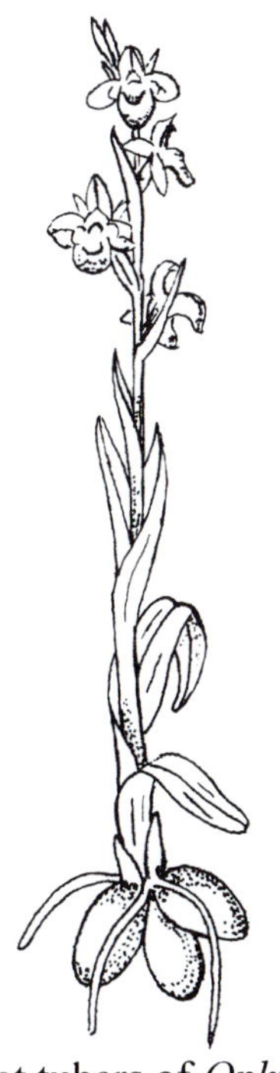

Root tubers of *Ophrys* sp.

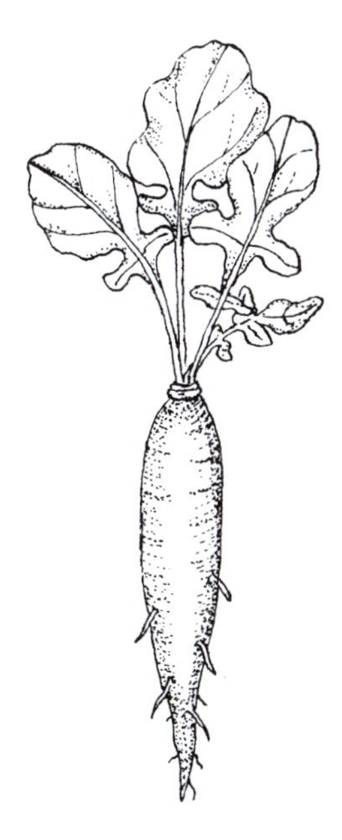

Taproot of *Raphanus sativus*

Roots 2

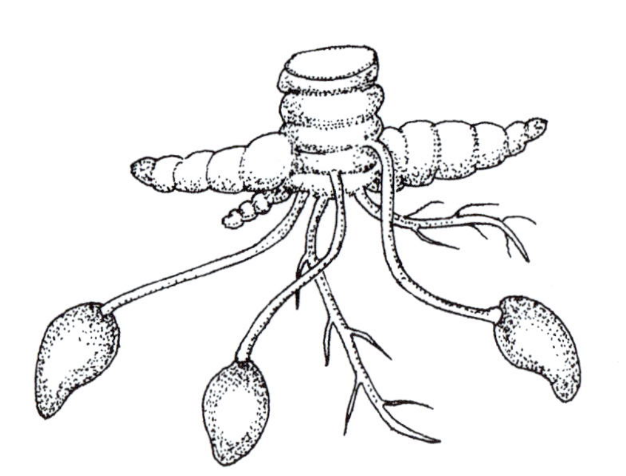

Nodulose roots of *Curcuma amada*

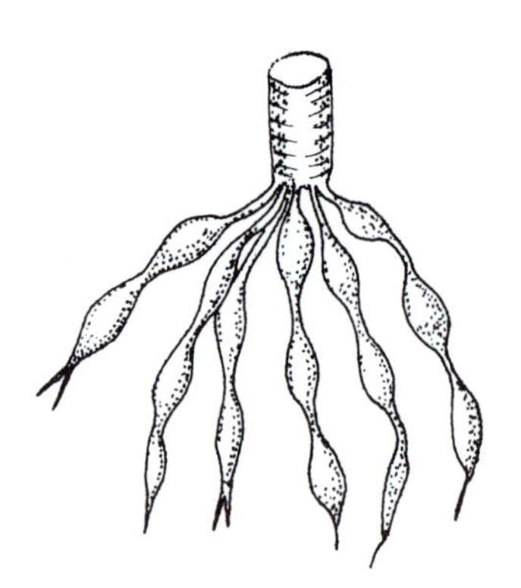

Moniliform roots of *Momordica* sp.

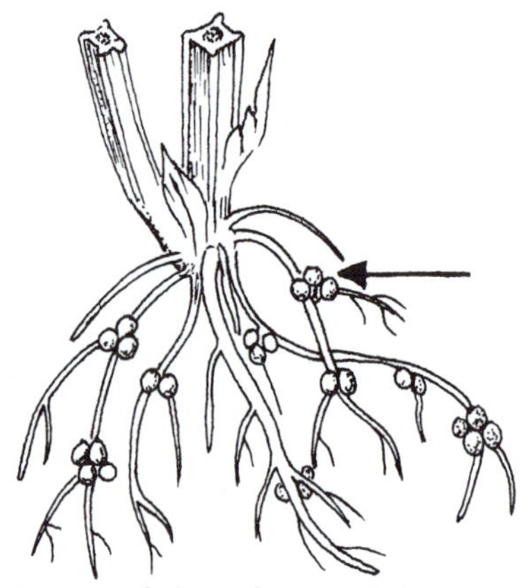

Root nodules of *Vicia faba*

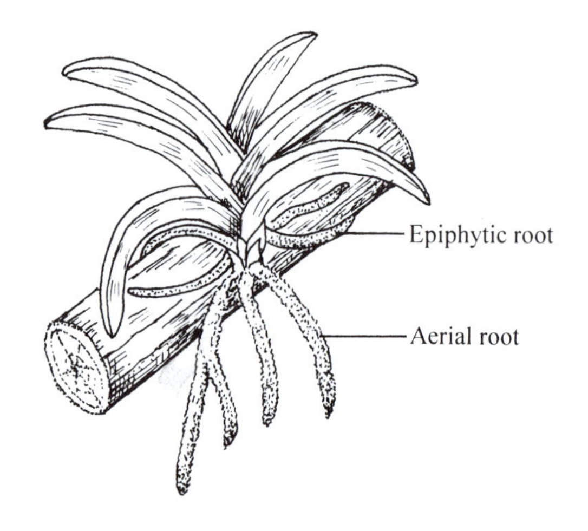

Aerial and epiphytic roots of *Vanda* sp.

Prop roots of *Rhizophora* sp.

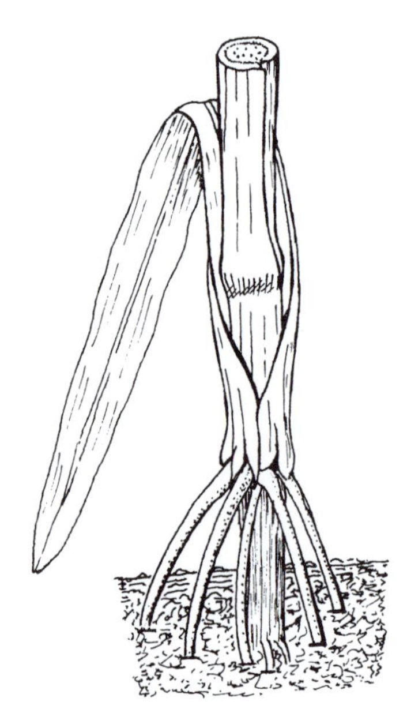

Prop roots of *Zea mays*

Roots 3

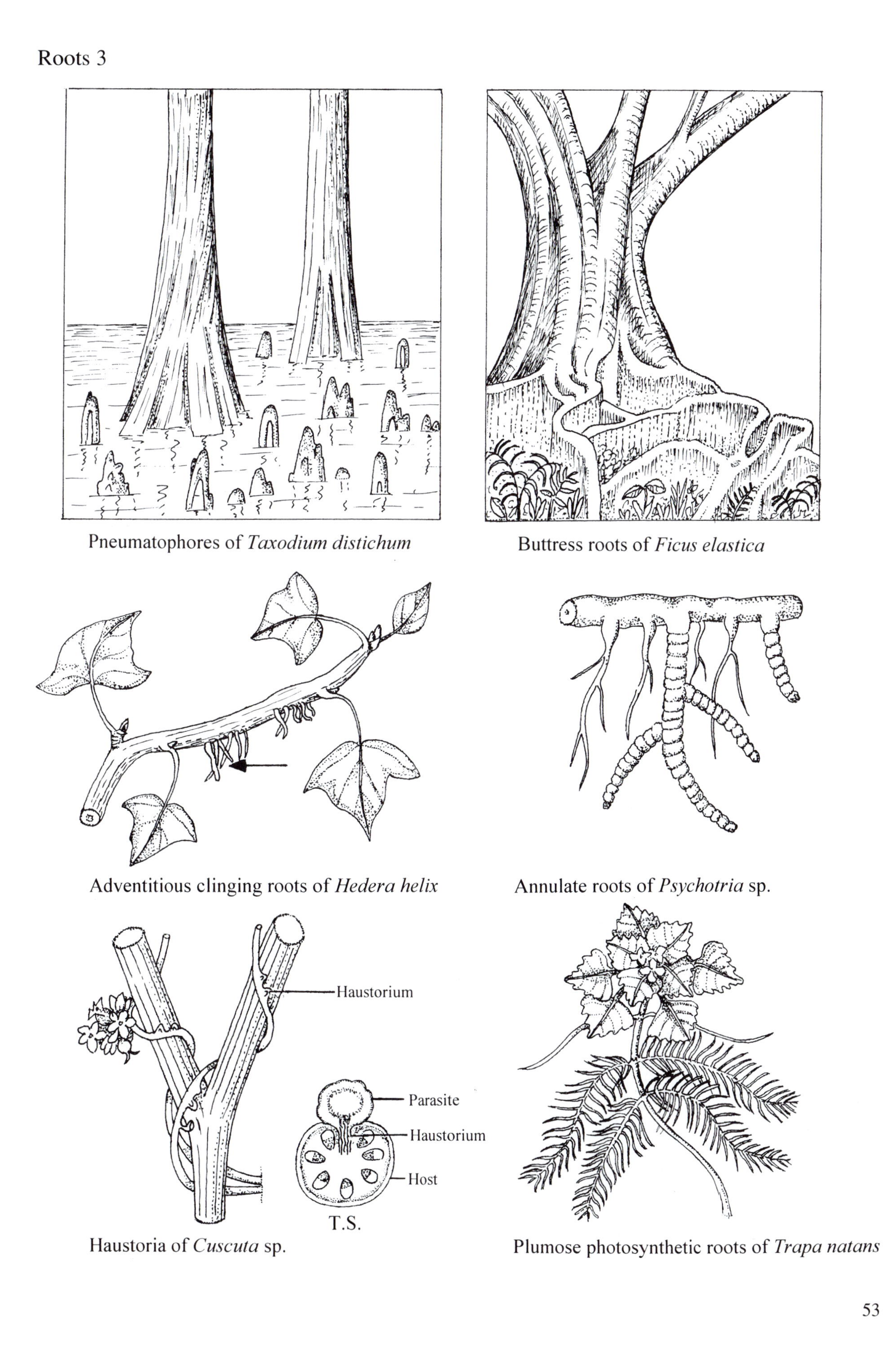

Pneumatophores of *Taxodium distichum*

Buttress roots of *Ficus elastica*

Adventitious clinging roots of *Hedera helix*

Annulate roots of *Psychotria* sp.

Haustoria of *Cuscuta* sp.

Plumose photosynthetic roots of *Trapa natans*

Roots 4

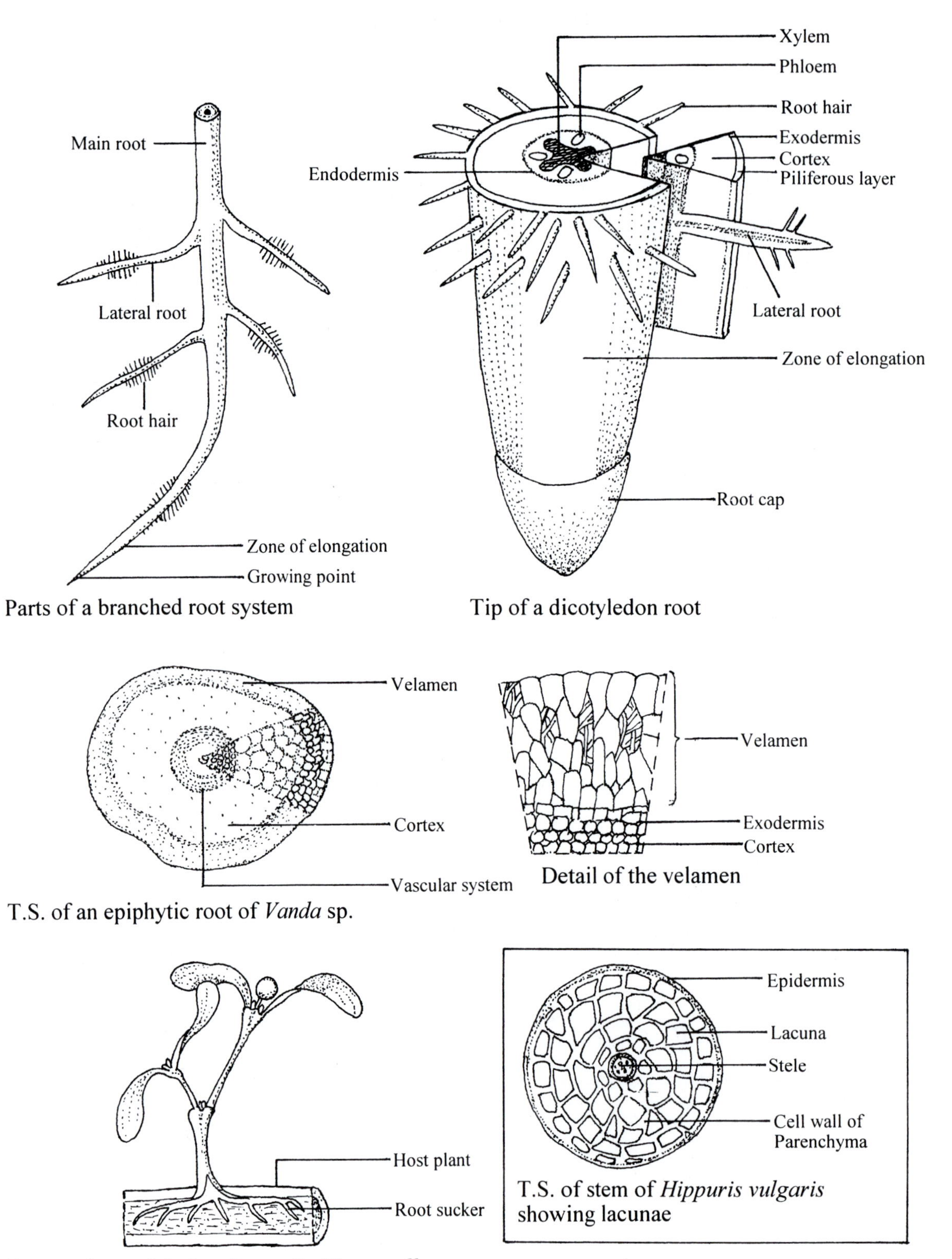

Parts of a branched root system

Tip of a dicotyledon root

T.S. of an epiphytic root of *Vanda* sp.

Detail of the velamen

Root suckers of the hemi-parasite *Viscum album*

T.S. of stem of *Hippuris vulgaris* showing lacunae

Roots 5

DIAGRAMS OF ROOT TYPES

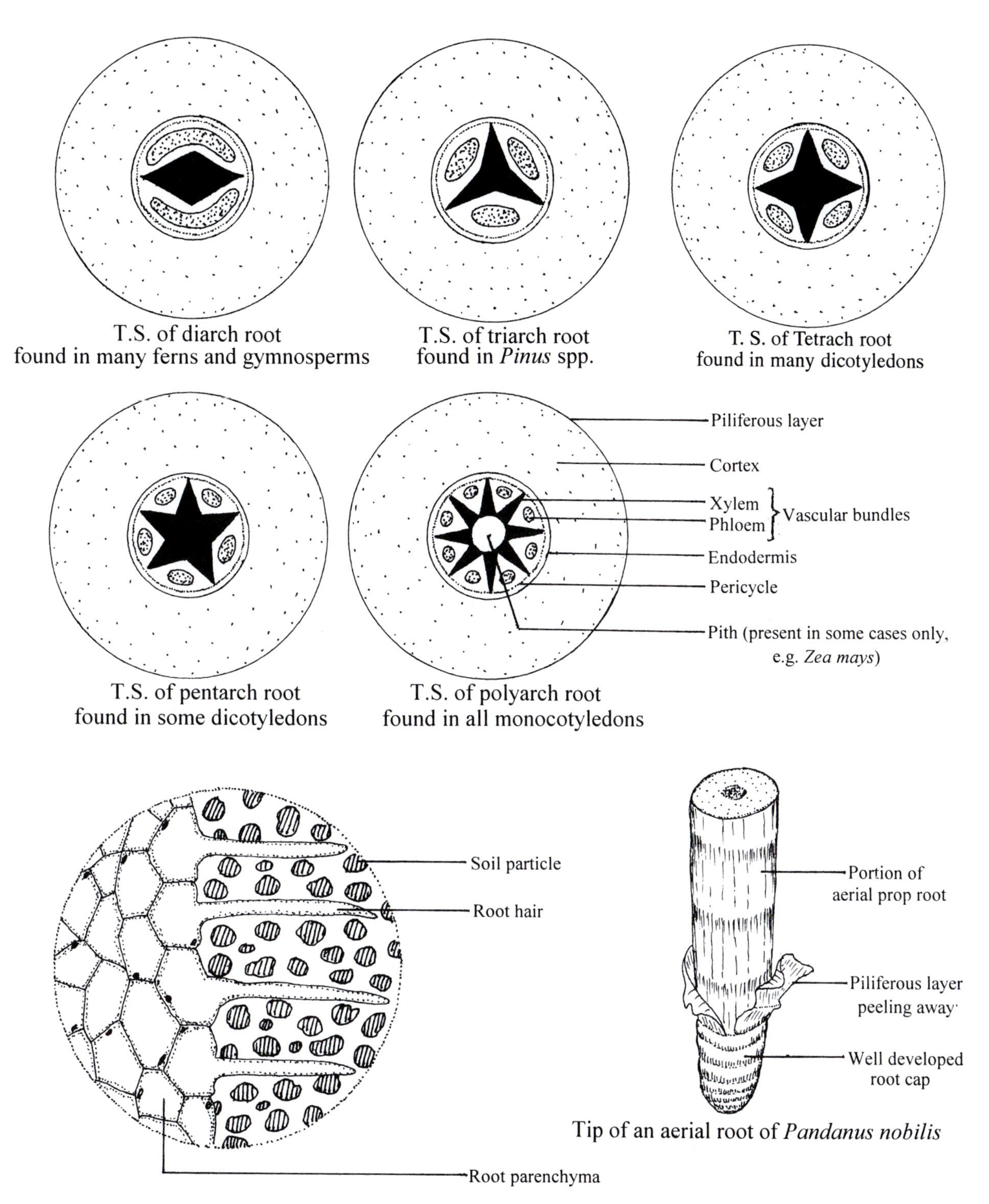

Detailed drawing of root hairs

Tip of an aerial root of *Pandanus nobilis*

Storage Organs 1

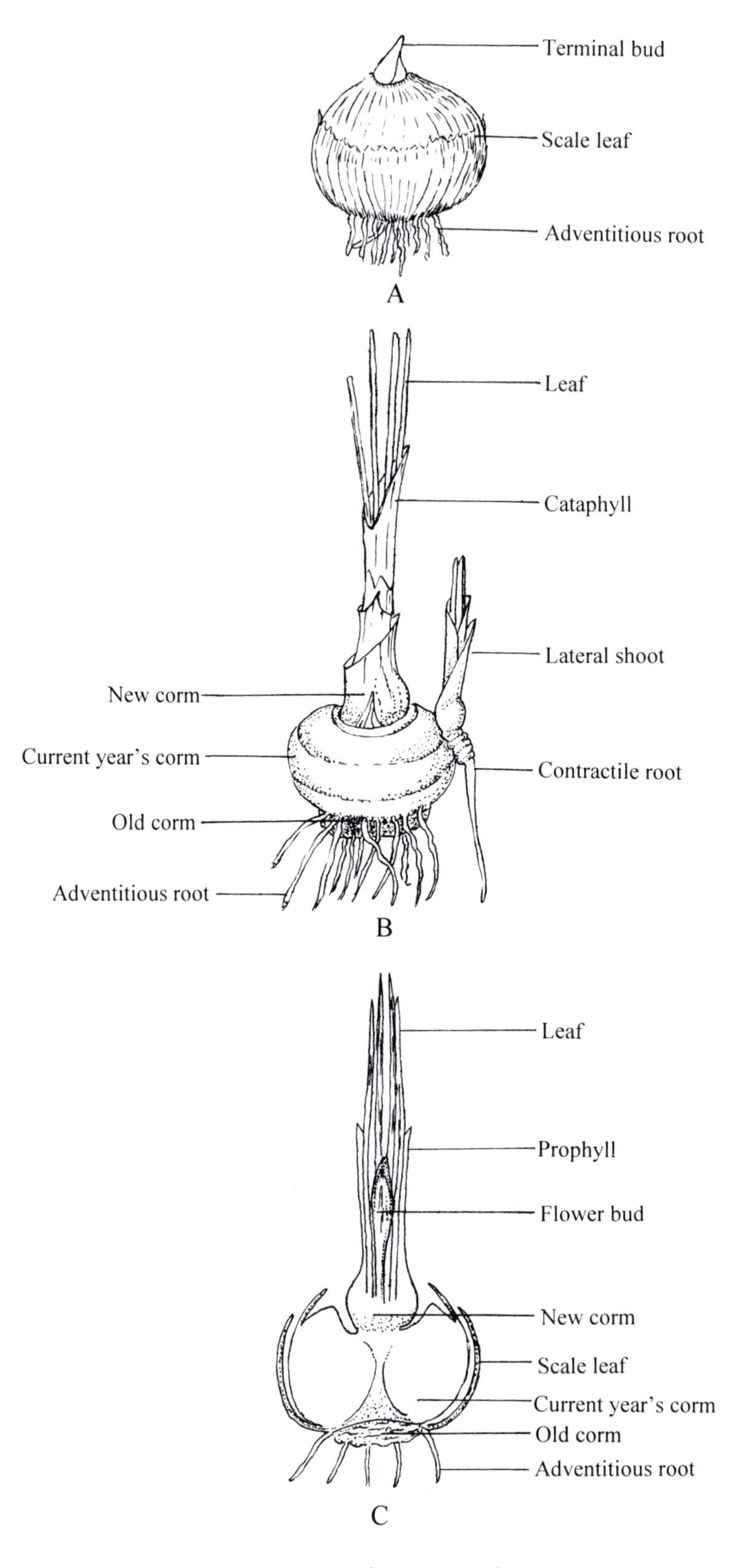

Corm of *Crocus* sp. A. Dormant stage B. Growing stage with scale leaves removed C. L.S. of corm

Storage Organs 2

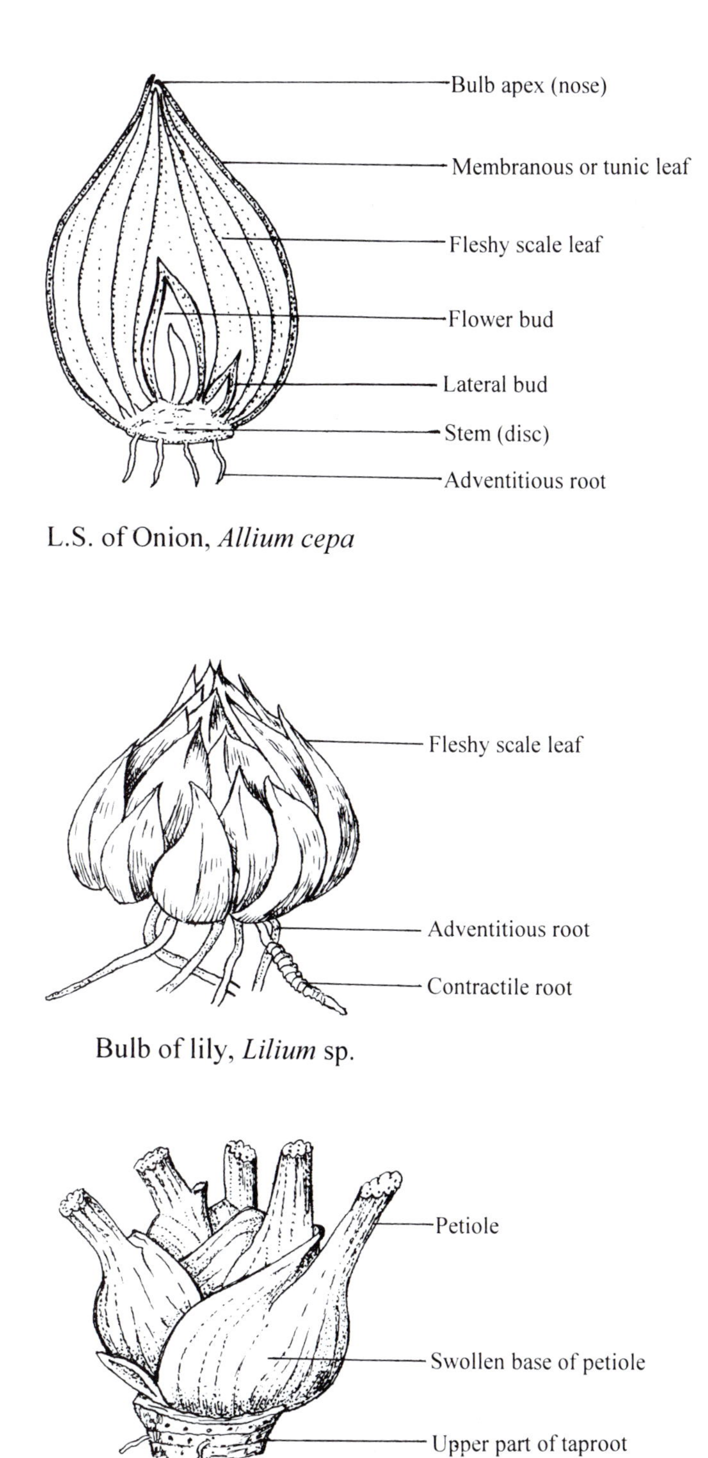

L.S. of Onion, *Allium cepa*

Bulb of lily, *Lilium* sp.

Swollen leaf-bases of *Foeniculum vulgare* var. *dulce*

Storage Organs 3

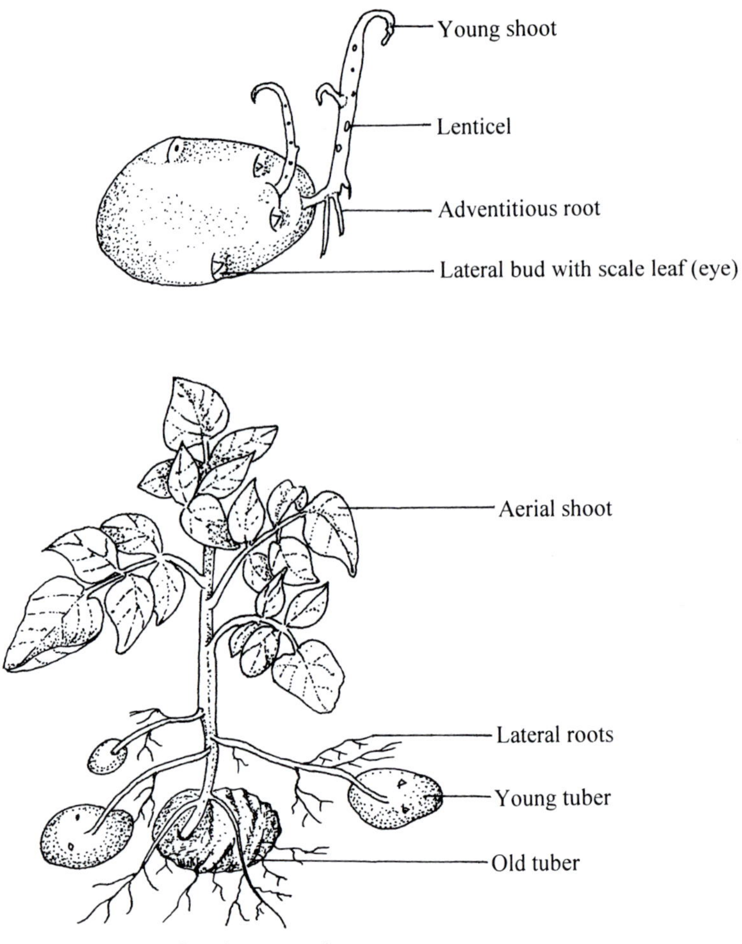

Stem tubers of *Solanum tuberosum*

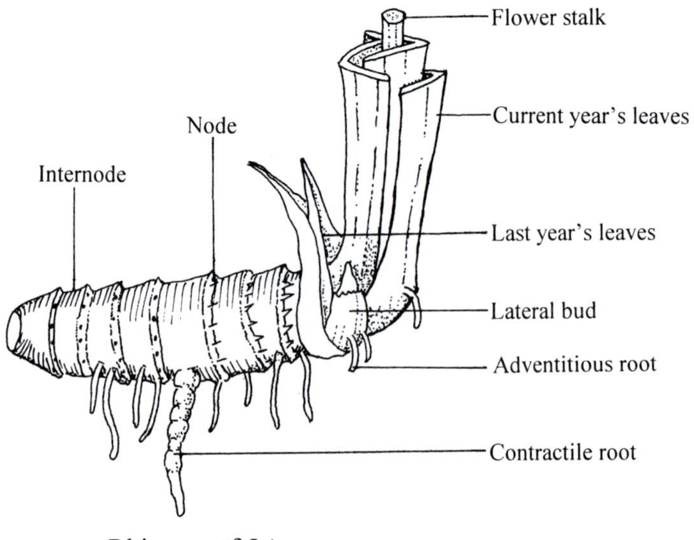

Rhizome of *Iris* sp.

Storage Organs 4

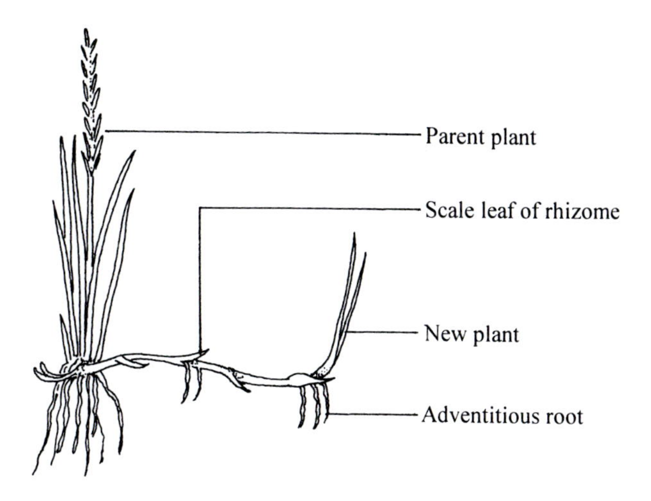

Rhizome of *Agropyron repens*

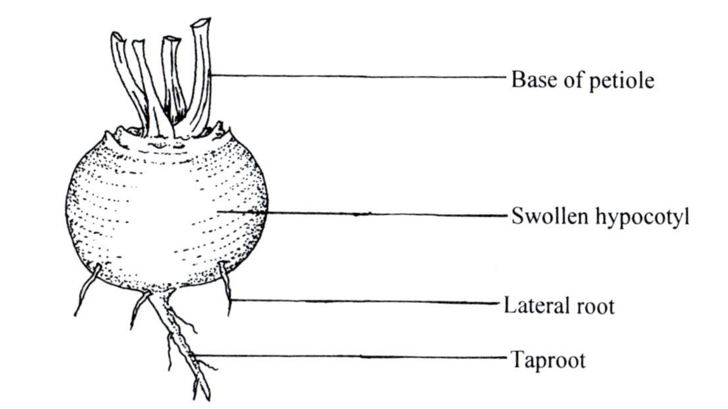

Swollen hypocotyl of *Brassica rapa*

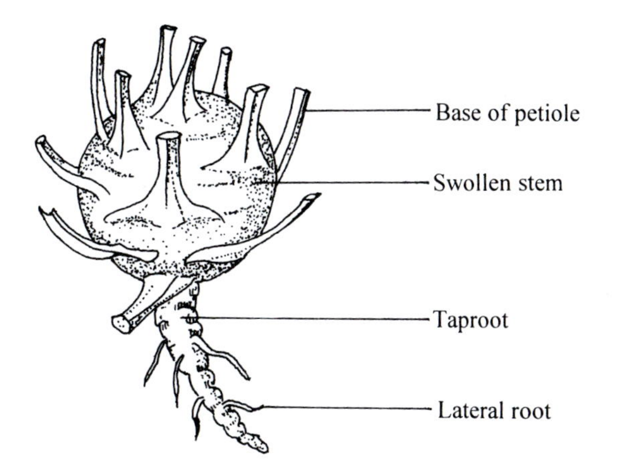

Swollen stem of *Brassica oleracea* var. *gongylodes*

Vegetative Reproduction 1

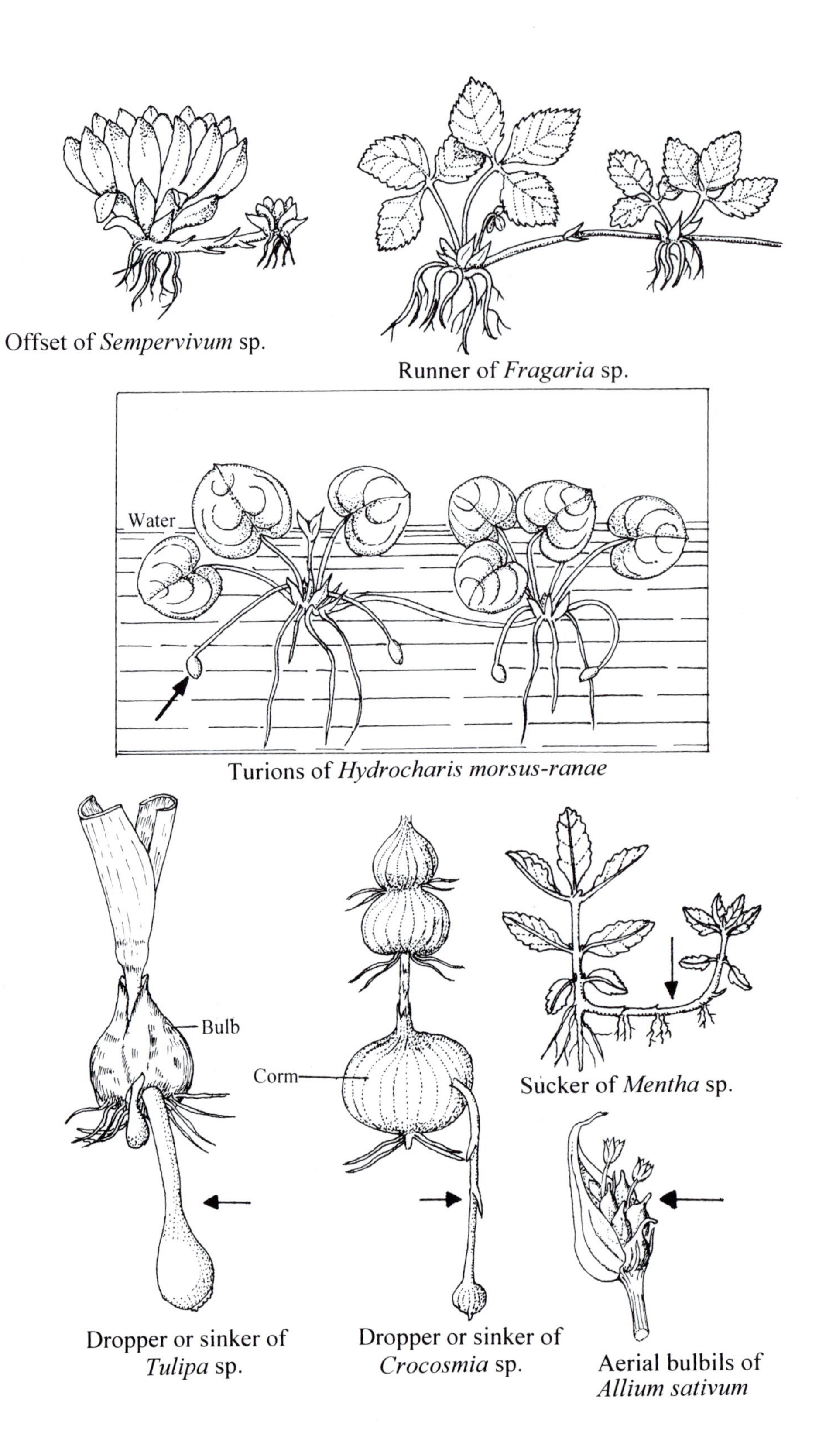

Offset of *Sempervivum* sp.

Runner of *Fragaria* sp.

Turions of *Hydrocharis morsus-ranae*

Dropper or sinker of *Tulipa* sp.

Dropper or sinker of *Crocosmia* sp.

Sucker of *Mentha* sp.

Aerial bulbils of *Allium sativum*

Vegetative Reproduction 2

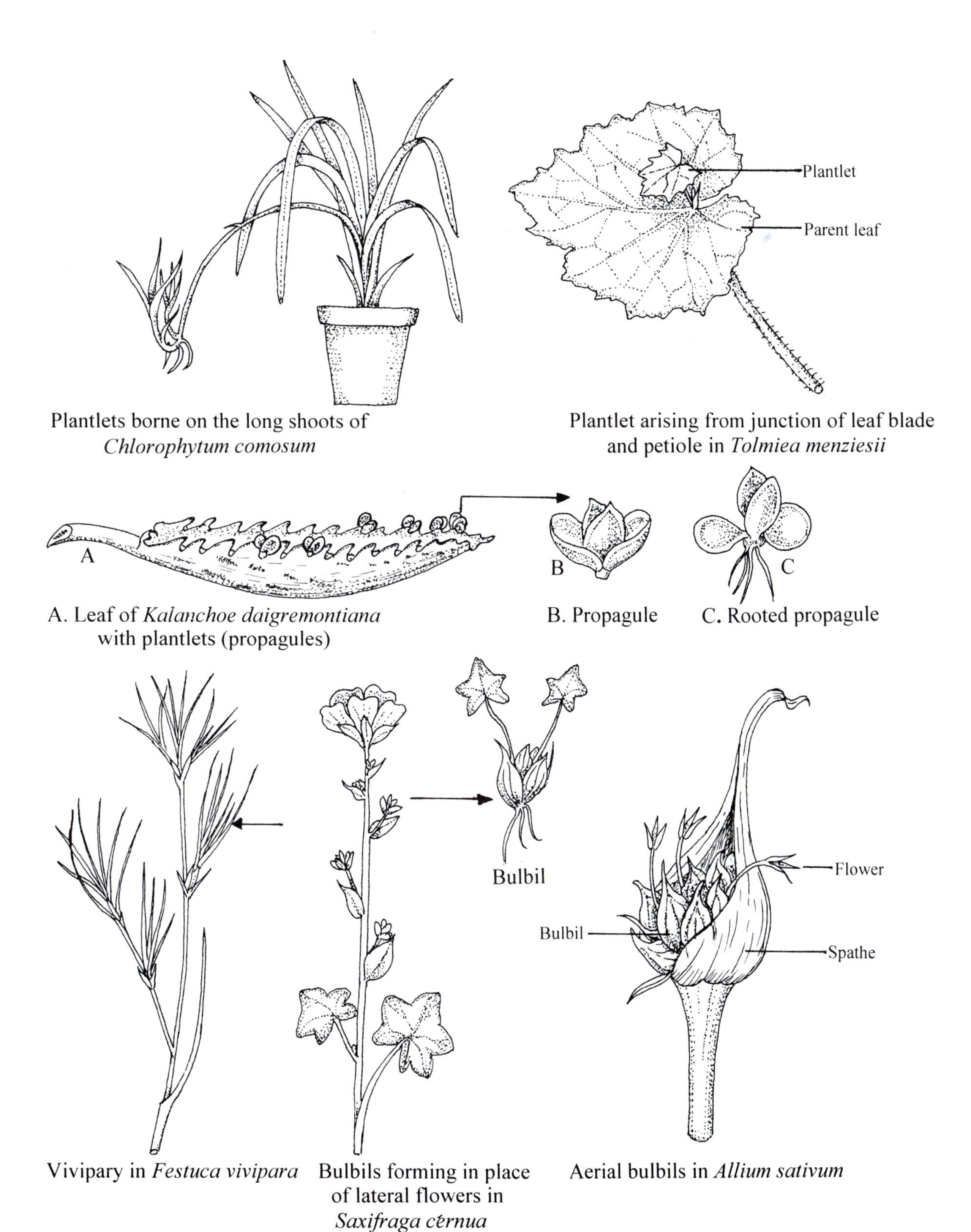

Plantlets borne on the long shoots of *Chlorophytum comosum*

Plantlet arising from junction of leaf blade and petiole in *Tolmiea menziesii*

A. Leaf of *Kalanchoe daigremontiana* with plantlets (propagules)

B. Propagule

C. Rooted propagule

Vivipary in *Festuca vivipara*

Bulbils forming in place of lateral flowers in *Saxifraga cernua*

Aerial bulbils in *Allium sativum*

Vegetative Reproduction 2

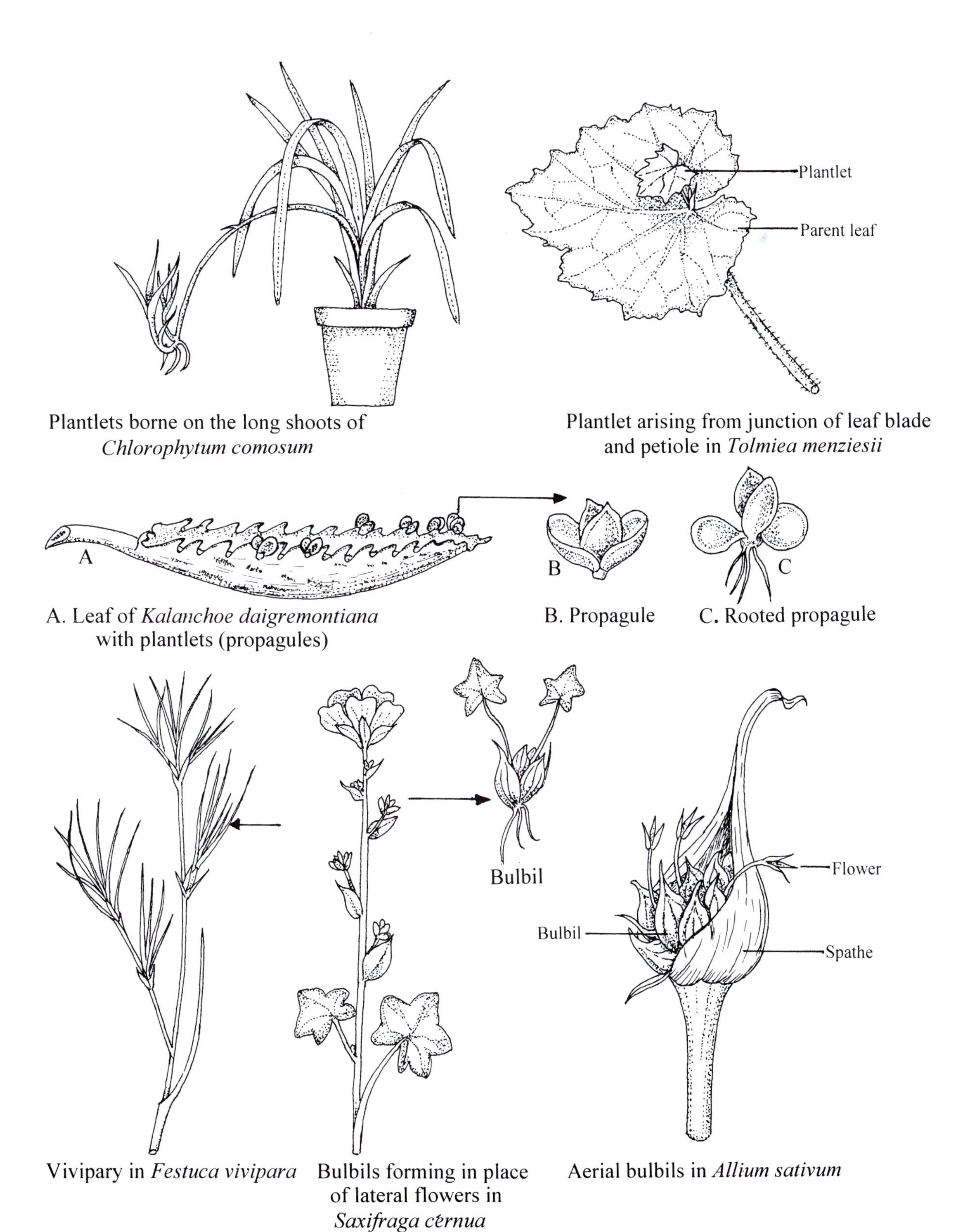

Plantlets borne on the long shoots of *Chlorophytum comosum*

Plantlet arising from junction of leaf blade and petiole in *Tolmiea menziesii*

A. Leaf of *Kalanchoe daigremontiana* with plantlets (propagules)

B. Propagule

C. Rooted propagule

Vivipary in *Festuca vivipara*

Bulbils forming in place of lateral flowers in *Saxifraga cernua*

Aerial bulbils in *Allium sativum*

2. Seeds and Seedlings

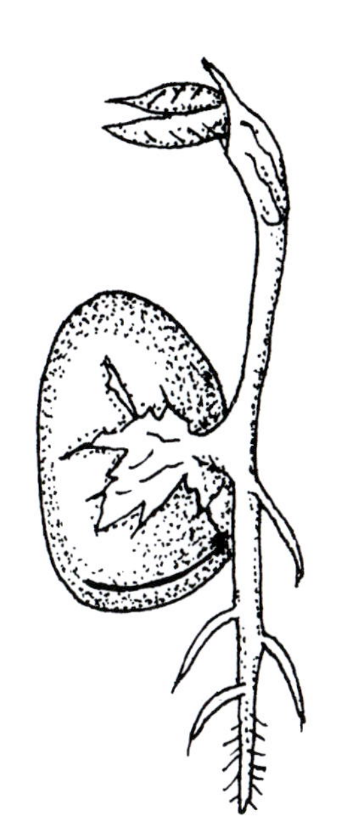

Seeds and Seedlings 1

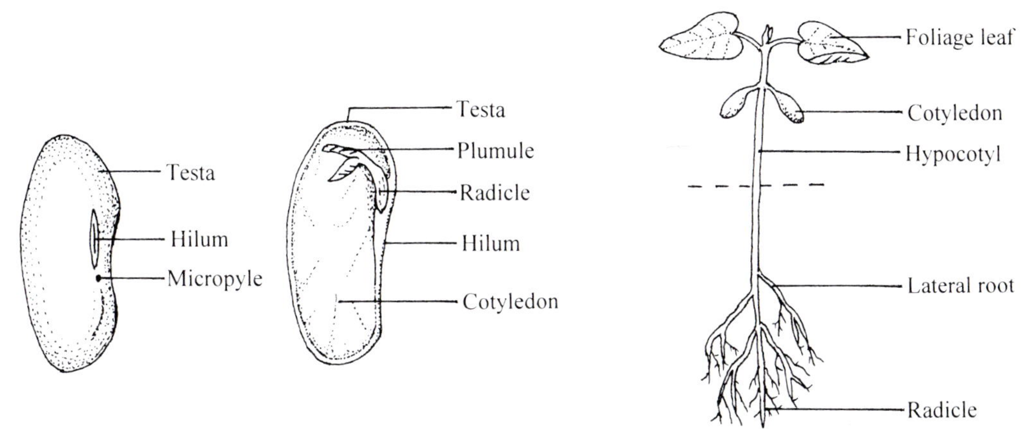

Epigeal germination and seed structure of *Phaseolus vulgaris*

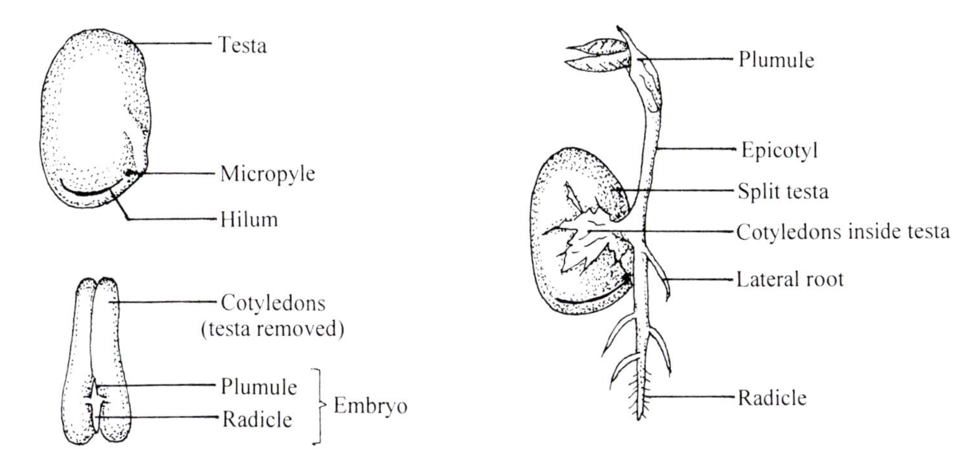

Hypogeal germination and seed structure of *Vicia faba*

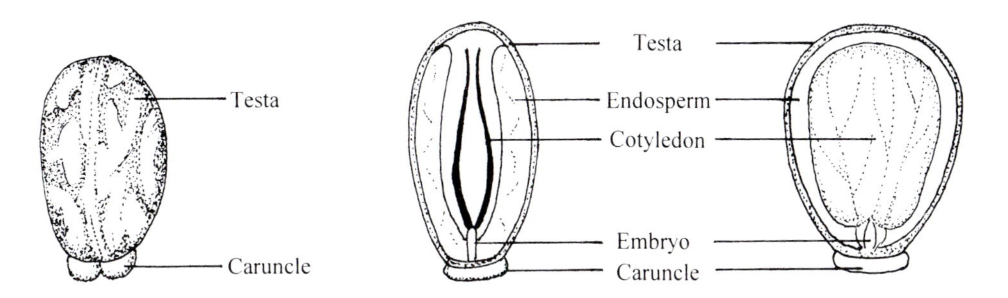

Seed structure of *Ricinus communis*

Seeds and Seedlings 2

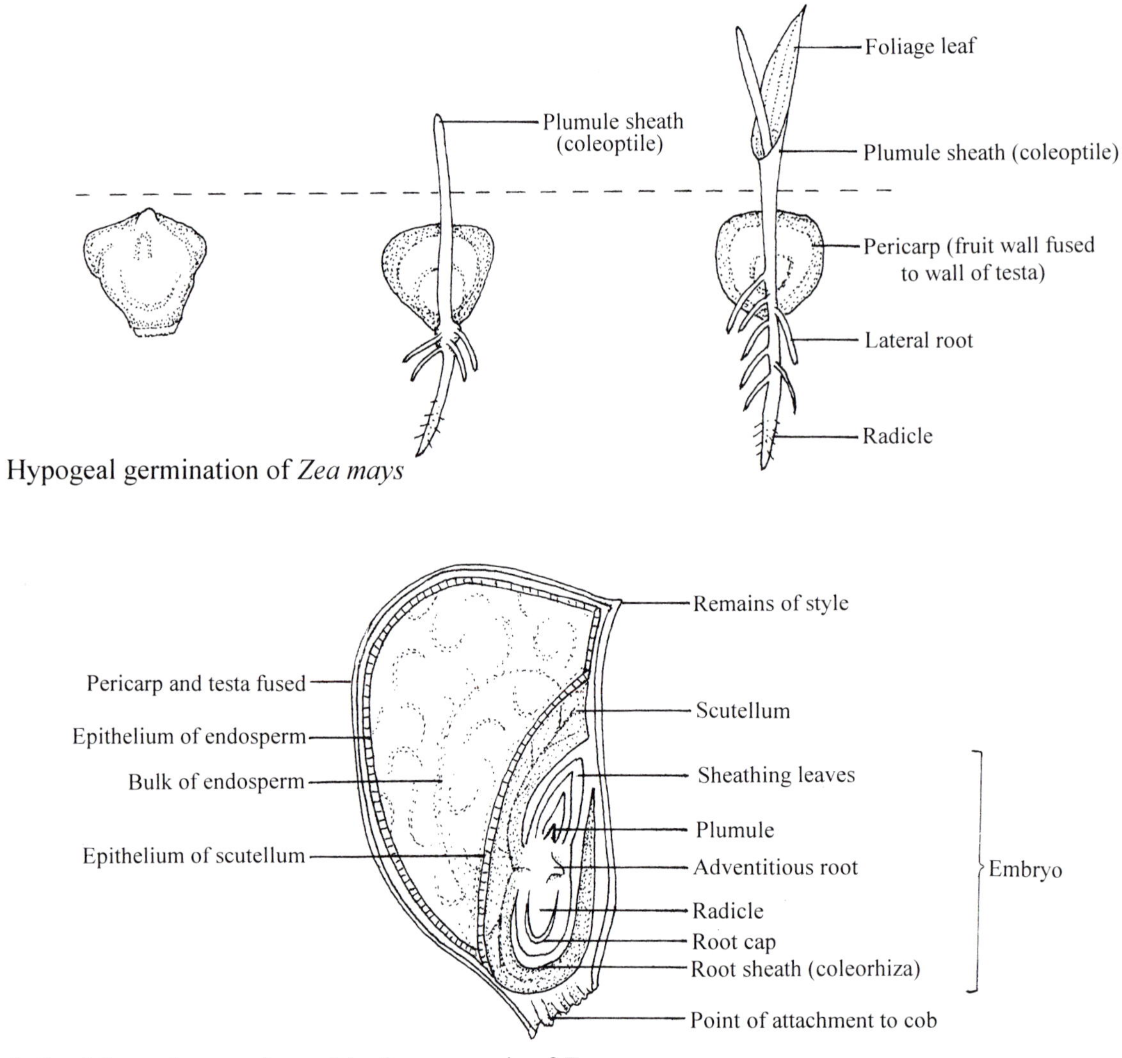

Hypogeal germination of *Zea mays*

L.S. of the endospermic seed in the caryopsis of *Zea mays*

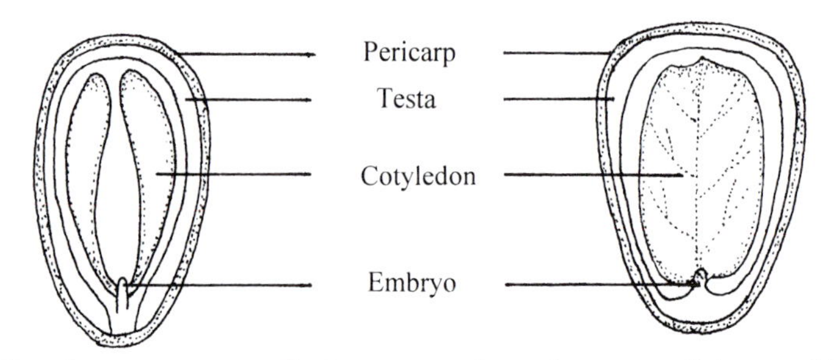

L.S. of fruit of *Helianthus annuus,* cut in 2 ways to show the testa of the seed joined to the pericarp

Seeds and Seedlings 3

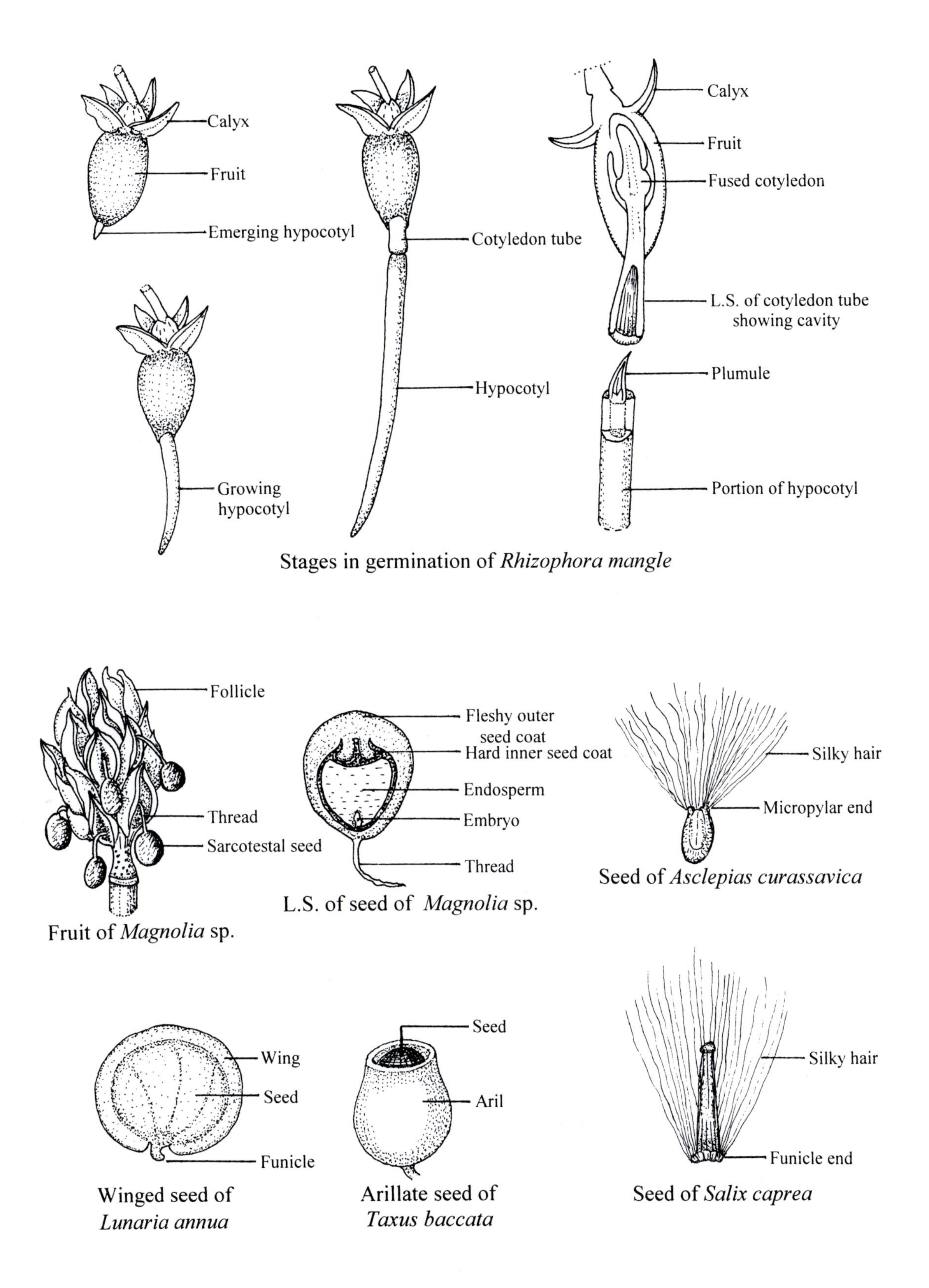

Stages in germination of *Rhizophora mangle*

Fruit of *Magnolia* sp.

L.S. of seed of *Magnolia* sp.

Seed of *Asclepias curassavica*

Winged seed of *Lunaria annua*

Arillate seed of *Taxus baccata*

Seed of *Salix caprea*

Seeds and Seedlings 4

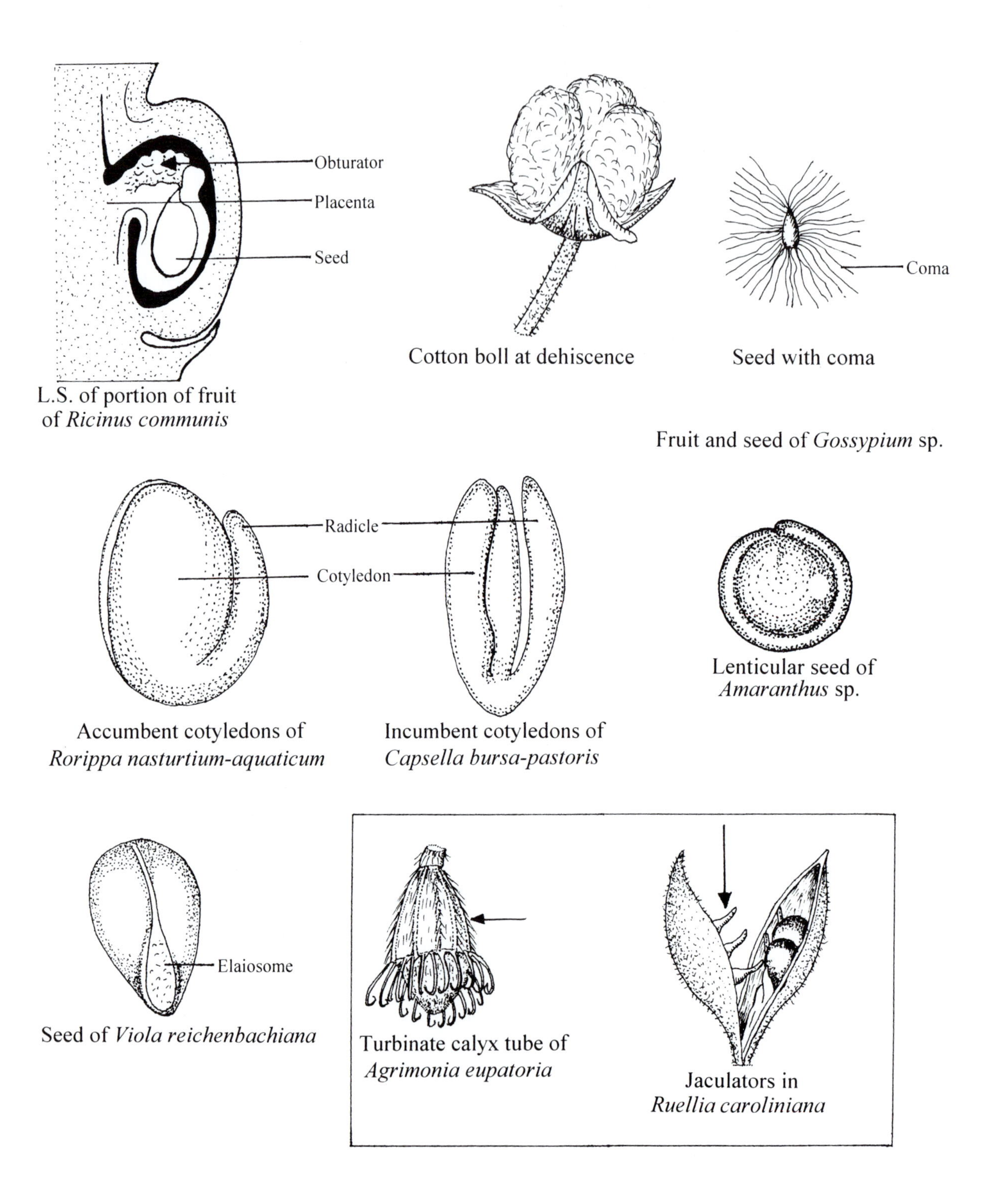

L.S. of portion of fruit of *Ricinus communis*

Cotton boll at dehiscence

Seed with coma

Fruit and seed of *Gossypium* sp.

Accumbent cotyledons of *Rorippa nasturtium-aquaticum*

Incumbent cotyledons of *Capsella bursa-pastoris*

Lenticular seed of *Amaranthus* sp.

Seed of *Viola reichenbachiana*

Turbinate calyx tube of *Agrimonia eupatoria*

Jaculators in *Ruellia caroliniana*

3. Growth and Life Forms

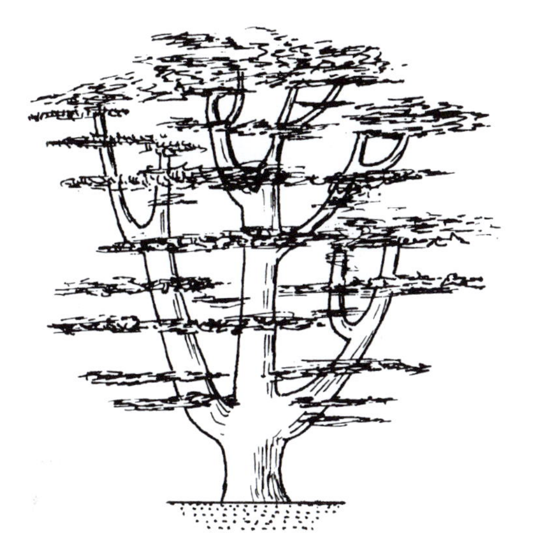

Growth Forms 1

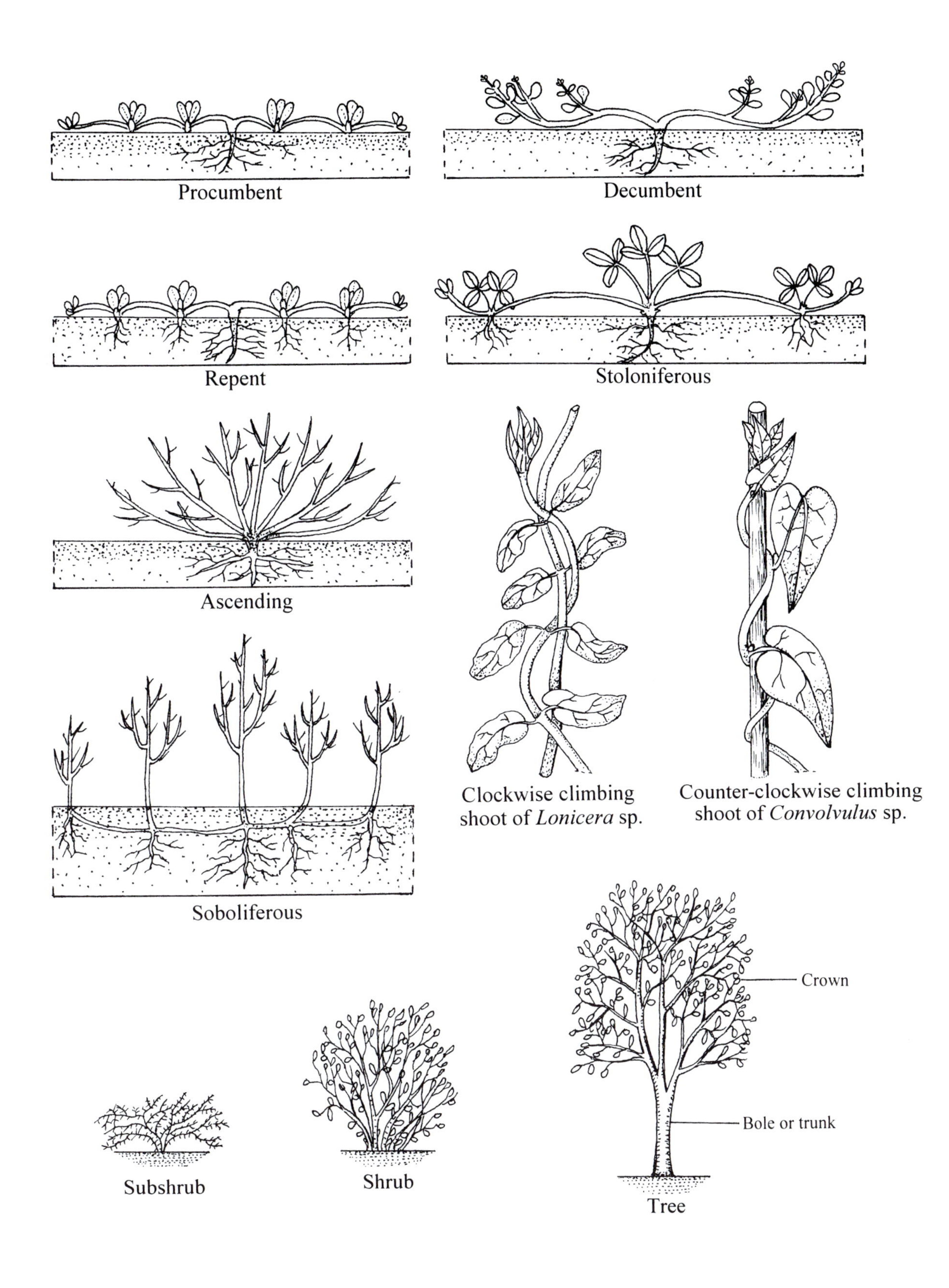

Growth Forms 2

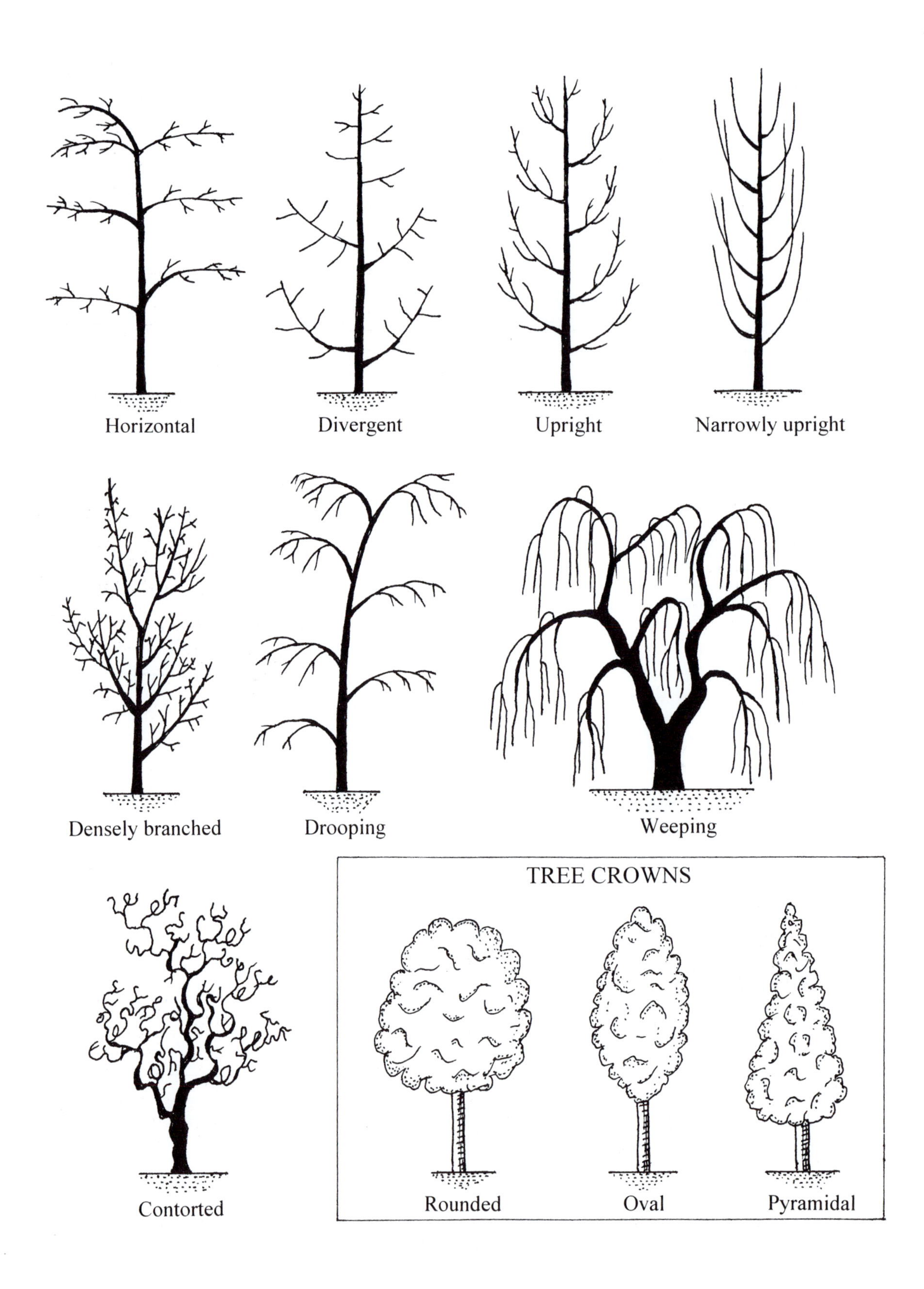

Growth Forms 3

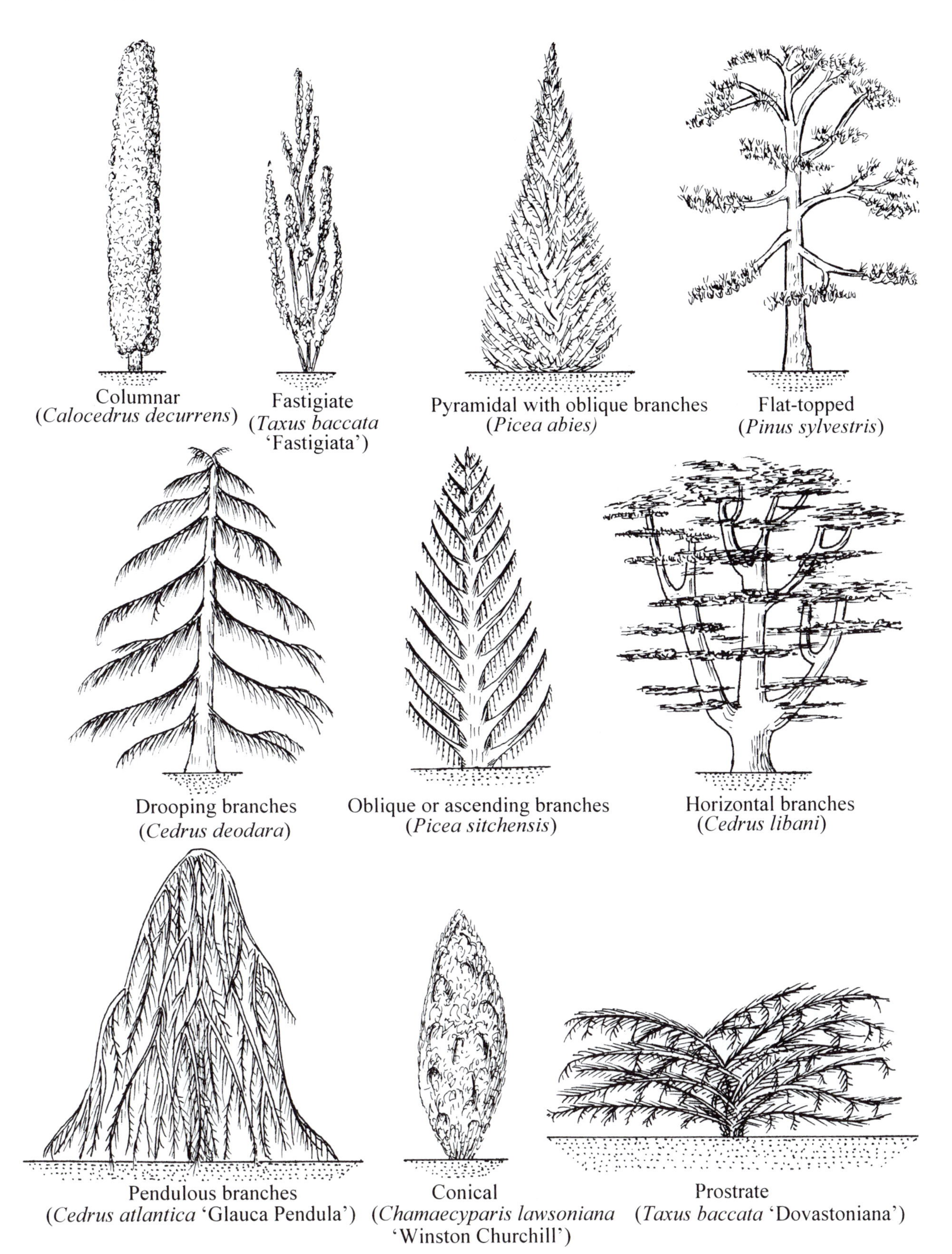

Columnar (*Calocedrus decurrens*)

Fastigiate (*Taxus baccata* 'Fastigiata')

Pyramidal with oblique branches (*Picea abies*)

Flat-topped (*Pinus sylvestris*)

Drooping branches (*Cedrus deodara*)

Oblique or ascending branches (*Picea sitchensis*)

Horizontal branches (*Cedrus libani*)

Pendulous branches (*Cedrus atlantica* 'Glauca Pendula')

Conical (*Chamaecyparis lawsoniana* 'Winston Churchill')

Prostrate (*Taxus baccata* 'Dovastoniana')

Growth Forms 4

Acaulescent (*Cirsium acaule*)

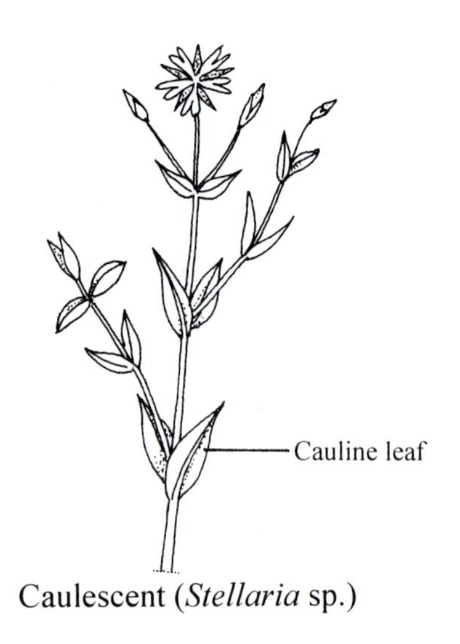

Caulescent (*Stellaria* sp.)

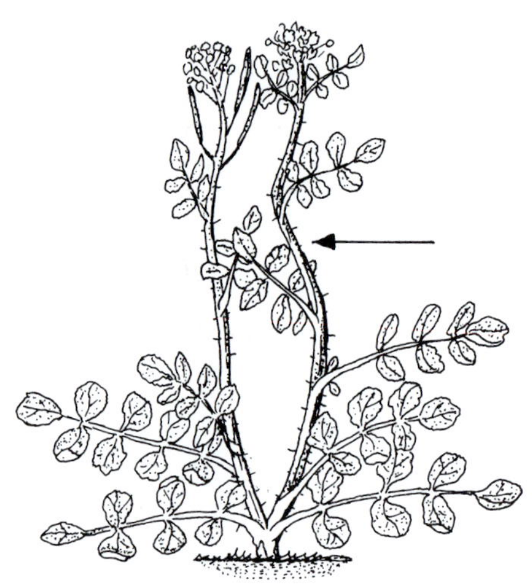

Flexuous (*Cardamine flexuosa*)

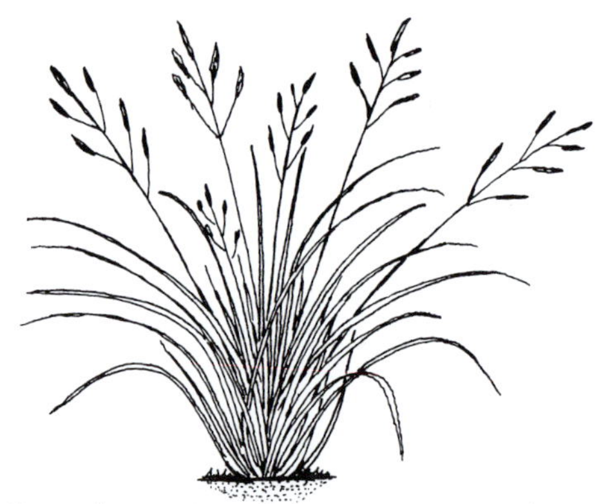

Caespitose (*Deschampsia cespitosa*)

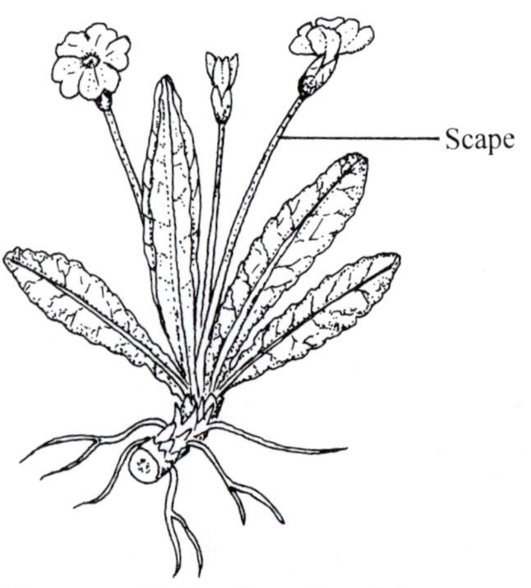

Scapose (*Primula vulgaris*)

Burs with epicormic shoots on *Tilia* x *europaea*

Life Forms (Raunkiaer's Classification)

The perennating parts of plants are shown in black

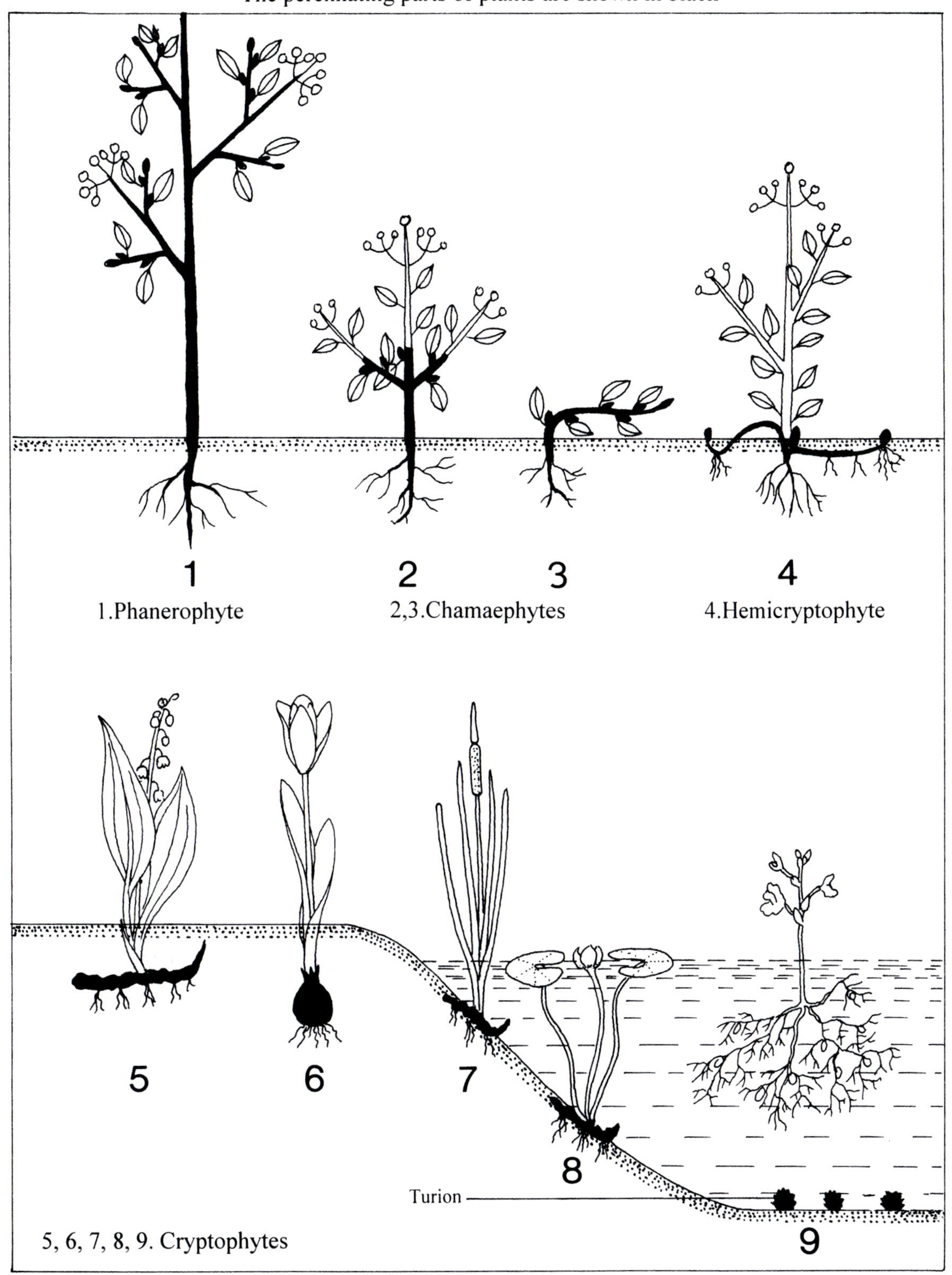

4. General Features of Flowering Plants

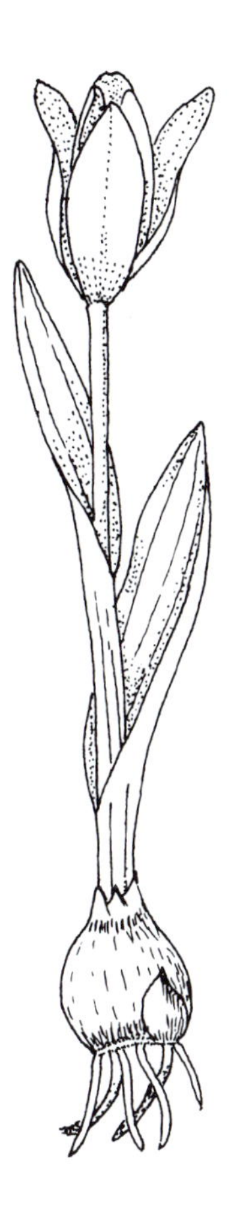

Parts of a Typical Monocotyledon

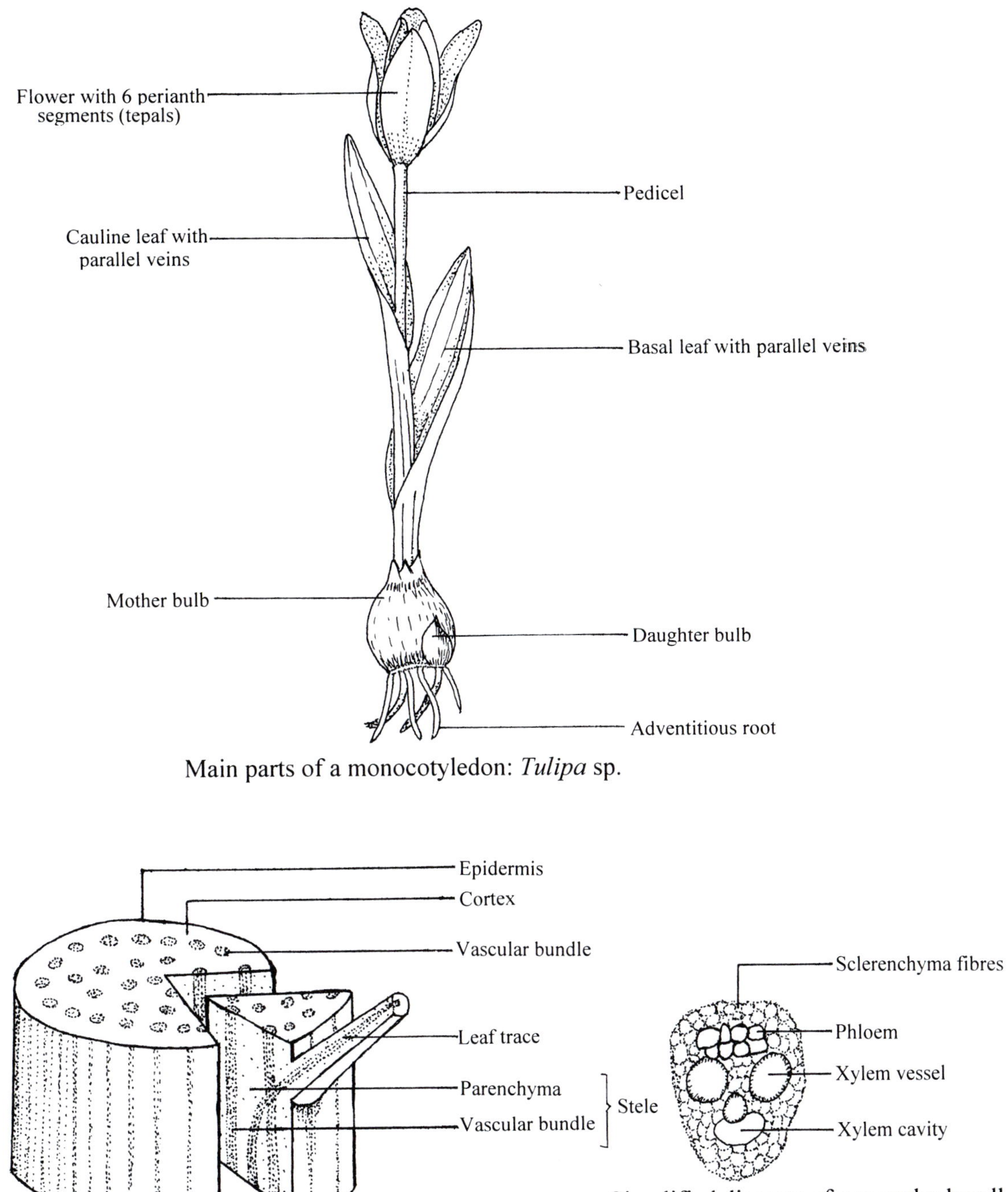

Main parts of a monocotyledon: *Tulipa* sp.

Structure of a monocotyledon stem: *Zea mays*

Simplified diagram of a vascular bundle of *Zea mays* (x 300)

Parts of a Typical Herbaceous Dicotyledon

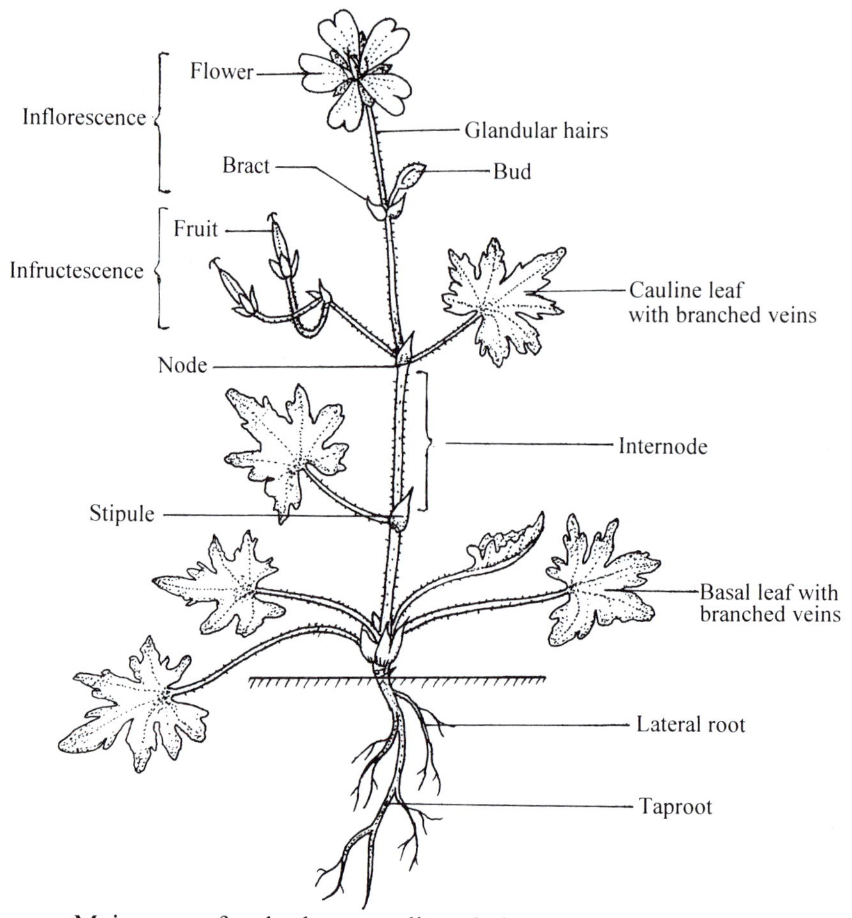

Main parts of an herbaceous dicotyledon: *Geranium* sp.

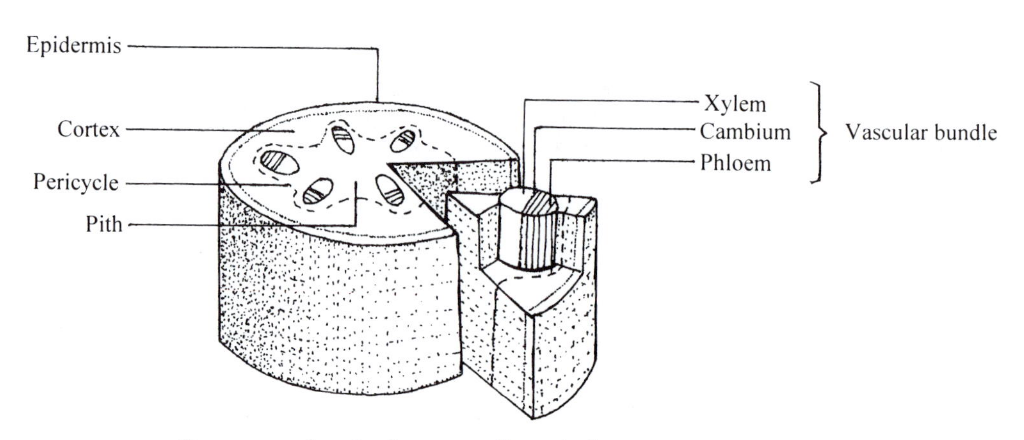

Structure of an herbaceous dicotyledon stem

Parts of a Winter Twig

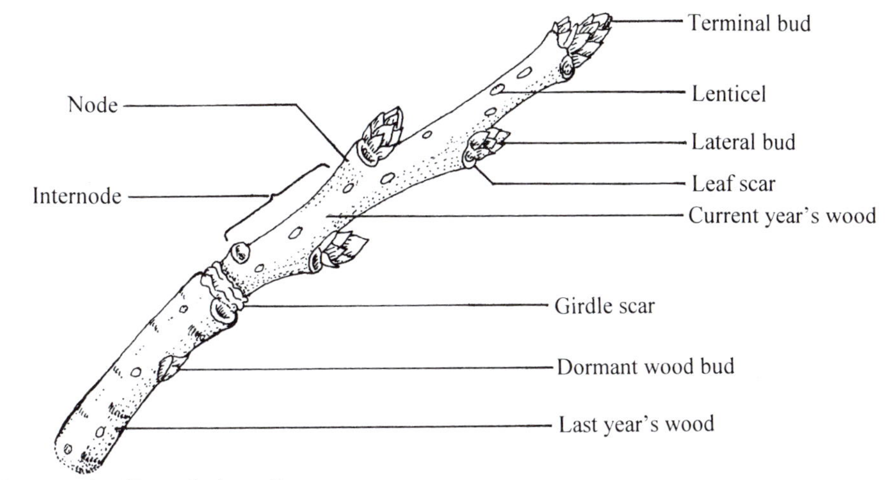

Stem of a woody dicotyledon: *Prunus* sp.

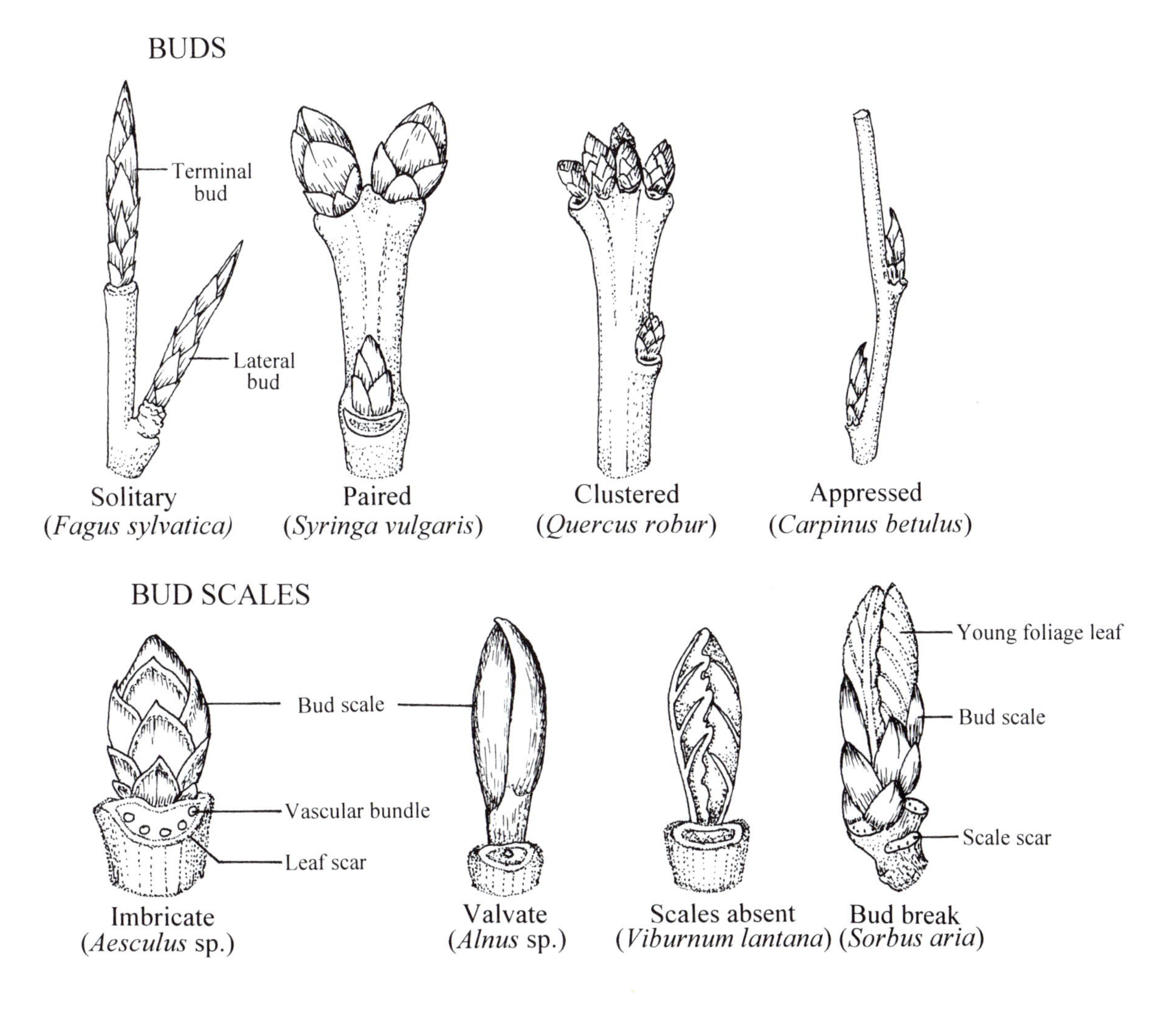

Structure of a Woody Dicotyledon Stem

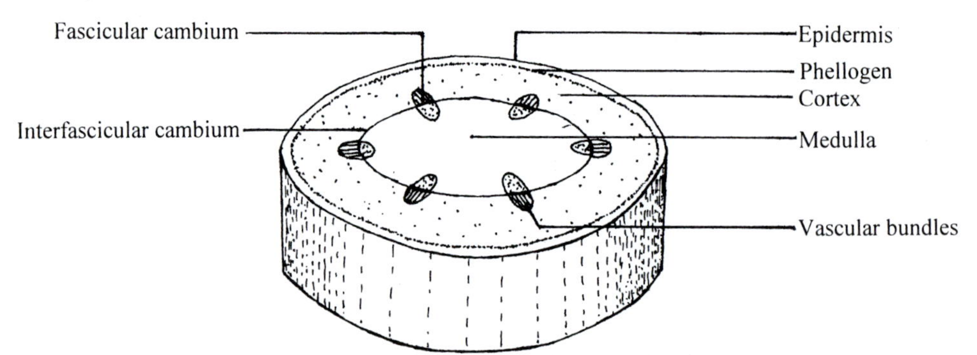

T.S. of a dicotyledon stem at the start of secondary thickening

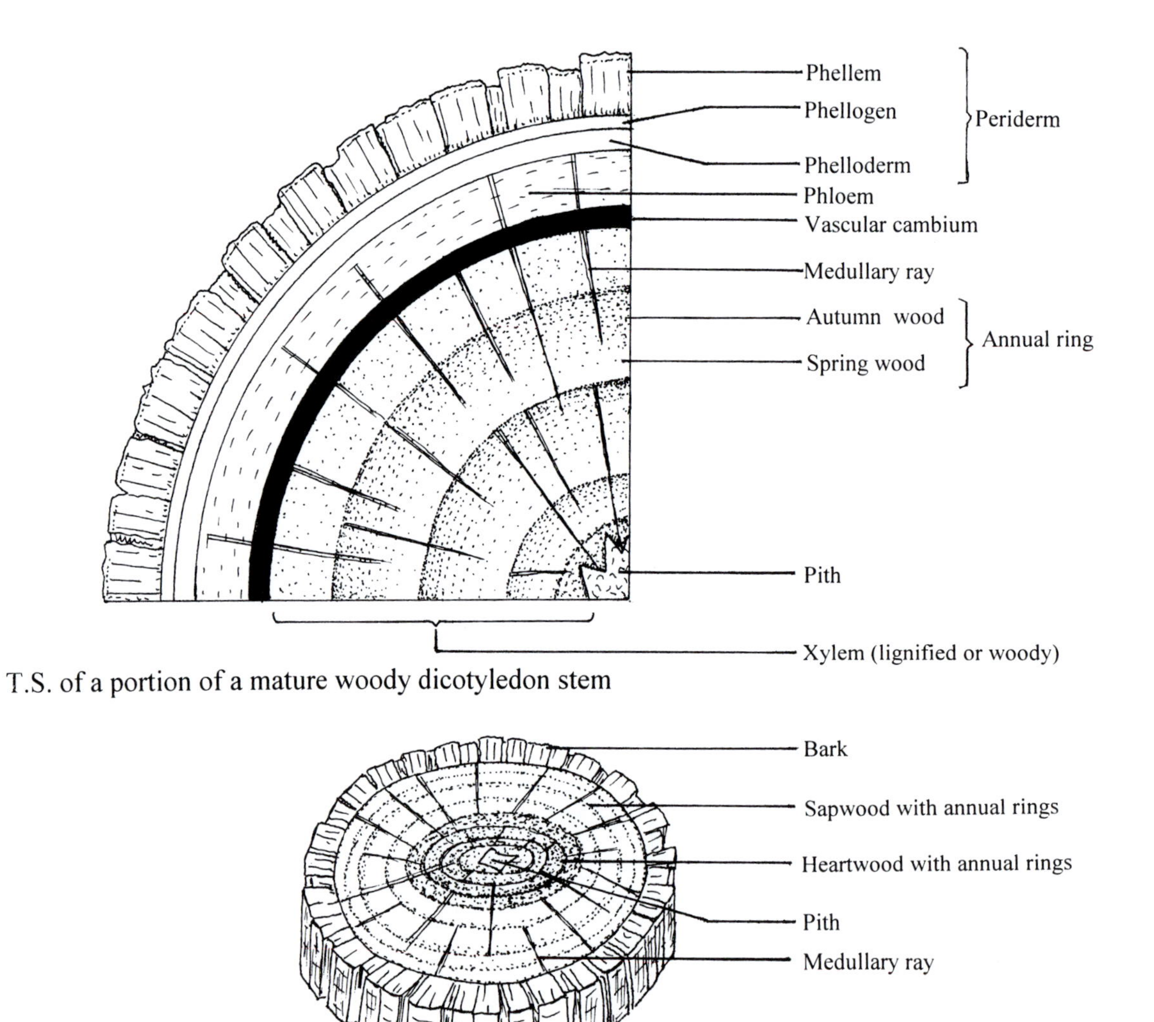

T.S. of a portion of a mature woody dicotyledon stem

Gross structure of a log of wood

5. Plant Features and Responses

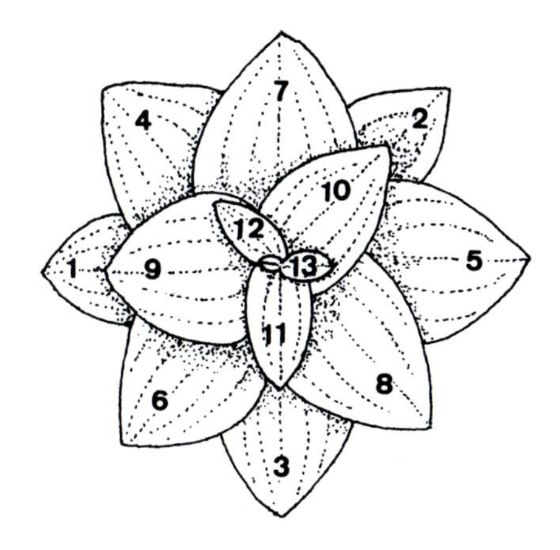

Stem Shapes, Leaf Fall, Apical Meristem, Phototropism

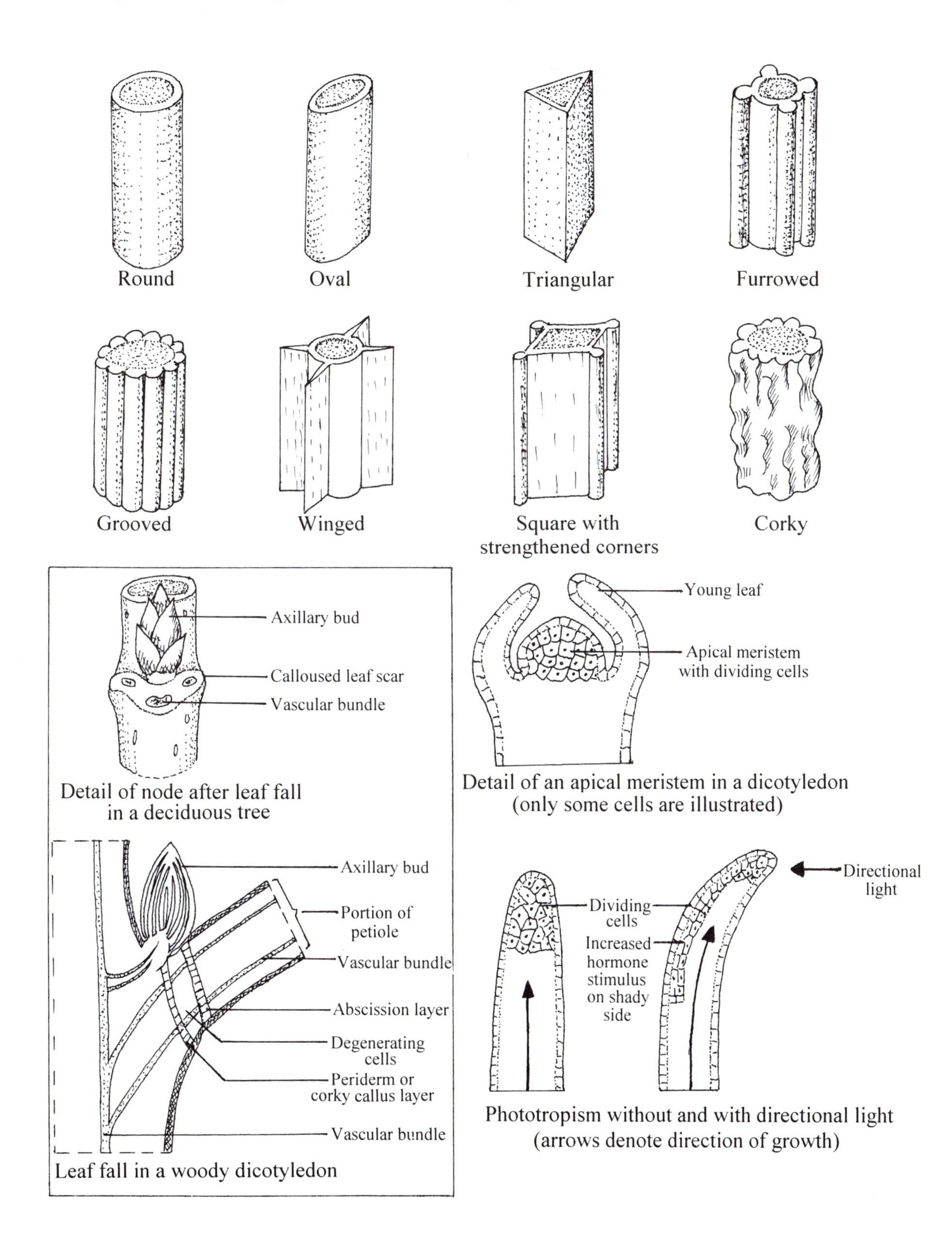

Plant Responses

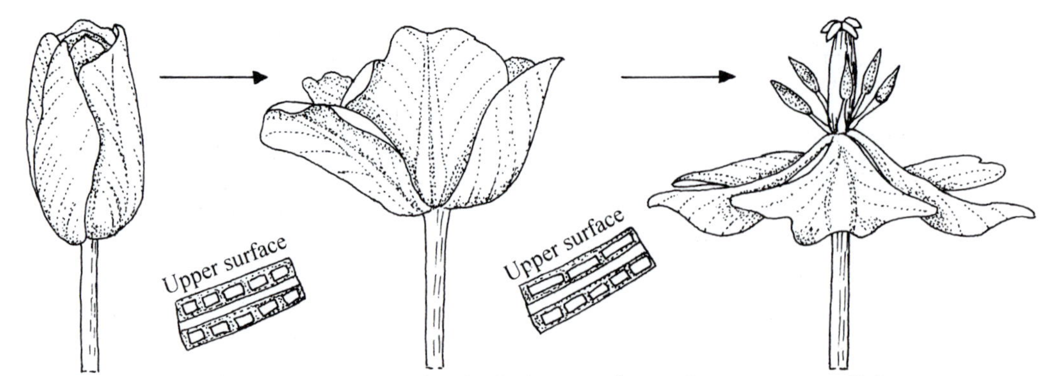

Thermonasty. Stages in flower opening in *Tulipa* sp. in cool to warm conditions

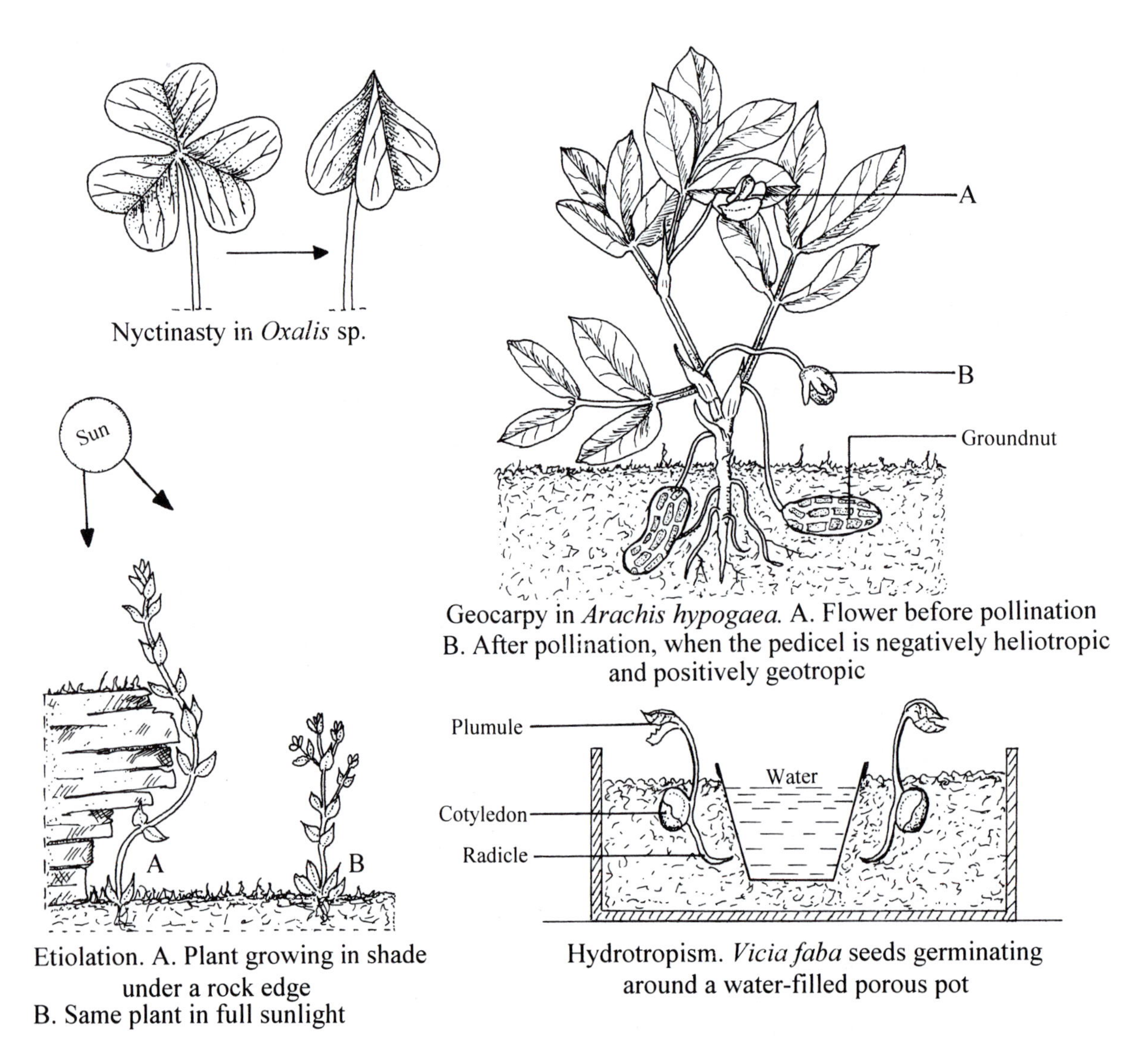

Nyctinasty in *Oxalis* sp.

Geocarpy in *Arachis hypogaea.* A. Flower before pollination
B. After pollination, when the pedicel is negatively heliotropic and positively geotropic

Etiolation. A. Plant growing in shade under a rock edge
B. Same plant in full sunlight

Hydrotropism. *Vicia faba* seeds germinating around a water-filled porous pot

Phyllotaxy, Fibonacci Series, Leaf Mosaic, Heterophylly

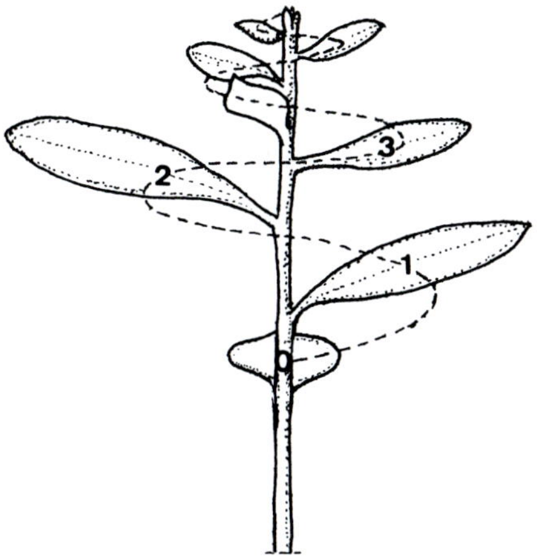

Phyllotaxy–the ideal spiral arrangement for leaves and branches (facial view)

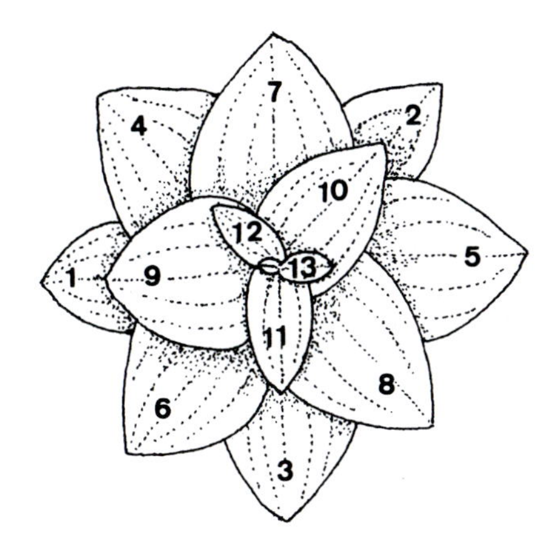

Leaf rosette of *Plantago major* showing phyllotaxy

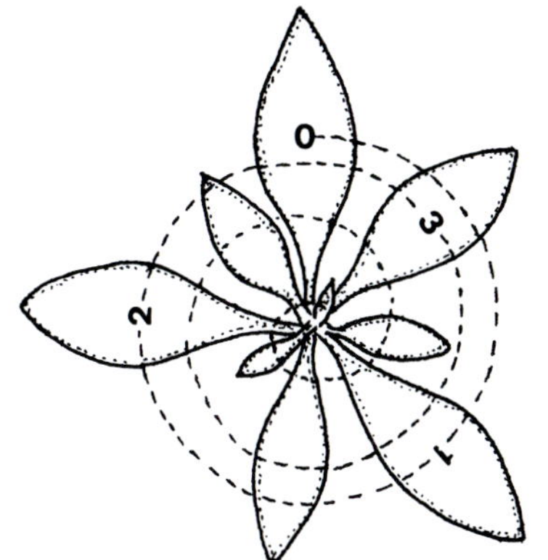

Phyllotaxy–the ideal spiral arrangement for leaves and branches (aerial view)

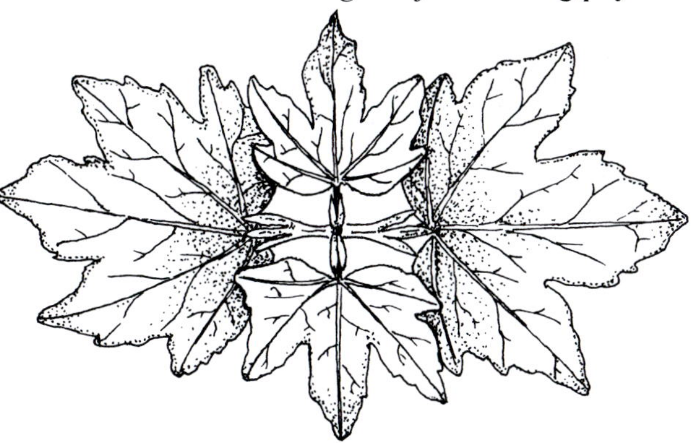

Leaf mosaic of *Acer pseudoplatanus*

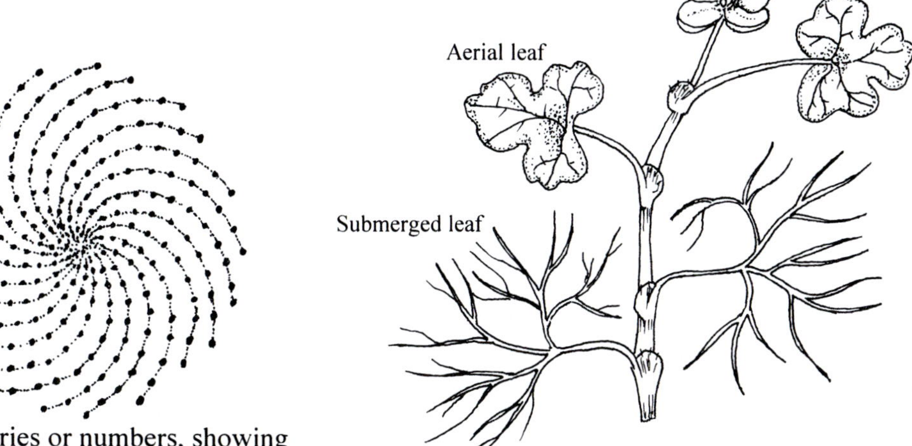

Fibonacci series or numbers, showing arrangement of disc florets in *Helianthus annuus* (each dot represents a floret)

Heterophylly in the aquatic plant *Ranunculus aquatilis*

6. Leaf-like Structures and Other Vegetative Features

Petioles, Stipules, Spines, Tendrils 1

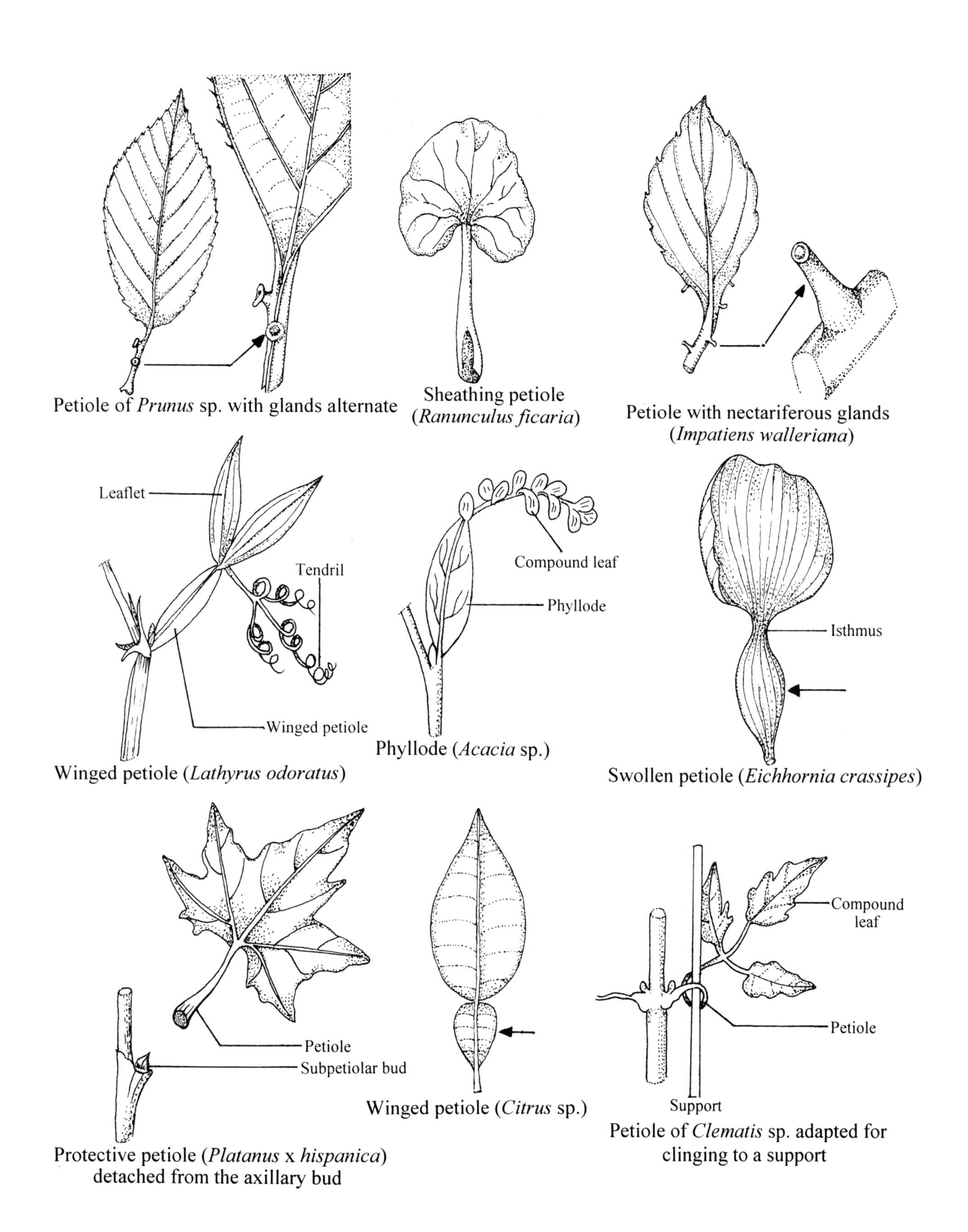

Petiole of *Prunus* sp. with glands alternate

Sheathing petiole (*Ranunculus ficaria*)

Petiole with nectariferous glands (*Impatiens walleriana*)

Winged petiole (*Lathyrus odoratus*)

Phyllode (*Acacia* sp.)

Swollen petiole (*Eichhornia crassipes*)

Protective petiole (*Platanus* x *hispanica*) detached from the axillary bud

Winged petiole (*Citrus* sp.)

Petiole of *Clematis* sp. adapted for clinging to a support

Petioles, Stipules, Spines, Tendrils 2

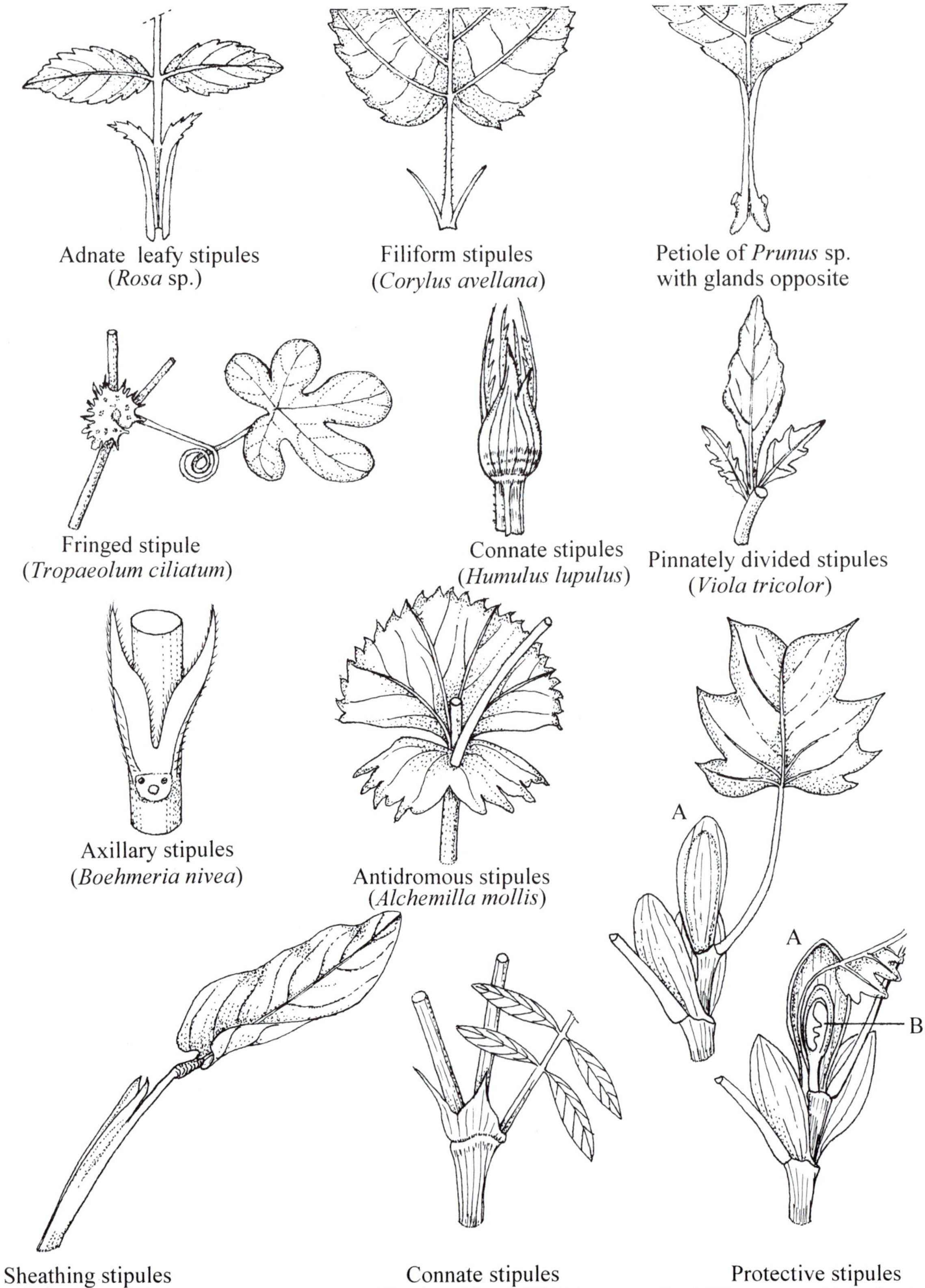

Adnate leafy stipules (*Rosa* sp.)

Filiform stipules (*Corylus avellana*)

Petiole of *Prunus* sp. with glands opposite

Fringed stipule (*Tropaeolum ciliatum*)

Connate stipules (*Humulus lupulus*)

Pinnately divided stipules (*Viola tricolor*)

Axillary stipules (*Boehmeria nivea*)

Antidromous stipules (*Alchemilla mollis*)

Sheathing stipules (*Maranta* sp.)

Connate stipules (*Onobrychis viciifolia*)

Protective stipules
Bud (B) protected by stipule (A)
(*Liriodendron tulipifera*)

Petioles, Stipules, Spines, Tendrils 3

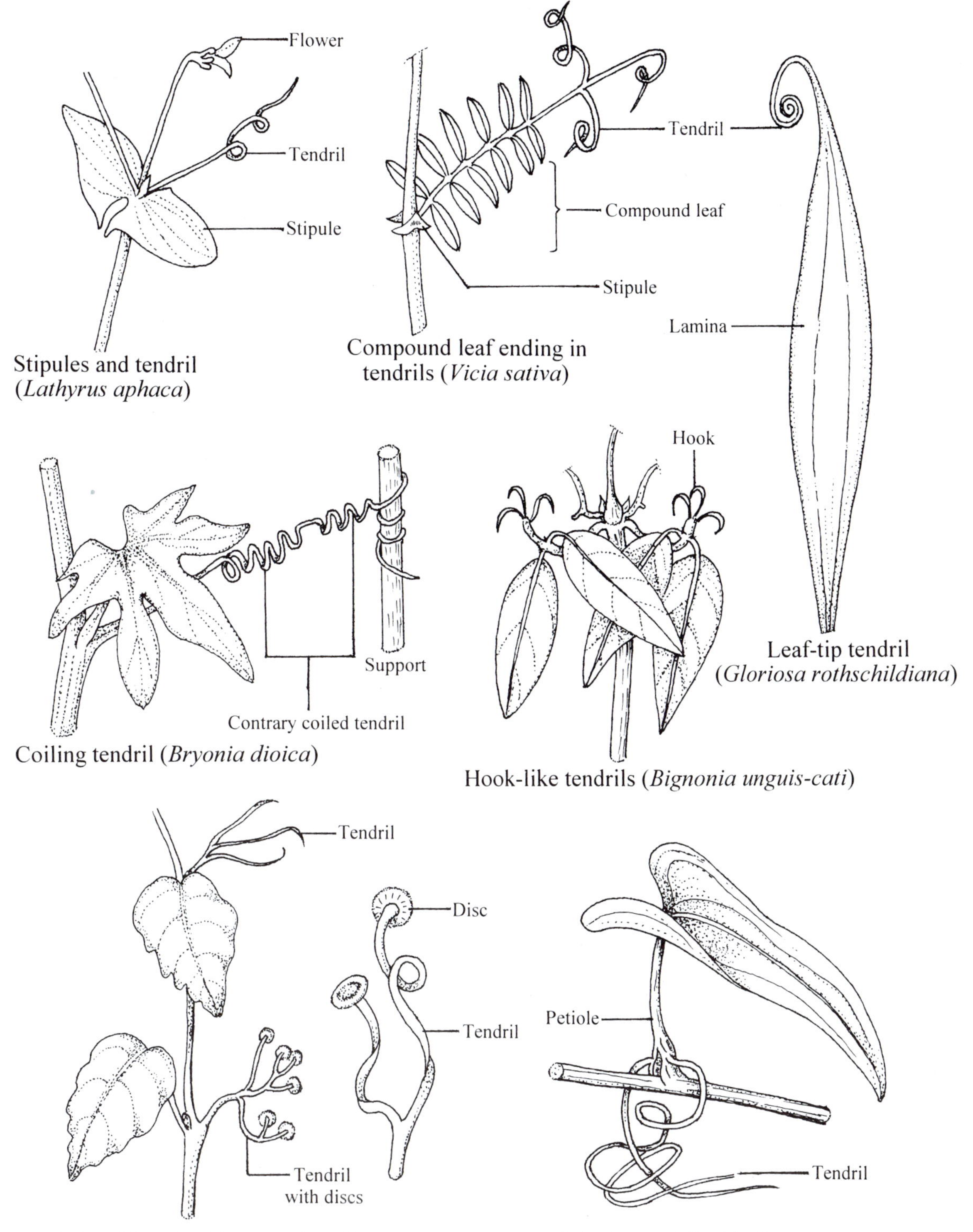

Stipules and tendril (*Lathyrus aphaca*)

Compound leaf ending in tendrils (*Vicia sativa*)

Leaf-tip tendril (*Gloriosa rothschildiana*)

Coiling tendril (*Bryonia dioica*)

Hook-like tendrils (*Bignonia unguis-cati*)

Tendrils with and without sucker discs (*Parthenocissus* sp.)

Stipular tendrils (*Smilax aspera*)

Petioles, Stipules, Spines, Tendrils 4

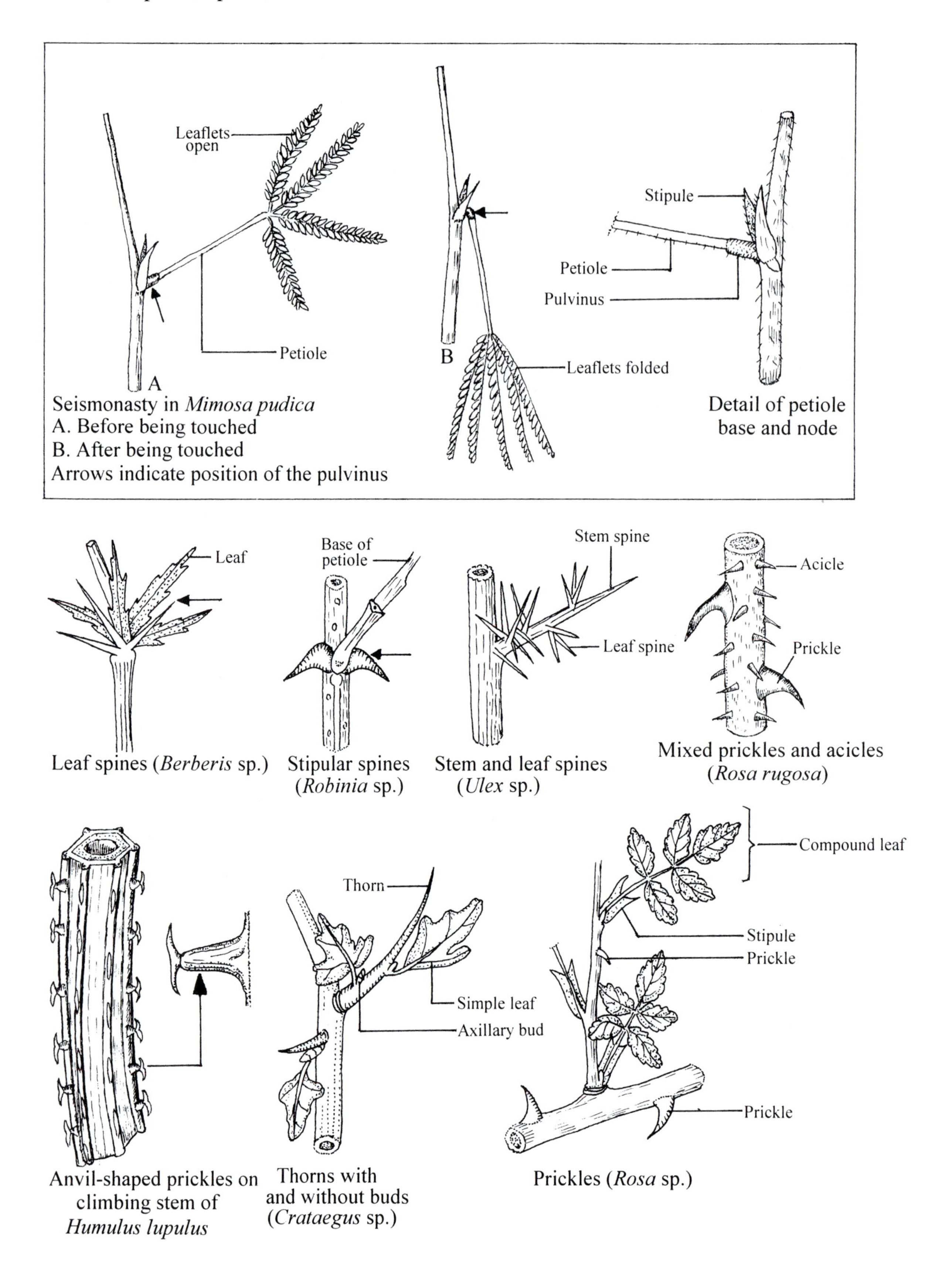

Seismonasty in *Mimosa pudica*
A. Before being touched
B. After being touched
Arrows indicate position of the pulvinus

Detail of petiole base and node

Leaf spines (*Berberis* sp.)

Stipular spines (*Robinia* sp.)

Stem and leaf spines (*Ulex* sp.)

Mixed prickles and acicles (*Rosa rugosa*)

Anvil-shaped prickles on climbing stem of *Humulus lupulus*

Thorns with and without buds (*Crataegus* sp.)

Prickles (*Rosa* sp.)

Other Leaf Features

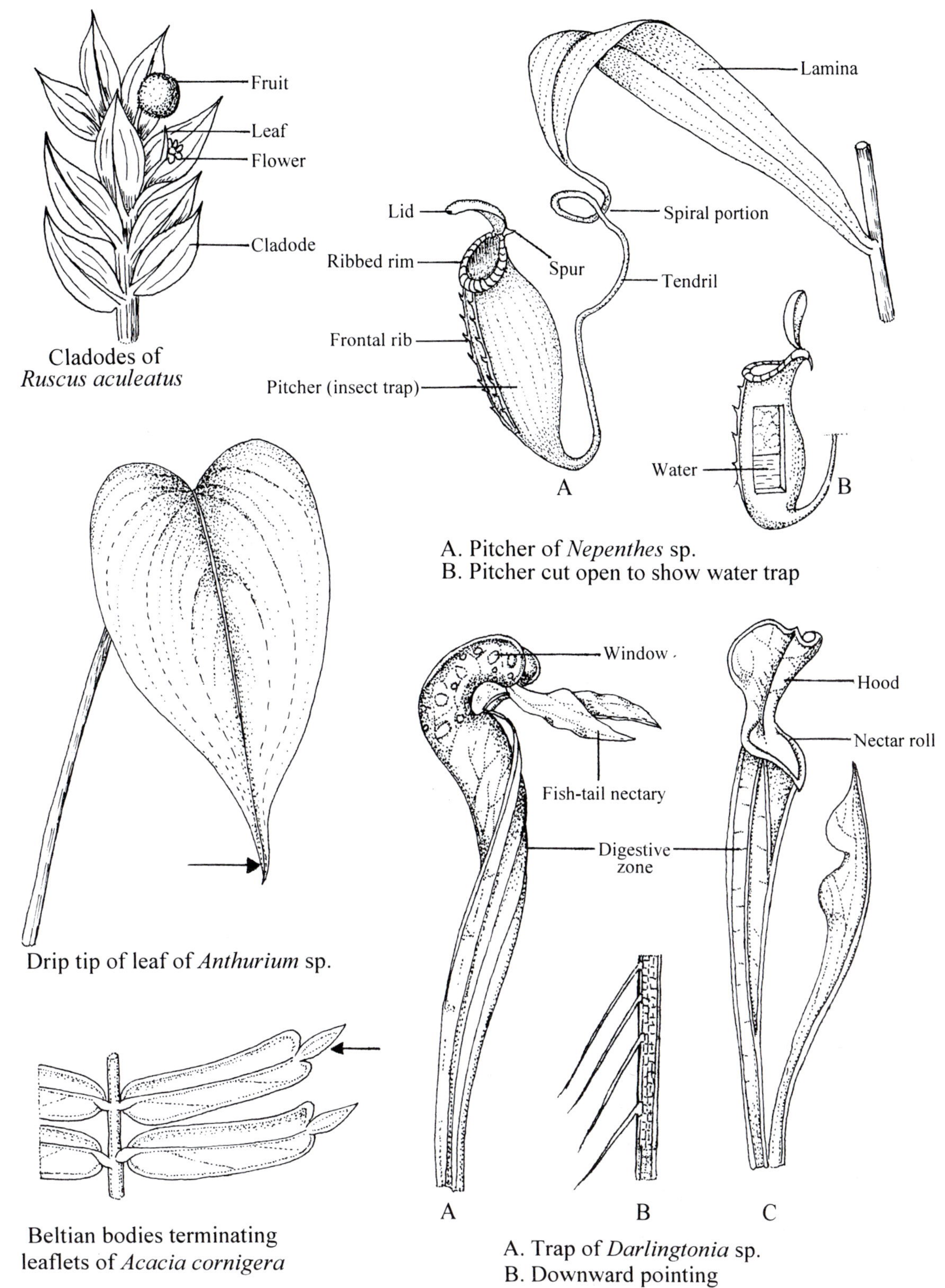

Cladodes of *Ruscus aculeatus*

A. Pitcher of *Nepenthes* sp.
B. Pitcher cut open to show water trap

Drip tip of leaf of *Anthurium* sp.

Beltian bodies terminating leaflets of *Acacia cornigera*

A. Trap of *Darlingtonia* sp.
B. Downward pointing hairs in trap of *Darlingtonia* sp.
C. Trap of *Sarracenia* sp.

Domatia and Galls

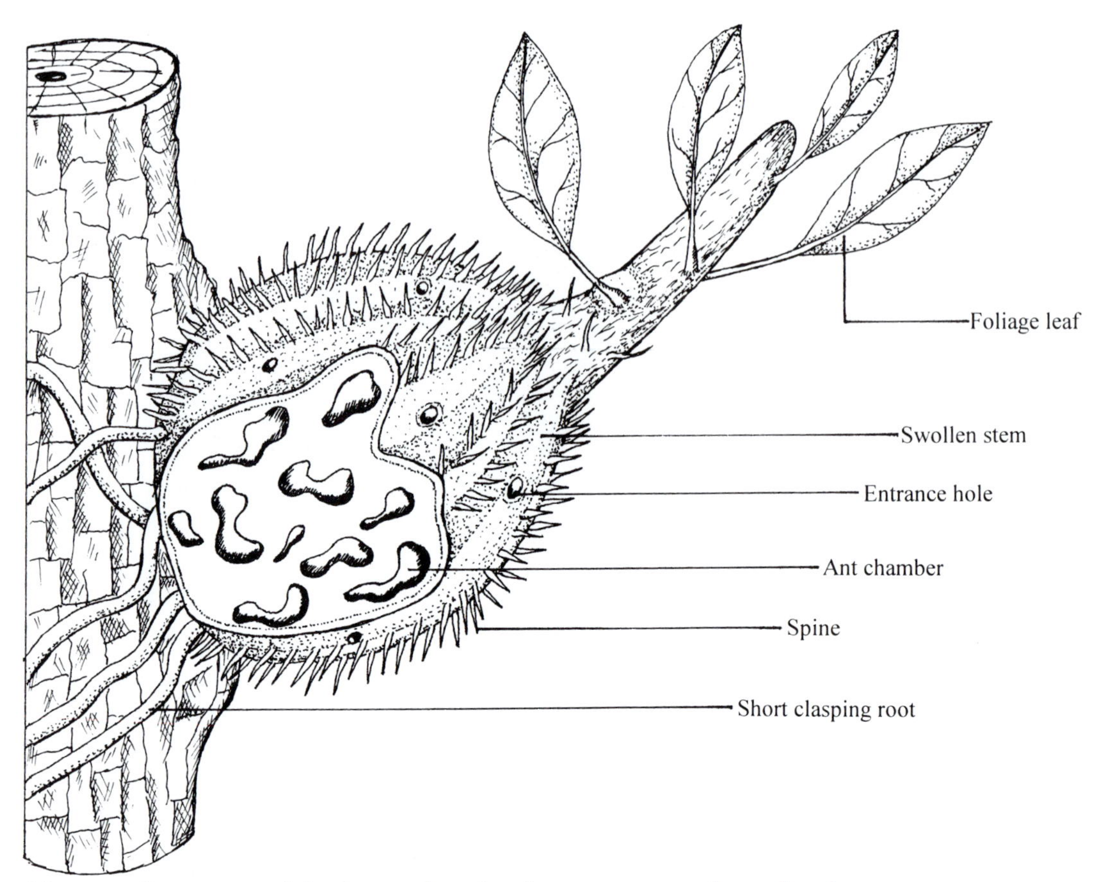

Myrmecodia sp., an epiphytic ant-plant. Portion cut open to show chambers

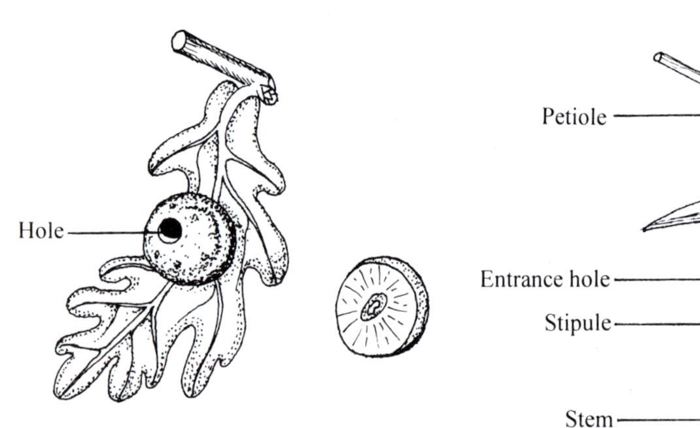

Insect-gall on *Quercus robur*, whole and in section

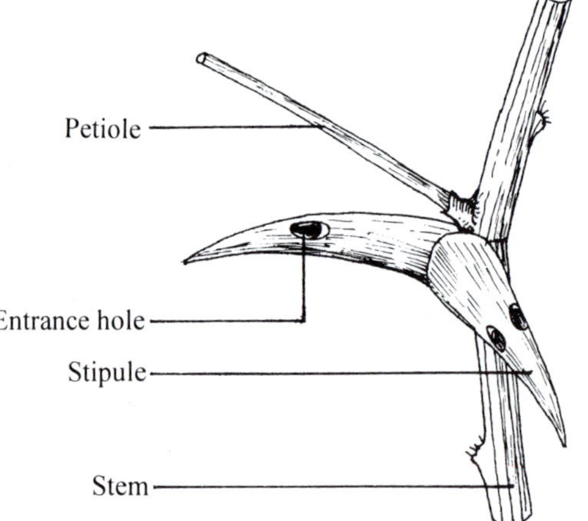

Hollow thorny stipule of myrmecophilous plant, *Acacia nicoyensis*

Abnormal Forms

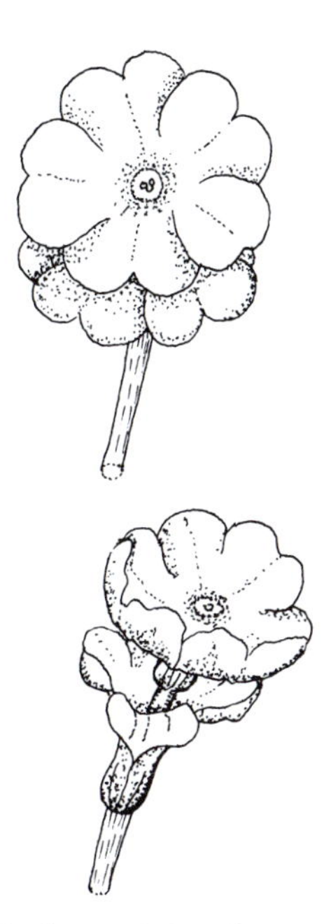

Hose-in-hose *Primula*

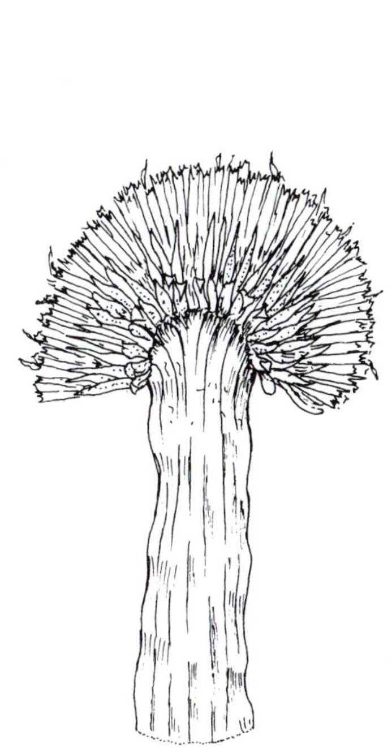

Fasciation in an inflorescence of *Taraxacum* sp.

Peloria in *Digitalis purpurea*

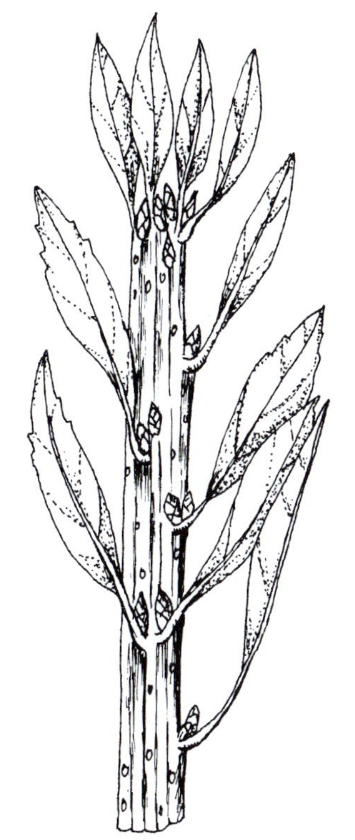

Fasciation in a *Forsythia* stem

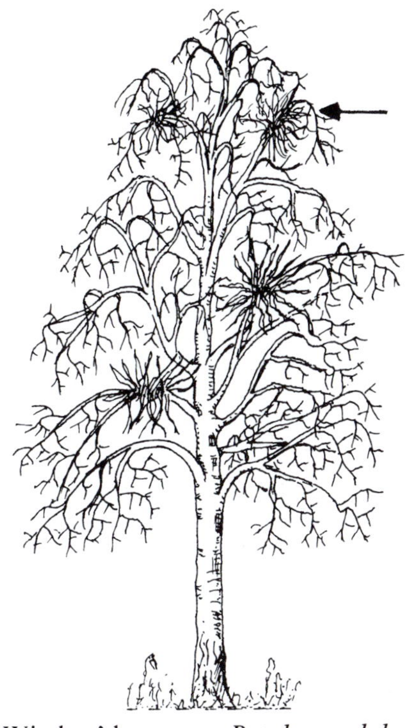

Witches' broom on *Betula pendula*

Bracts

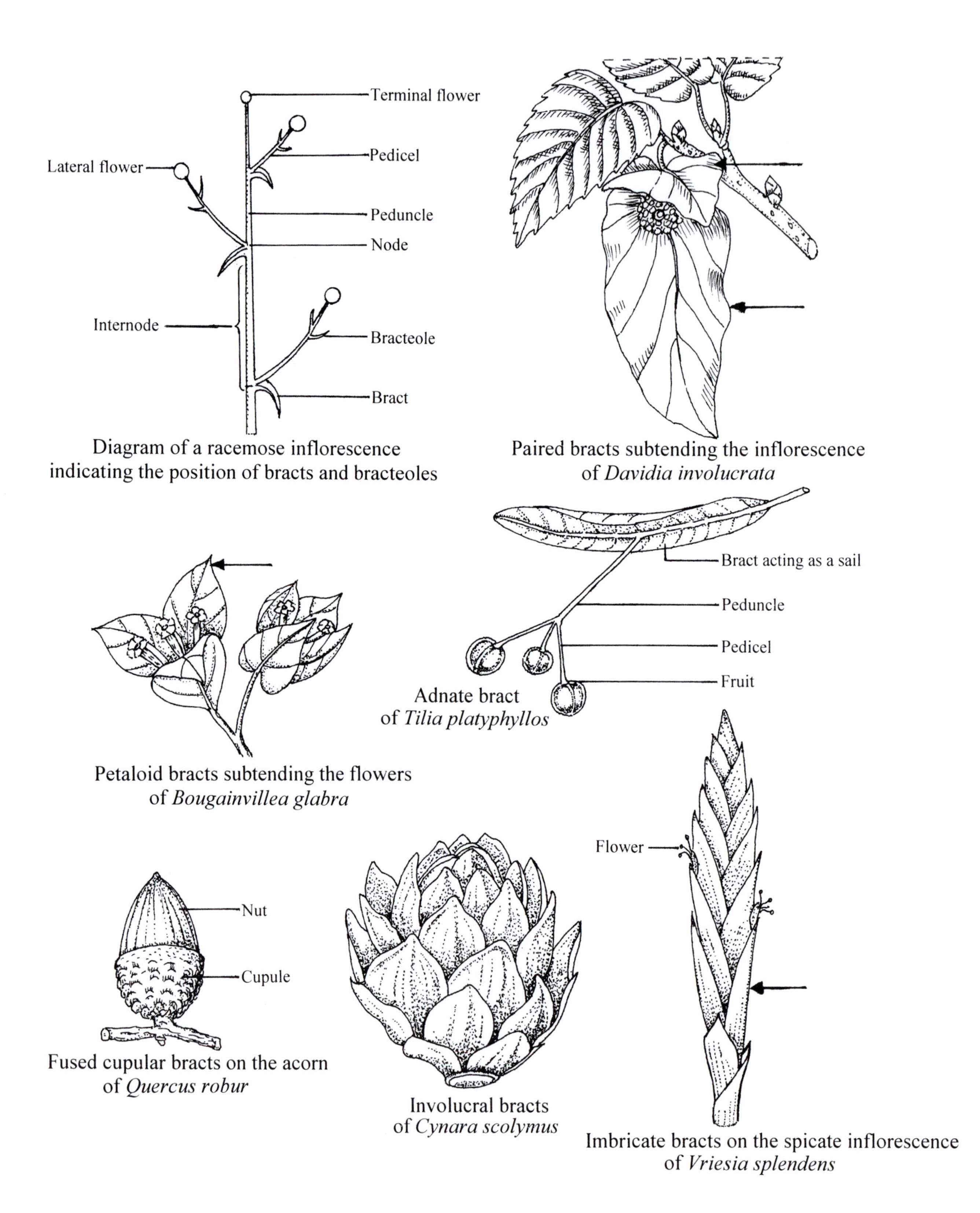

Diagram of a racemose inflorescence indicating the position of bracts and bracteoles

Paired bracts subtending the inflorescence of *Davidia involucrata*

Adnate bract of *Tilia platyphyllos*

Petaloid bracts subtending the flowers of *Bougainvillea glabra*

Fused cupular bracts on the acorn of *Quercus robur*

Involucral bracts of *Cynara scolymus*

Imbricate bracts on the spicate inflorescence of *Vriesia splendens*

7. Leaves

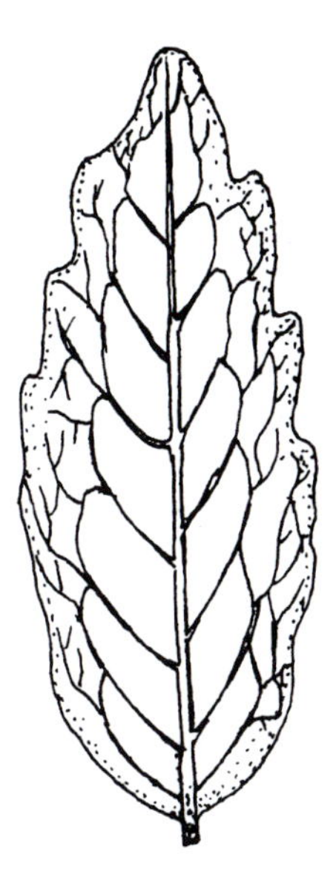

Leaf Arrangements

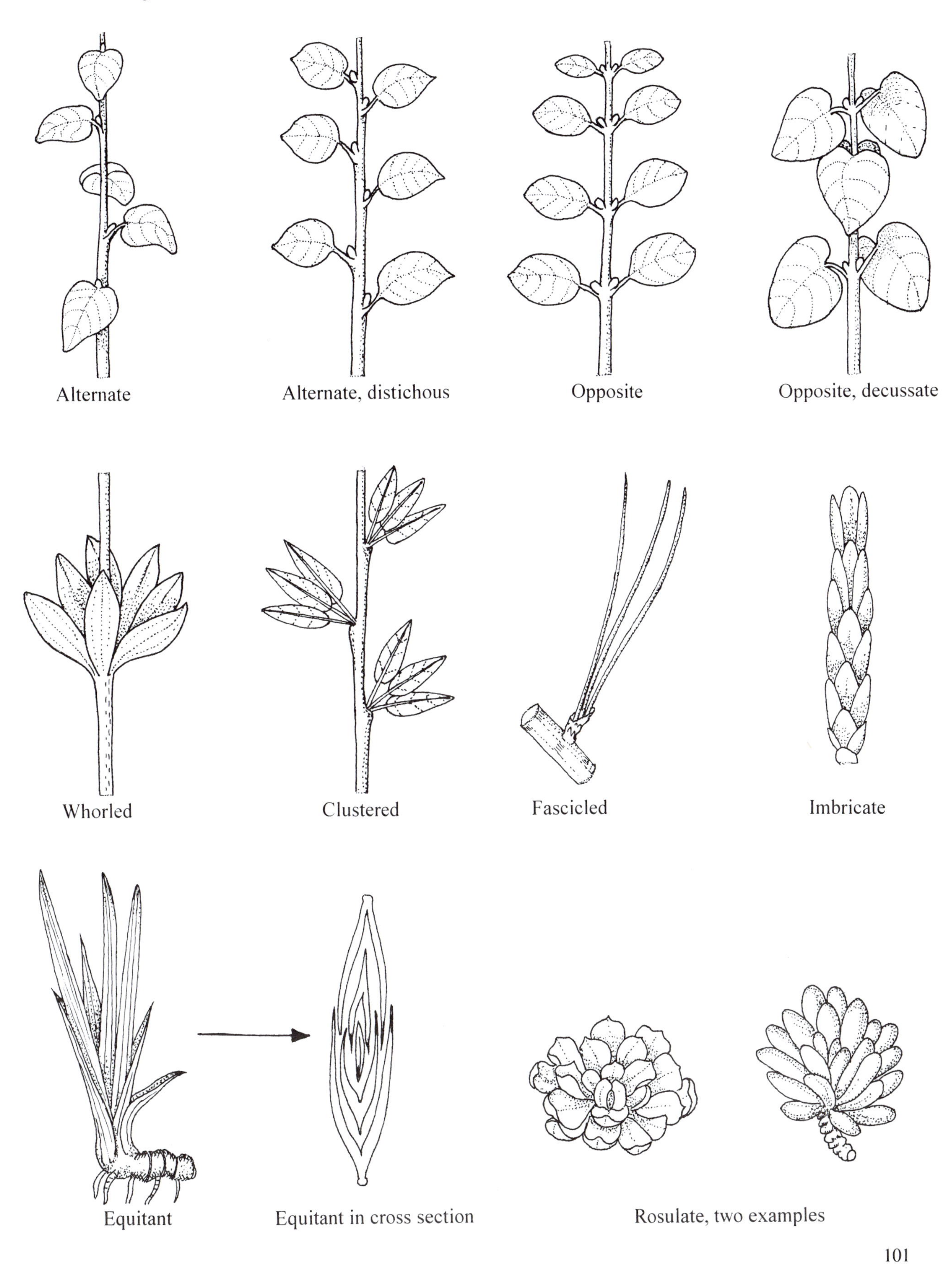

Points of Attachment

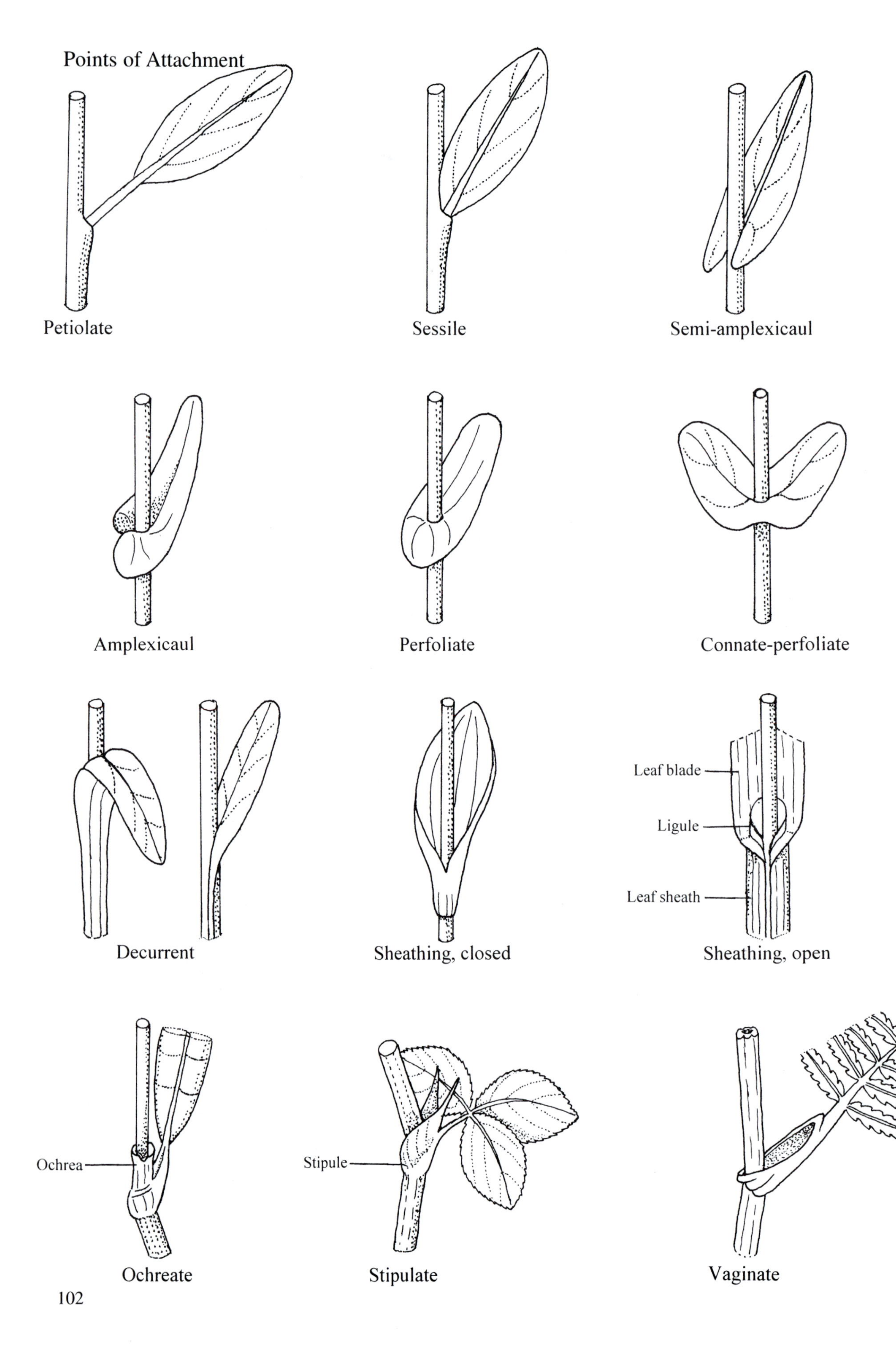

A Typical Leaf and its Attachment

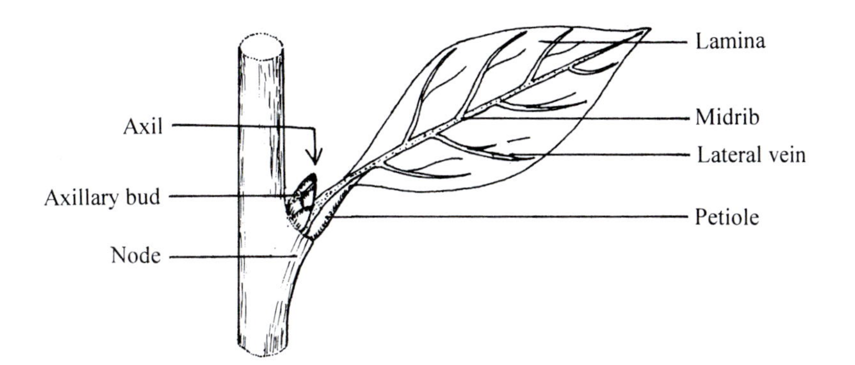

Leaf Folding including Ptyxis

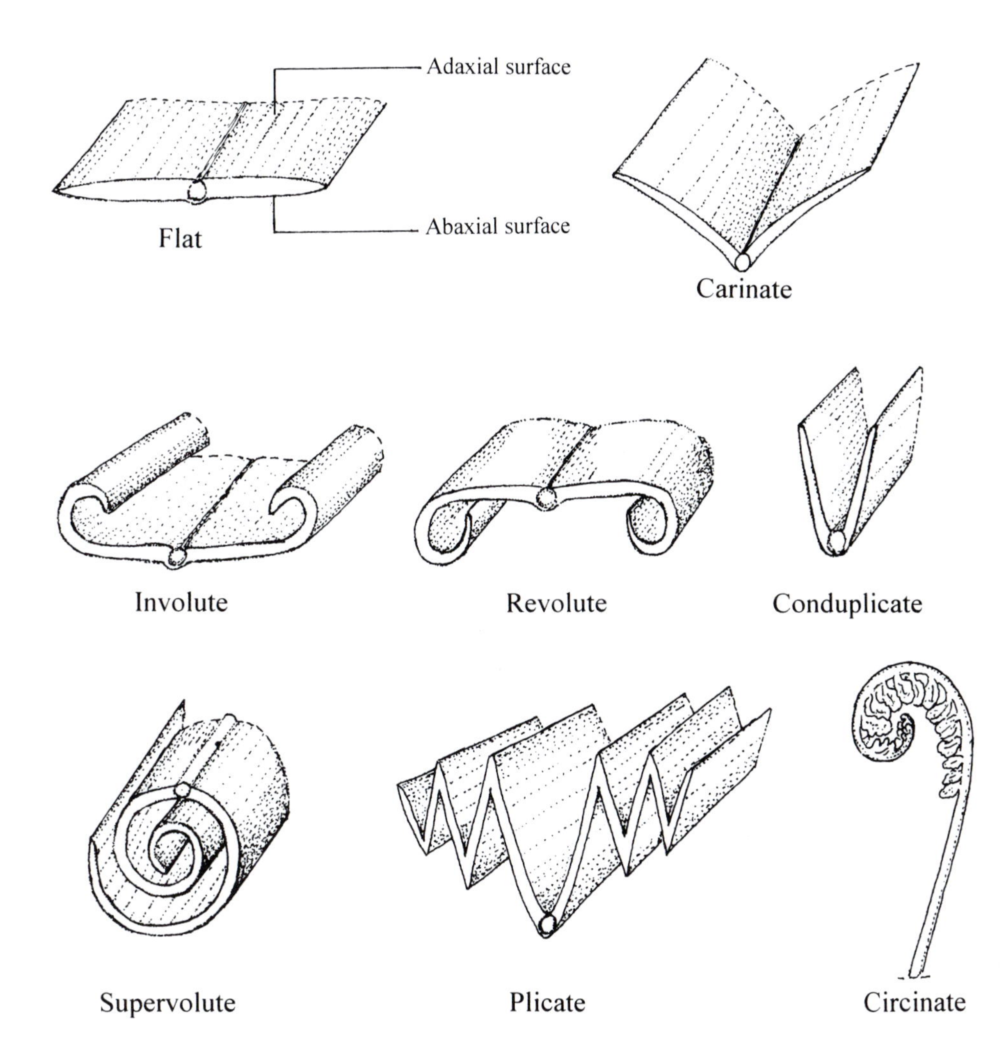

Vernation

Imbricate

Equitant

Convolute

Venation

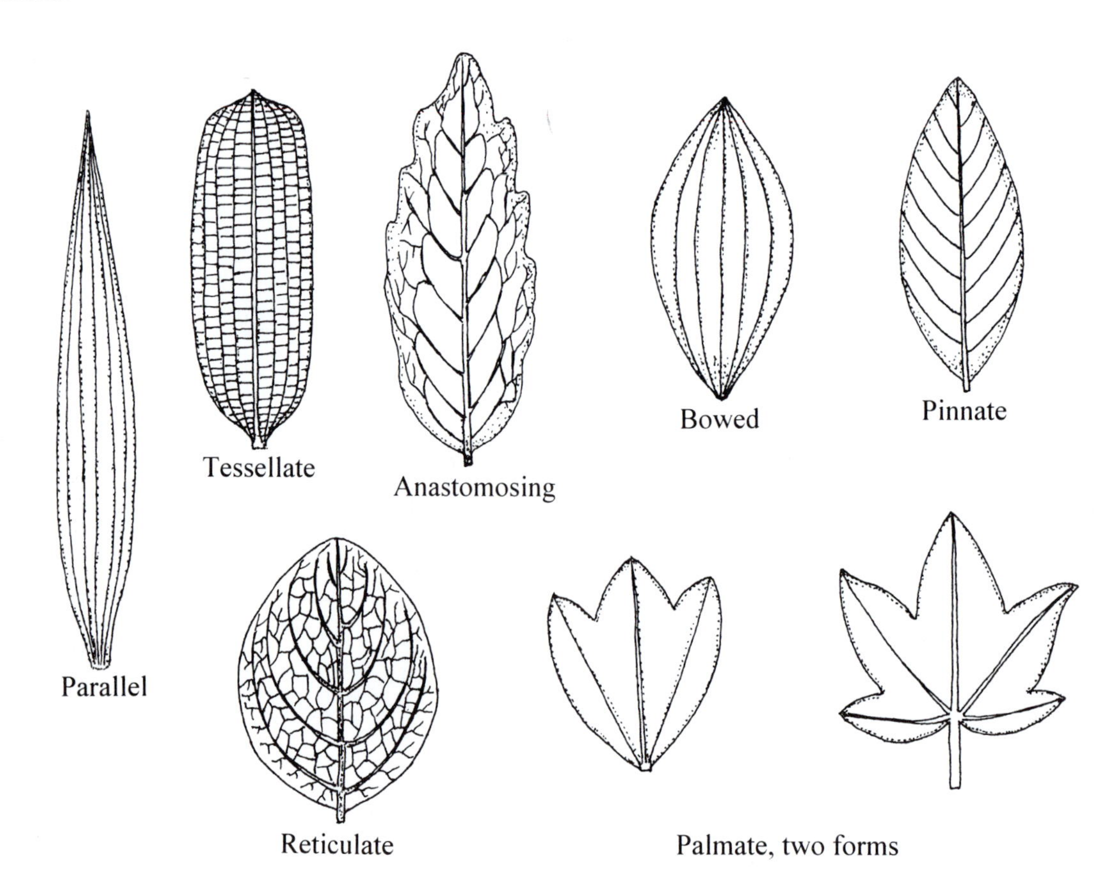

Simple Leaves 1

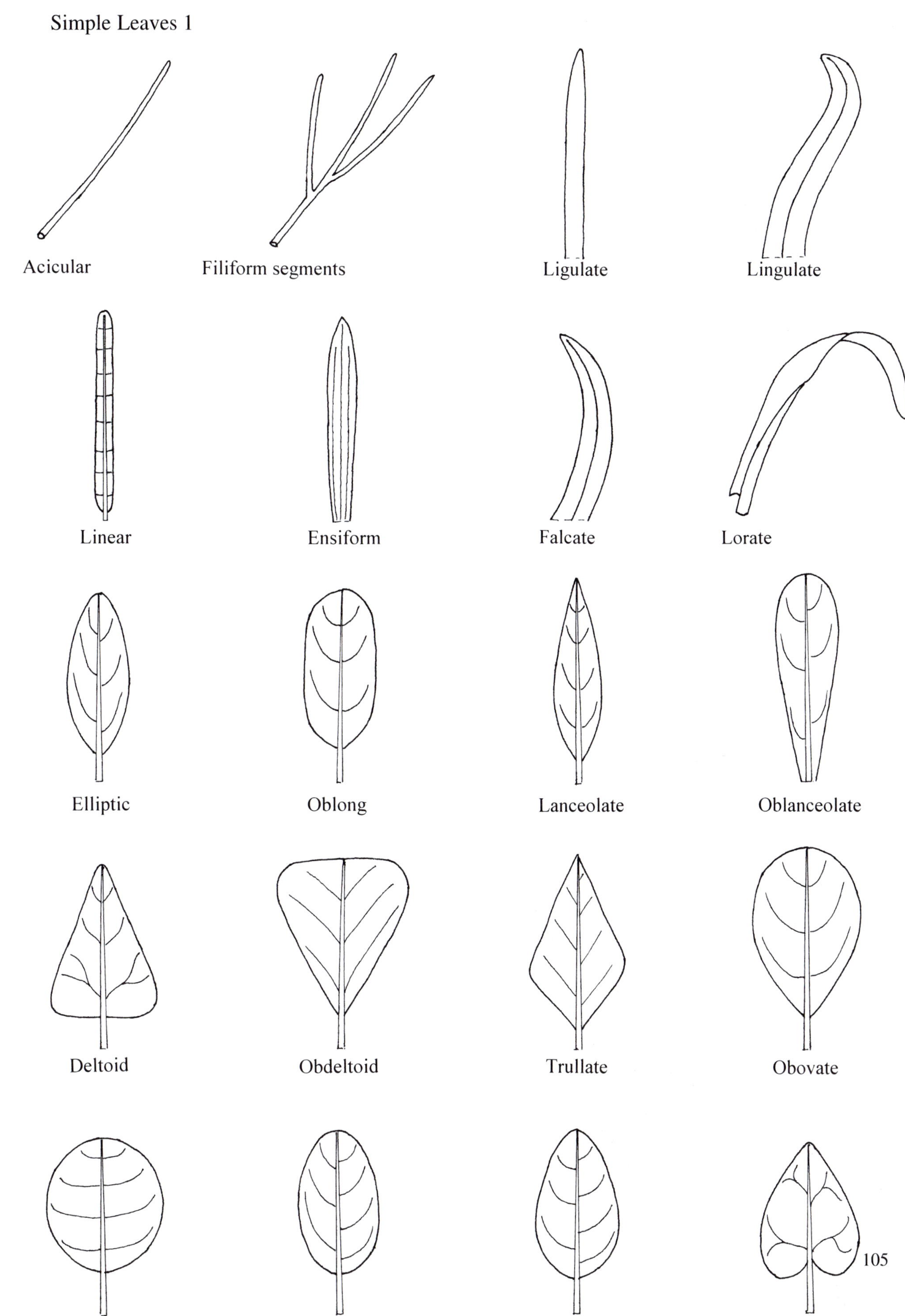

Simple Leaves 2

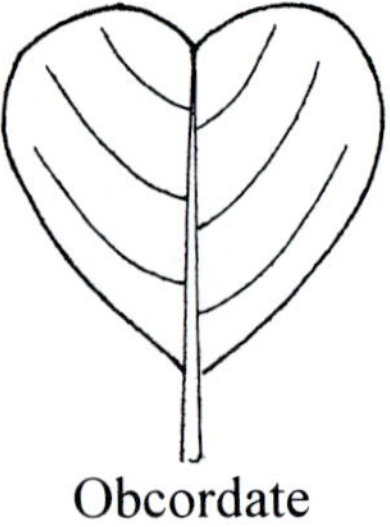

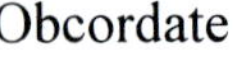

Obcordate

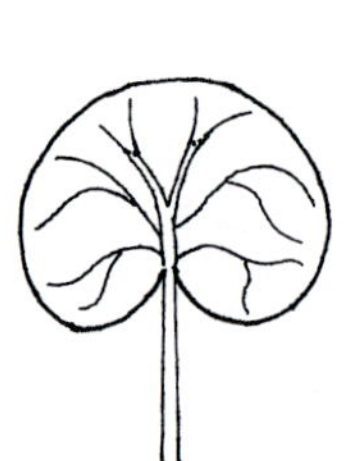

Reniform

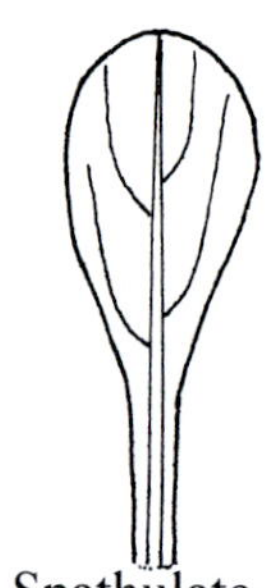

Spathulate

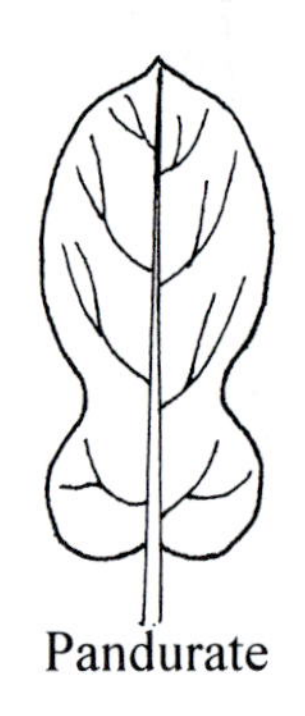

Pandurate

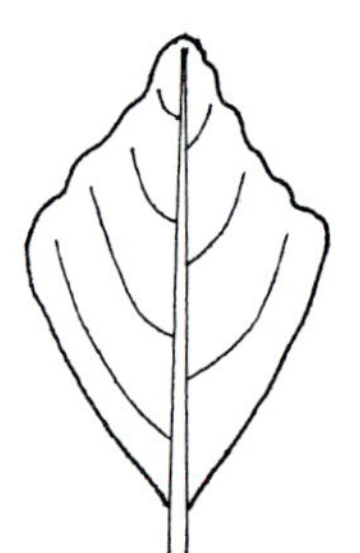

Rhomboid

Lunate

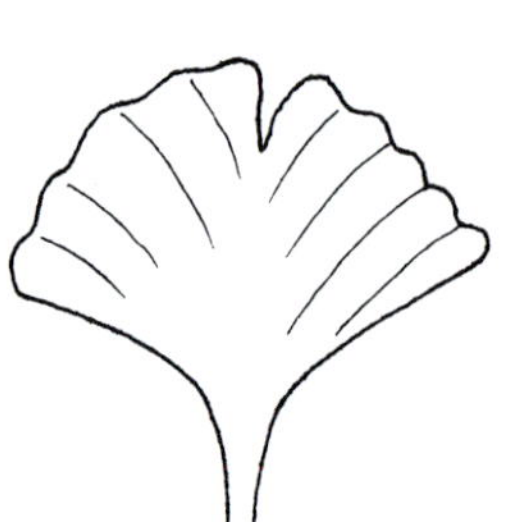

Flabellate

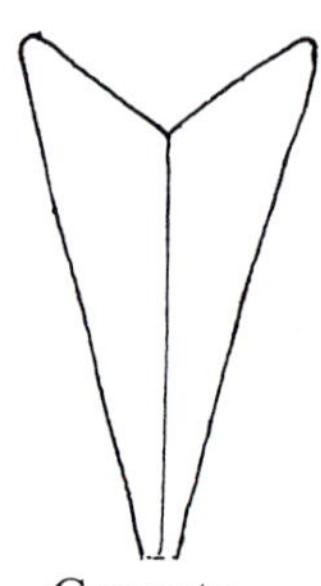

Cuneate

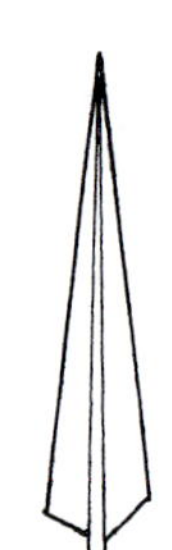

Subulate

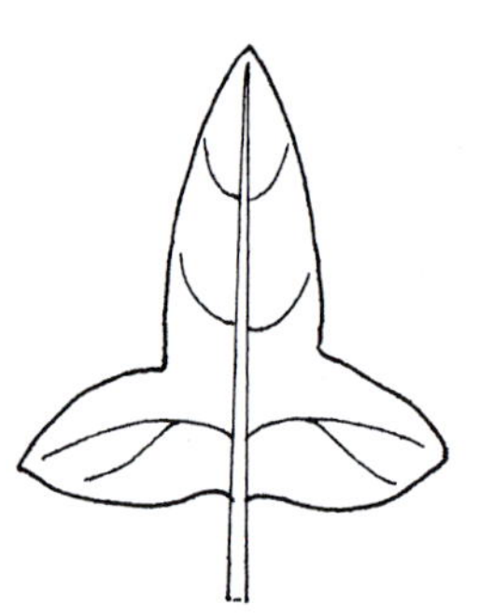

Hastate

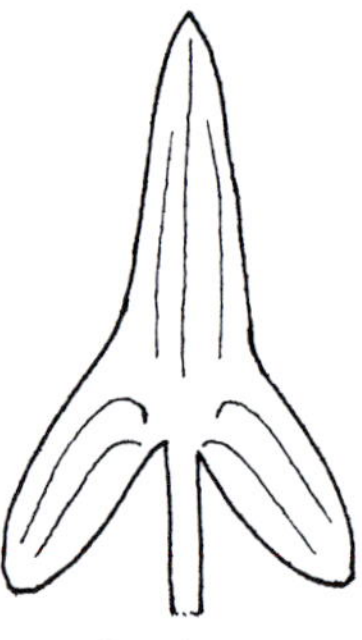

Sagittate

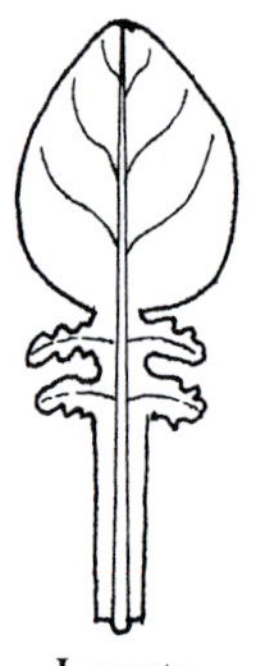

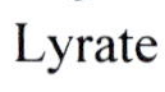

Lyrate

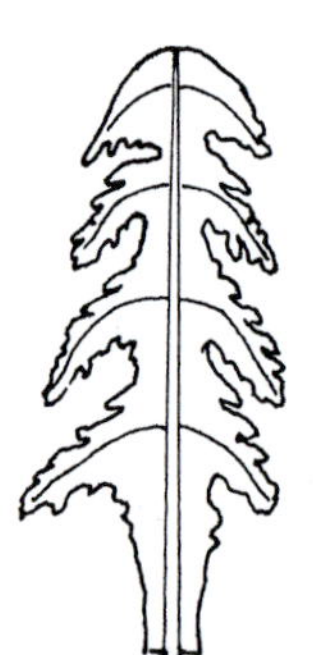

Runcinate

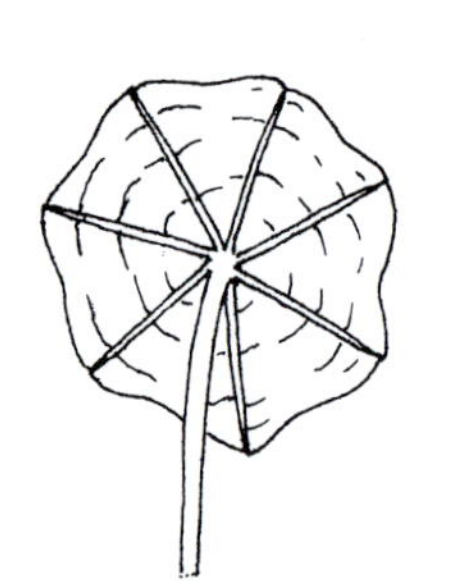

Peltate

Palmatifid

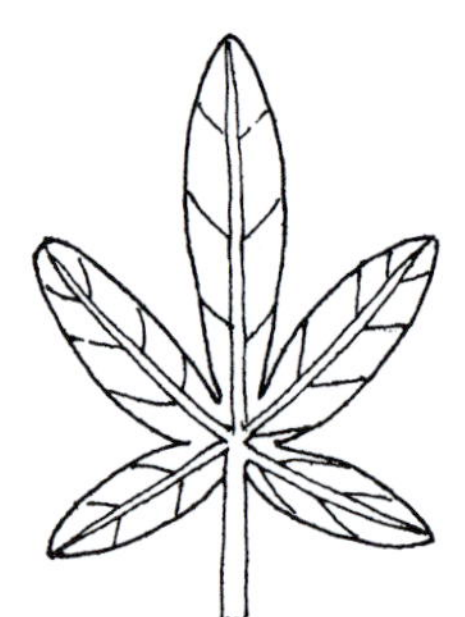

Palmatisect

Pinnatifid

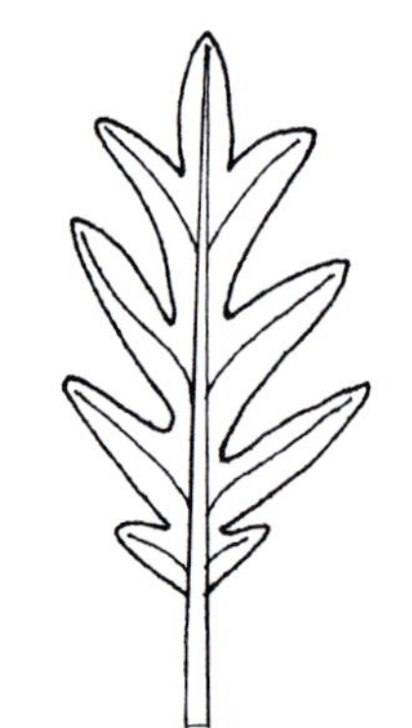

Pinnatisect

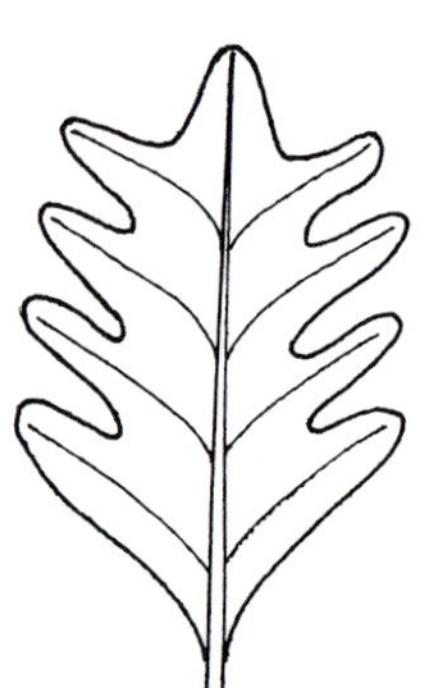

Pinnatipartite

Fan-shaped

Compound Leaves

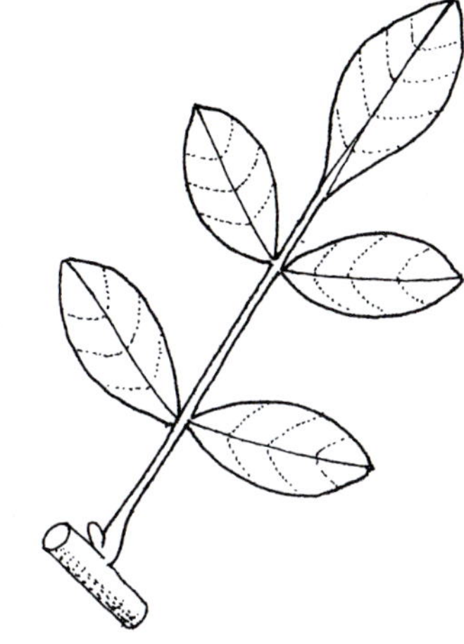

Imparipinnate (odd-pinnate)

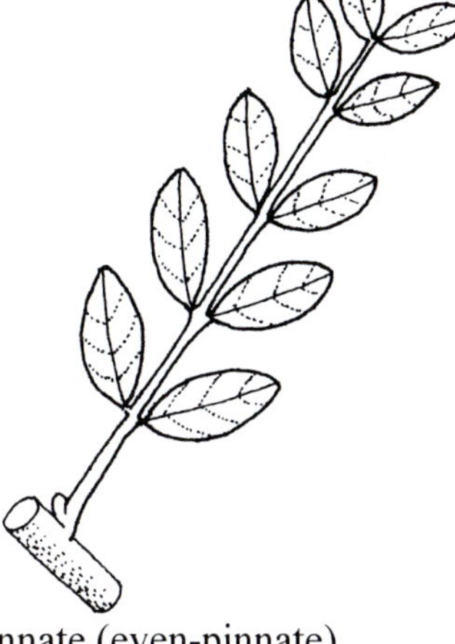

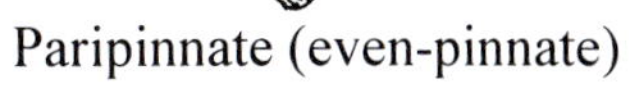

Paripinnate (even-pinnate)

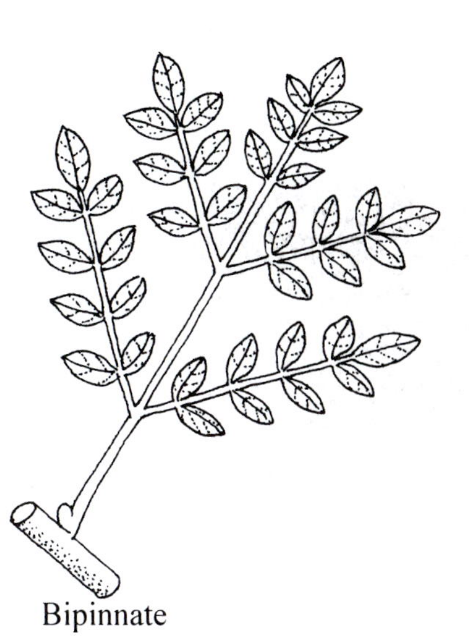

Bipinnate

Tripinnate

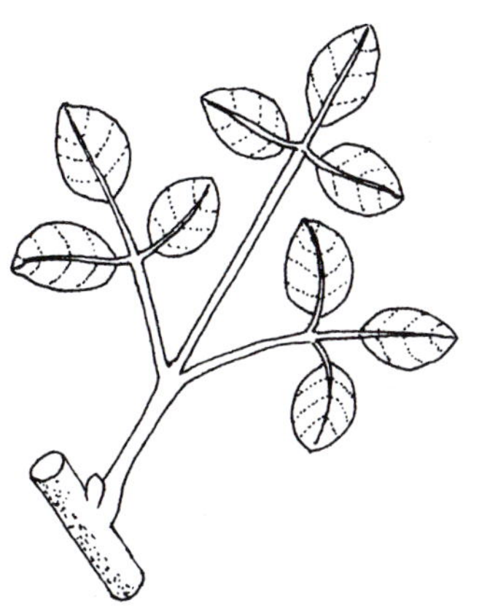

Biternate

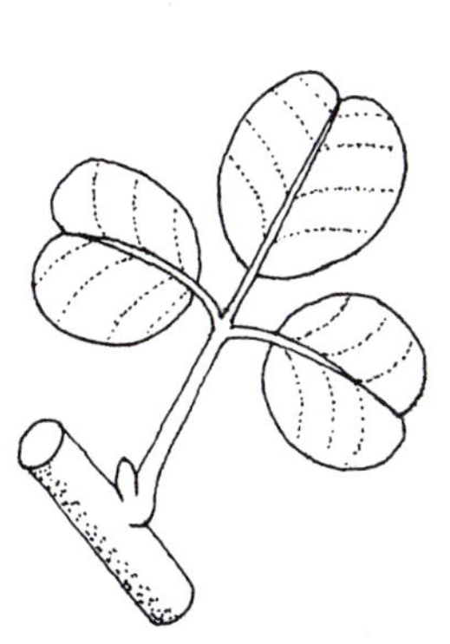

Trifoliolate

Pedate

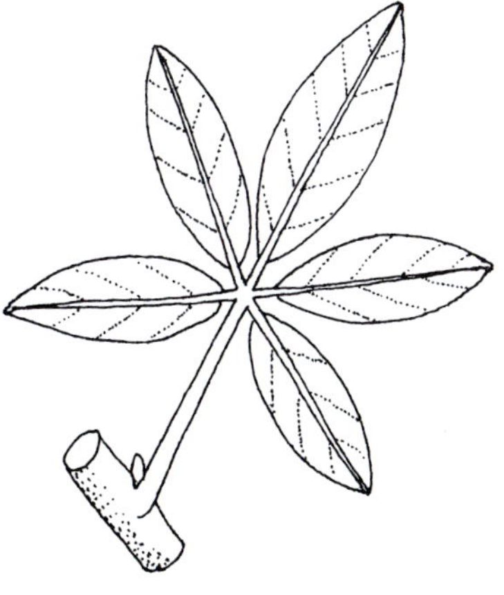

Palmate

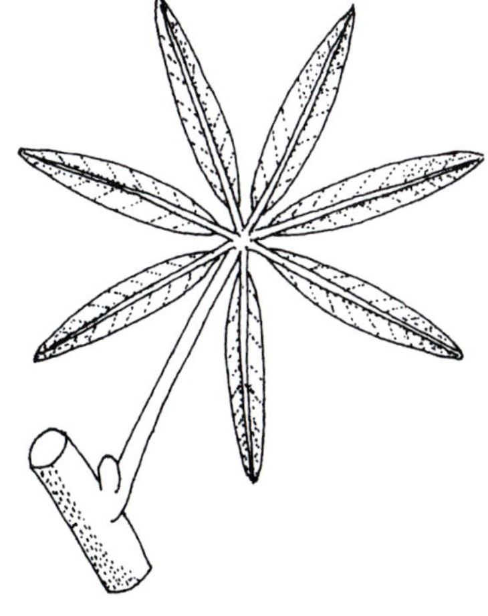

Digitate

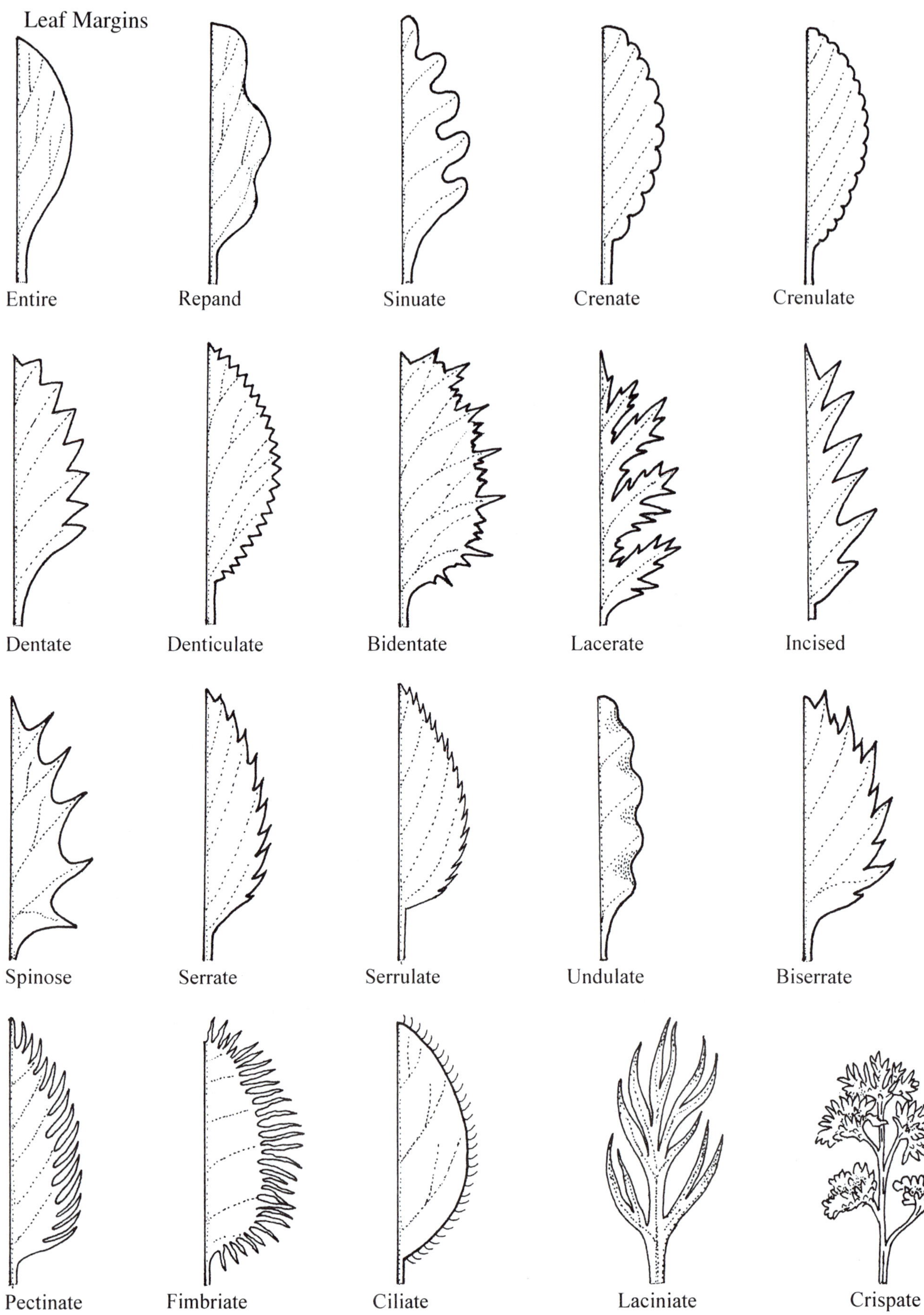
Leaf Margins
Entire
Repand
Sinuate
Crenate
Crenulate
Dentate
Denticulate
Bidentate
Lacerate
Incised
Spinose
Serrate
Serrulate
Undulate
Biserrate
Pectinate
Fimbriate
Ciliate
Laciniate
Crispate

Leaf Bases

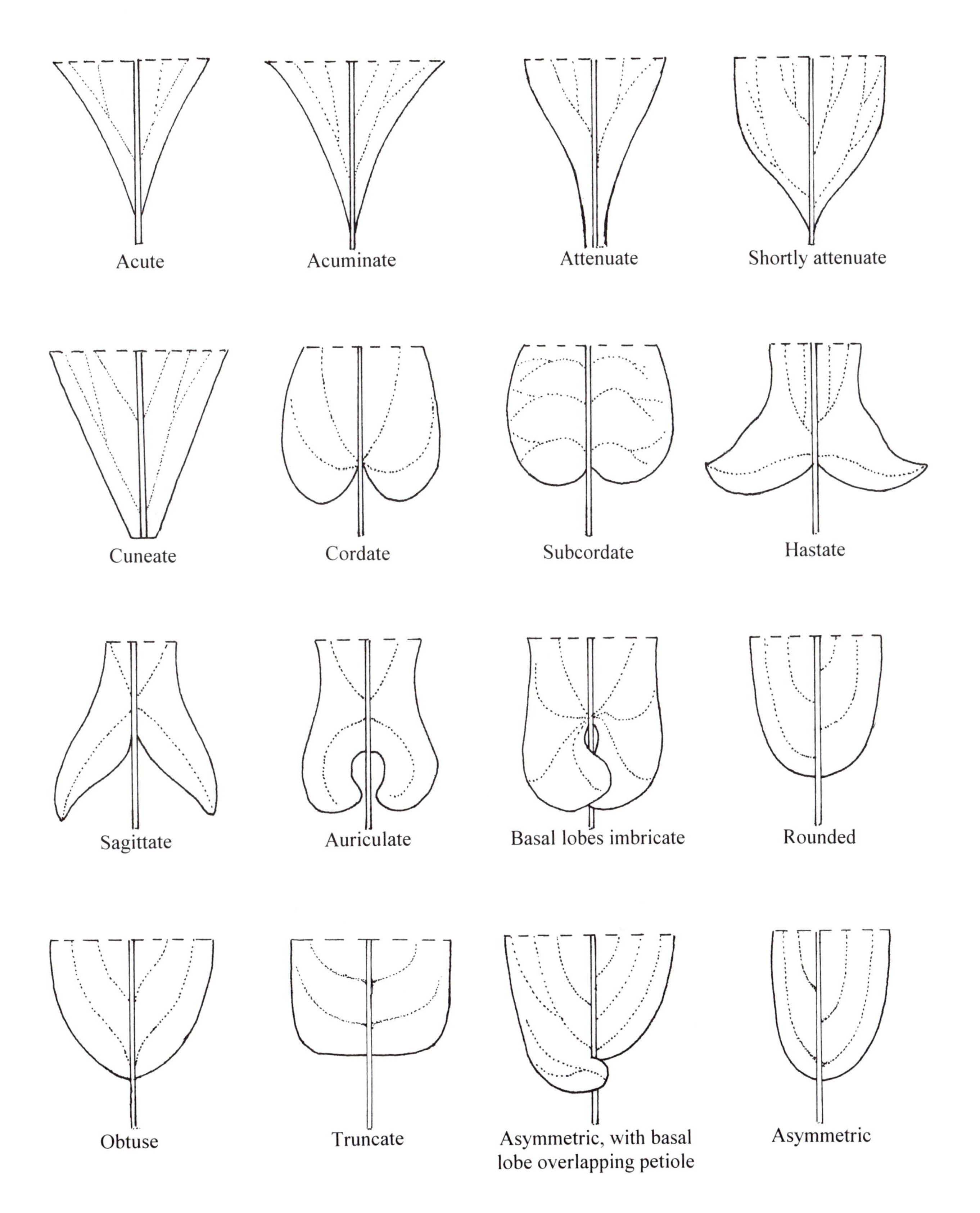
Acute
Acuminate
Attenuate
Shortly attenuate
Cuneate
Cordate
Subcordate
Hastate
Sagittate
Auriculate
Basal lobes imbricate
Rounded
Obtuse
Truncate
Asymmetric, with basal lobe overlapping petiole
Asymmetric

Leaf Apices

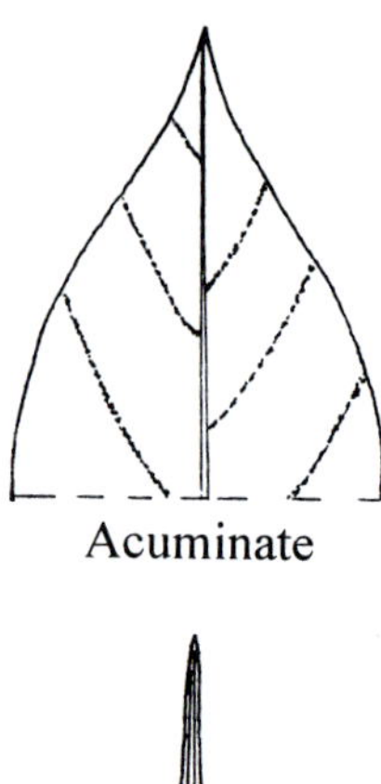
Acuminate

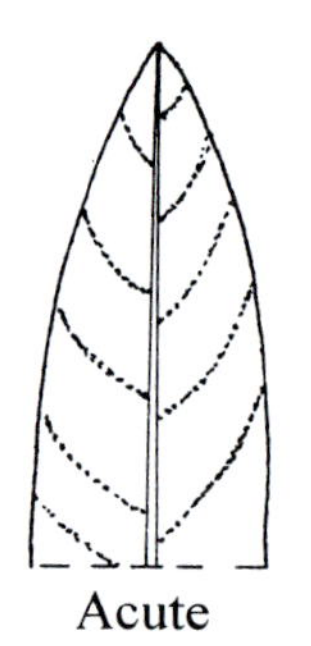
Acute

Abruptly acute

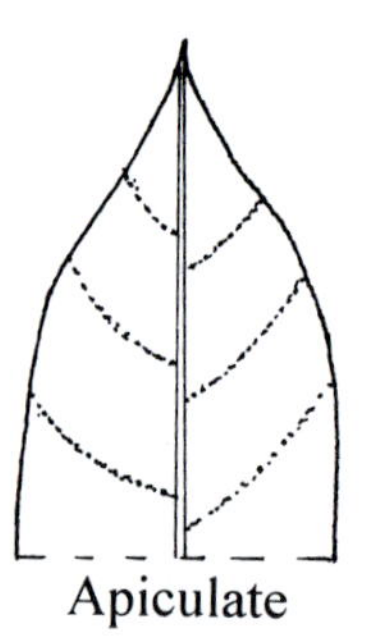
Apiculate

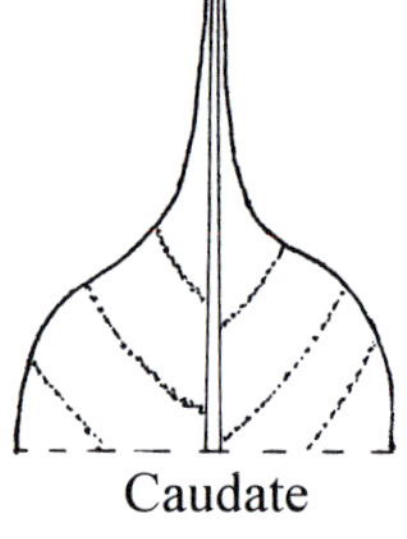
Caudate

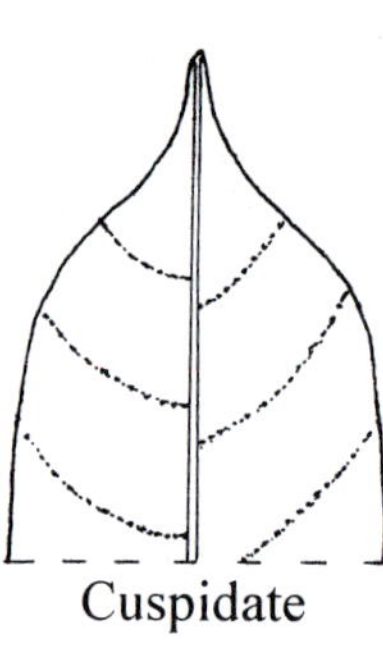
Cuspidate

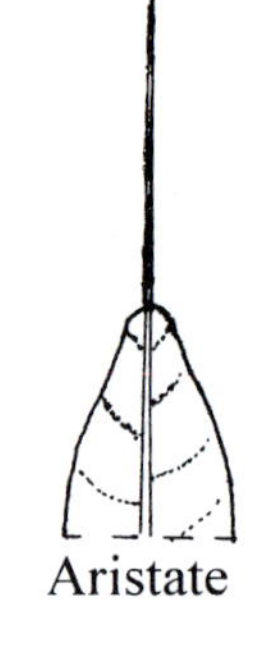
Aristate

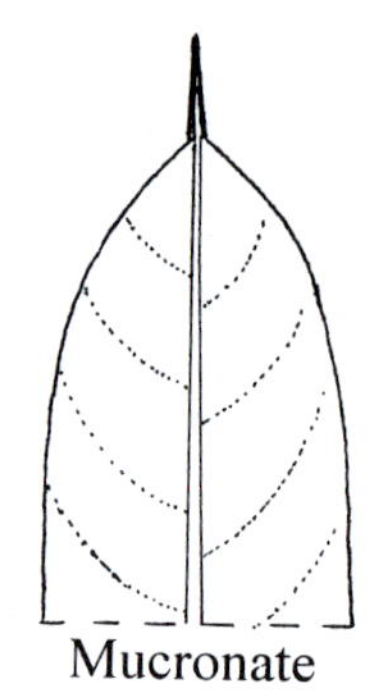
Mucronate

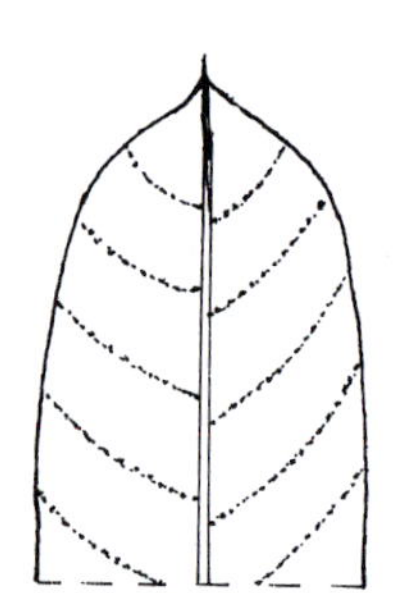
Mucronulate

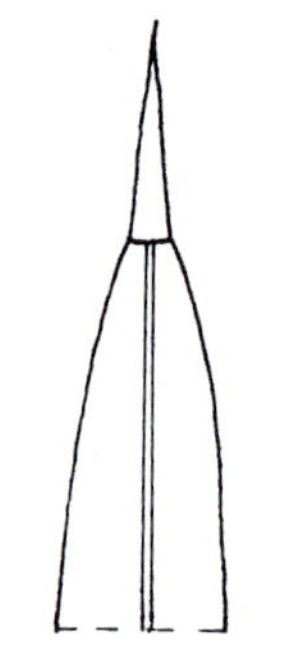
Pungent

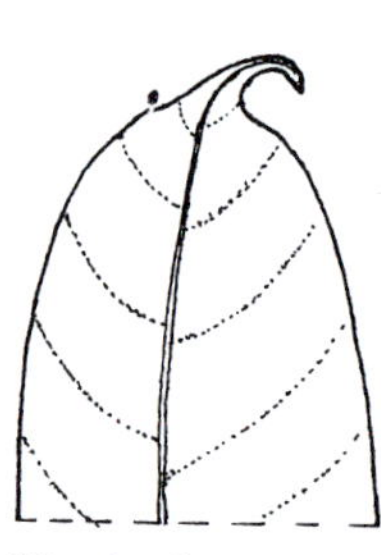
Hooked-truncate

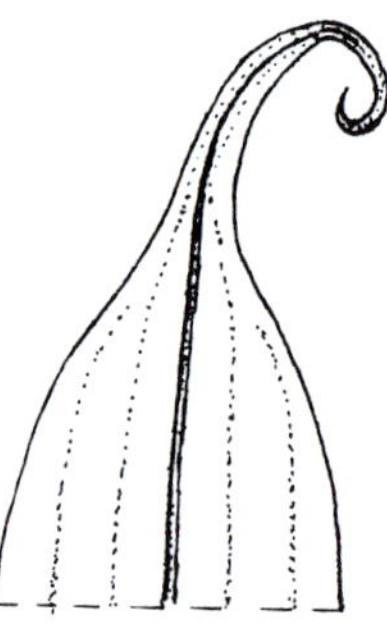
Cirrhose

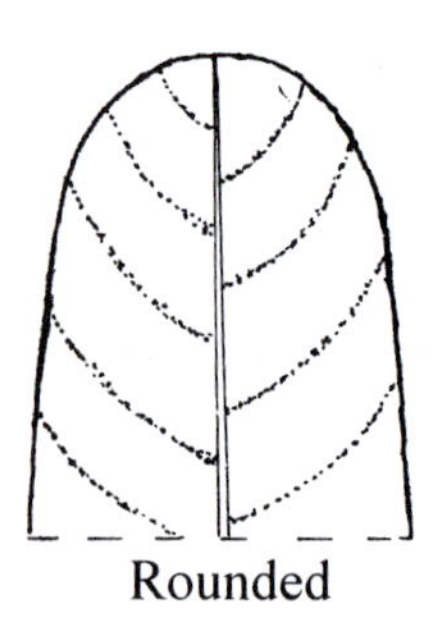
Rounded

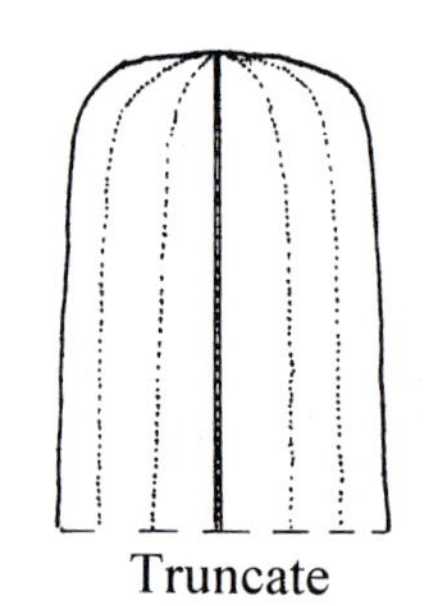
Truncate

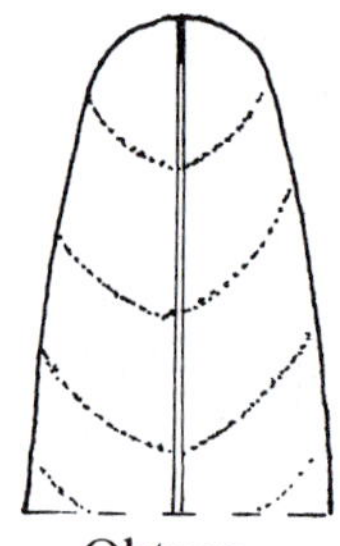
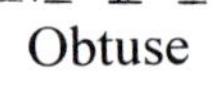
Obtuse

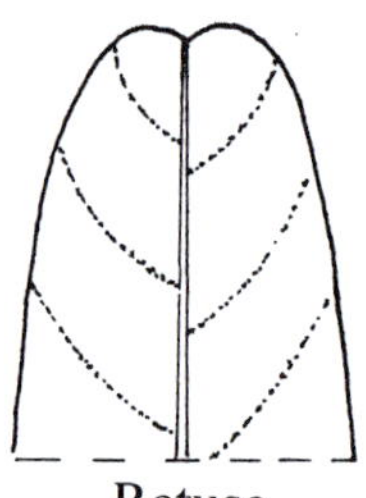
Retuse

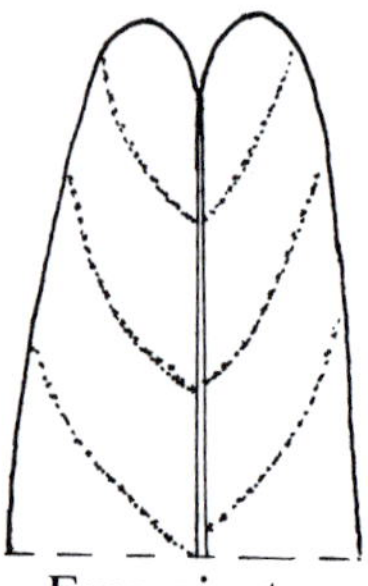
Emarginate

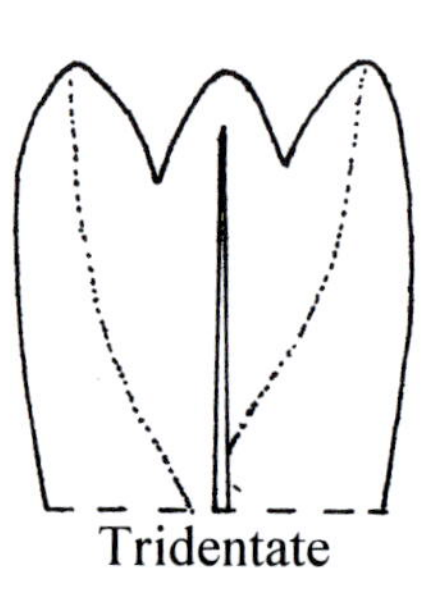
Tridentate

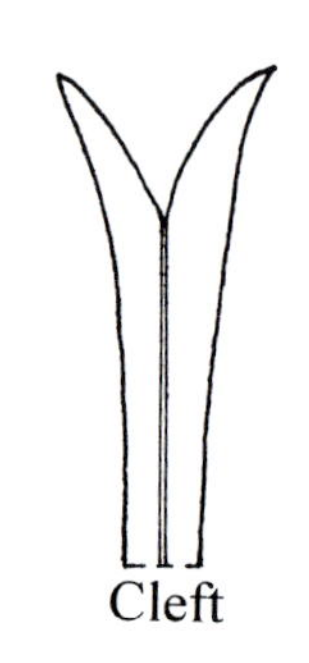
Cleft

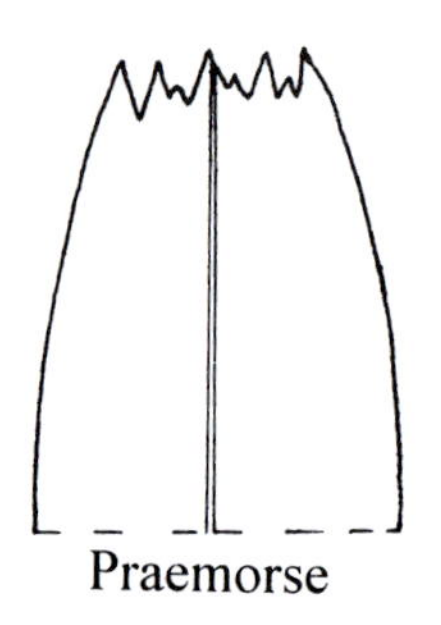
Praemorse

Leaf Structure 1

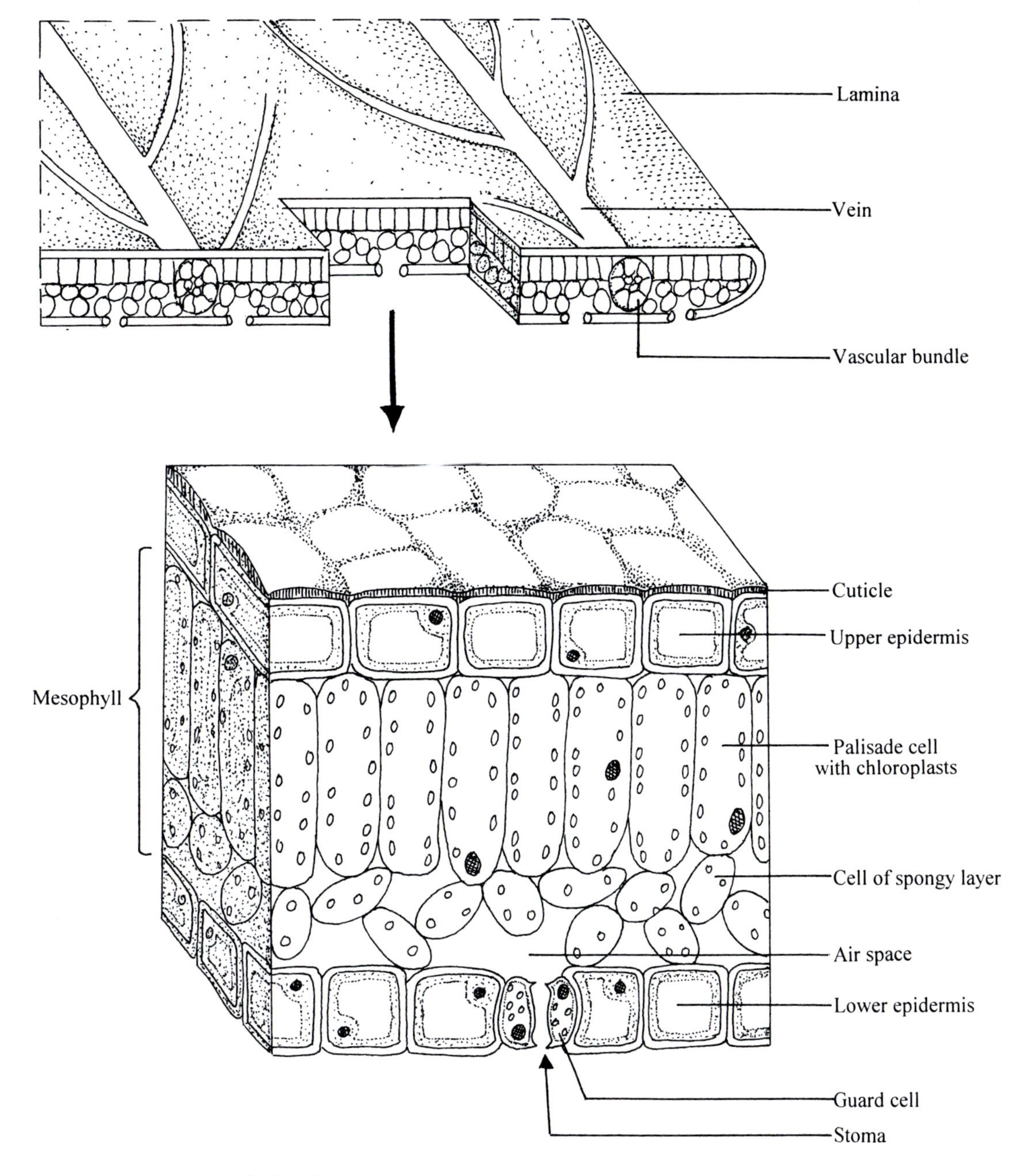

Internal structure of a leaf

Leaf Structure 2

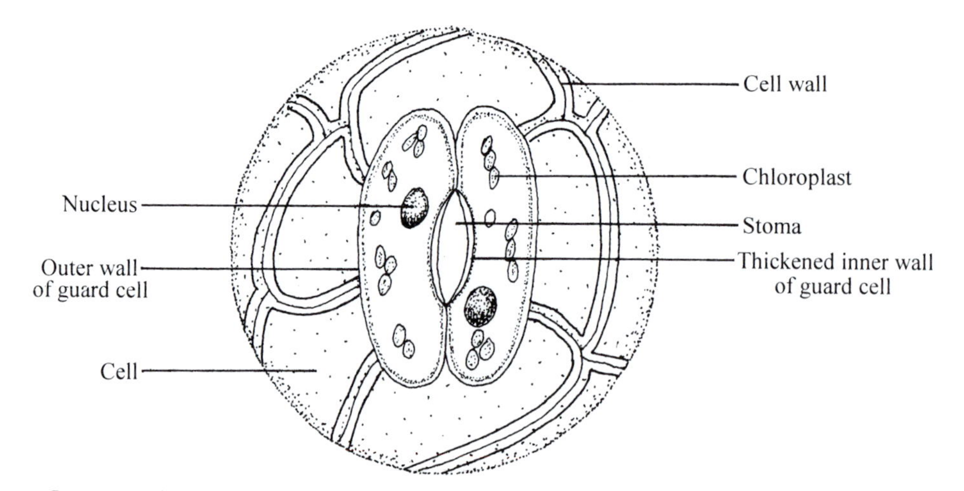

Lower epidermis of a leaf showing details of a typical stoma and guard cells

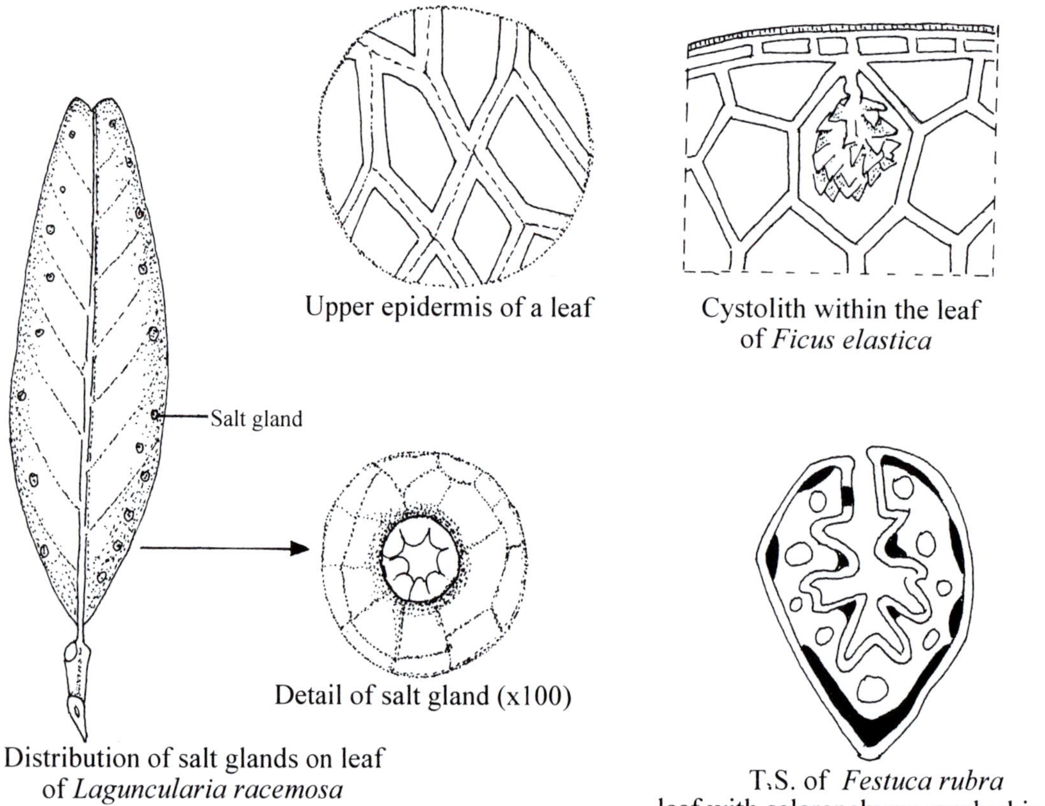

Upper epidermis of a leaf

Cystolith within the leaf of *Ficus elastica*

Detail of salt gland (x100)

Distribution of salt glands on leaf of *Laguncularia racemosa*

T.S. of *Festuca rubra* leaf with sclerenchyma marked in black

8. Hairs and Scales

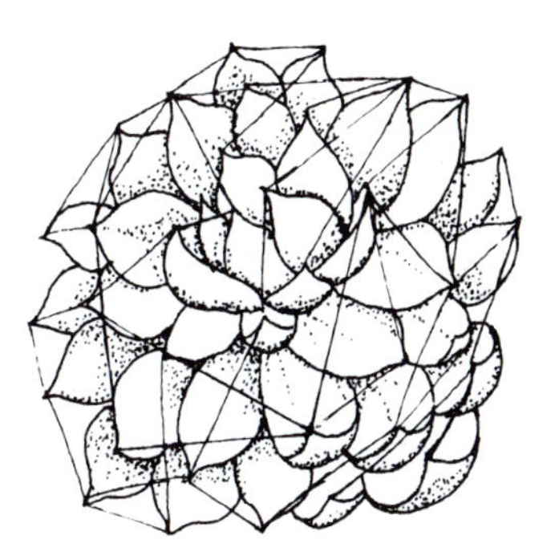

Hairs and Scales 1

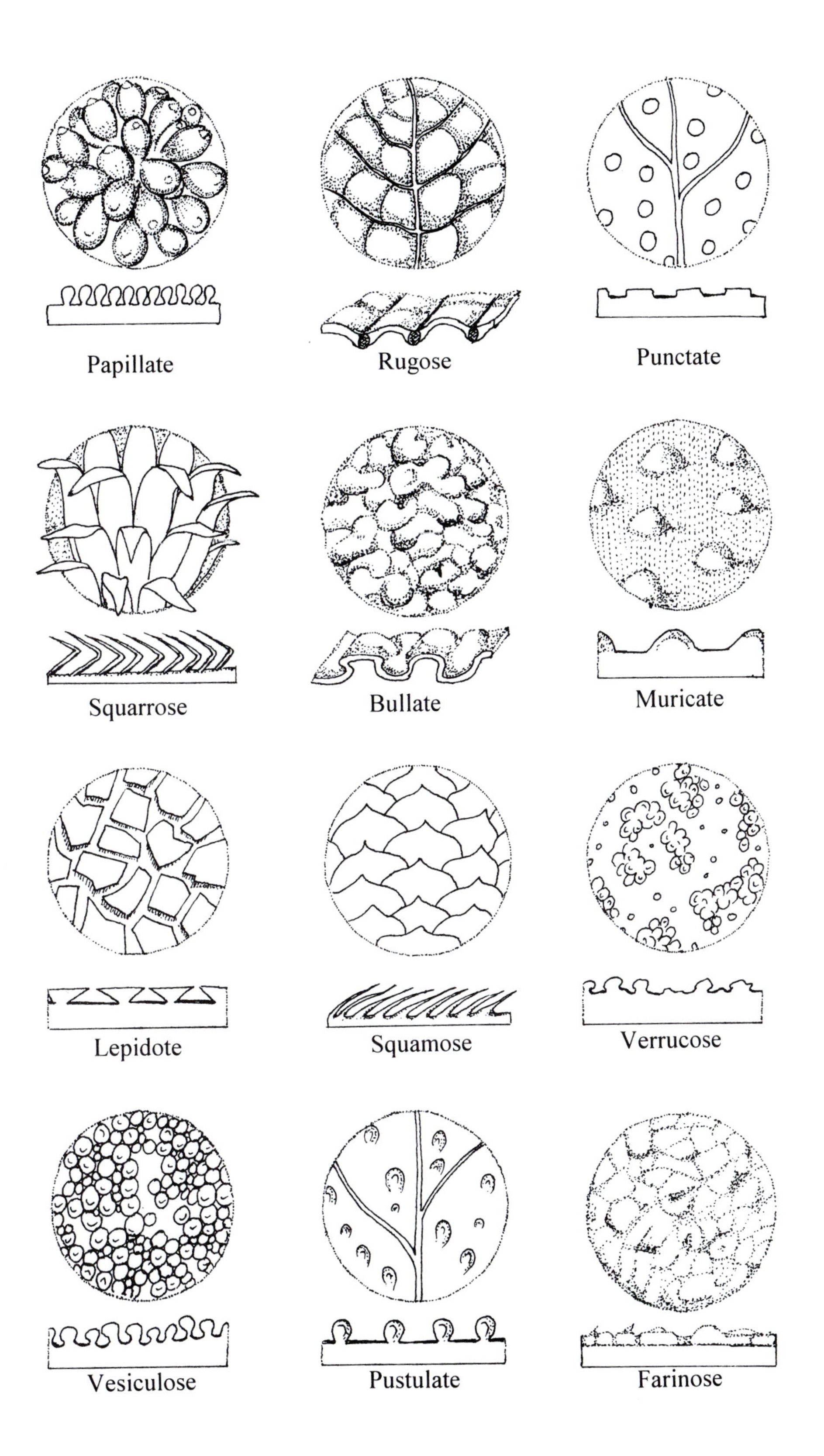

Hairs and Scales 2

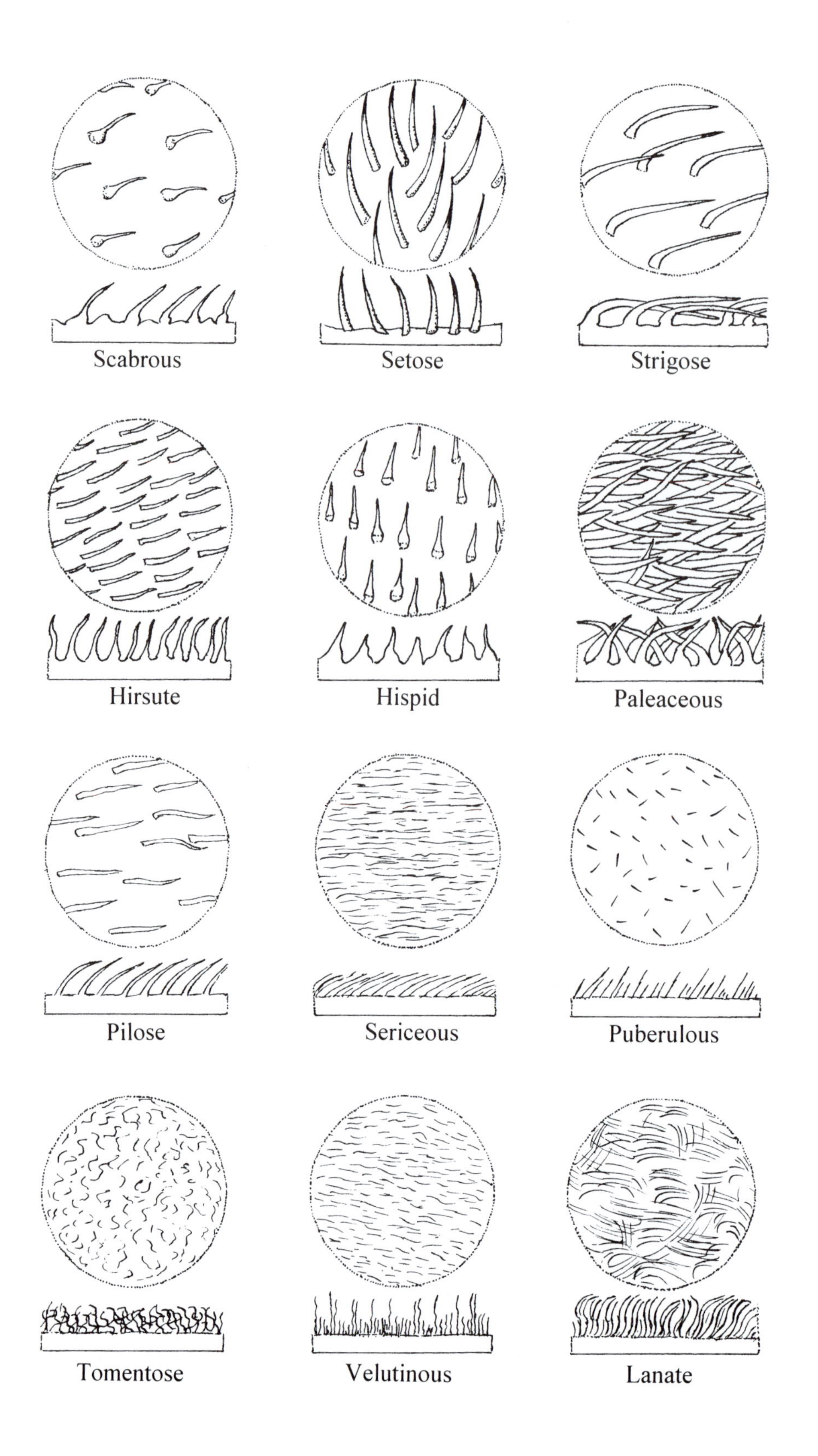

Hairs and Scales 3

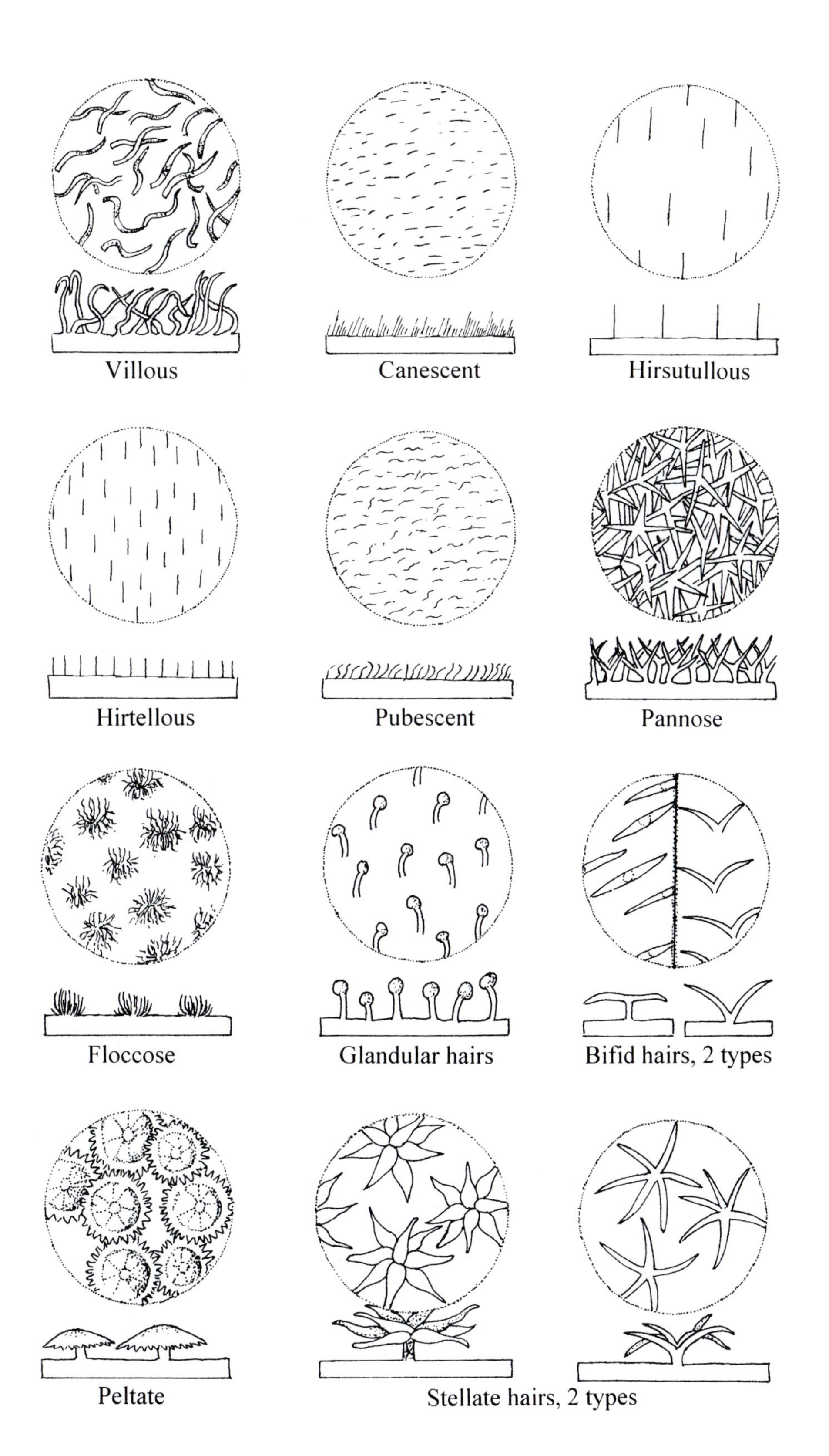

Hairs and Scales 4

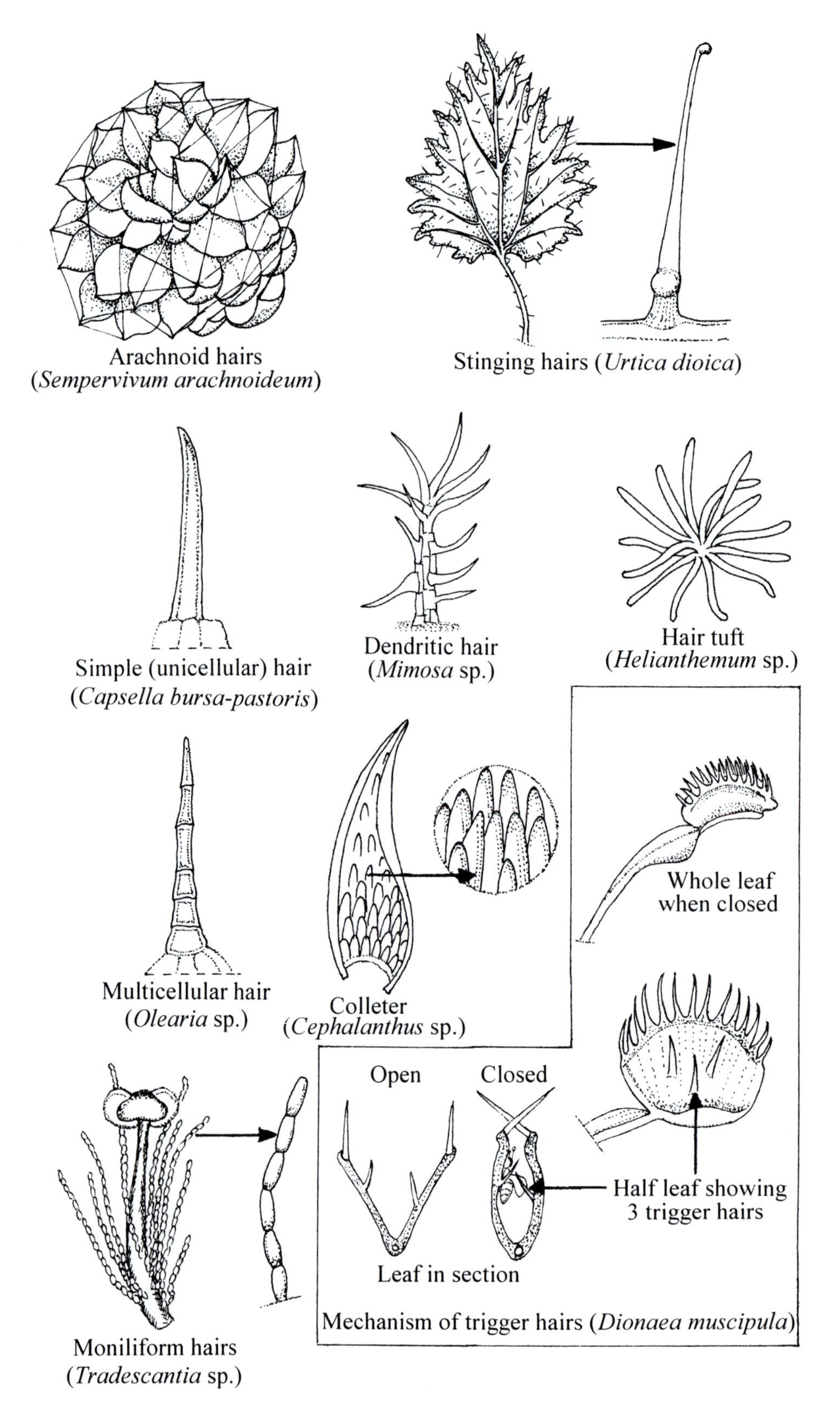

9. Floral Features

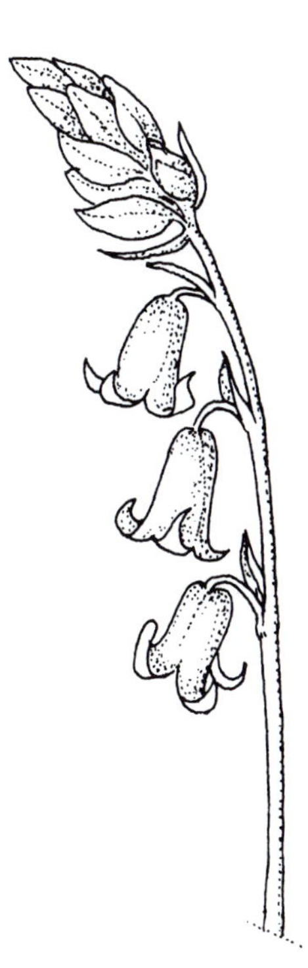

Inflorescences 1

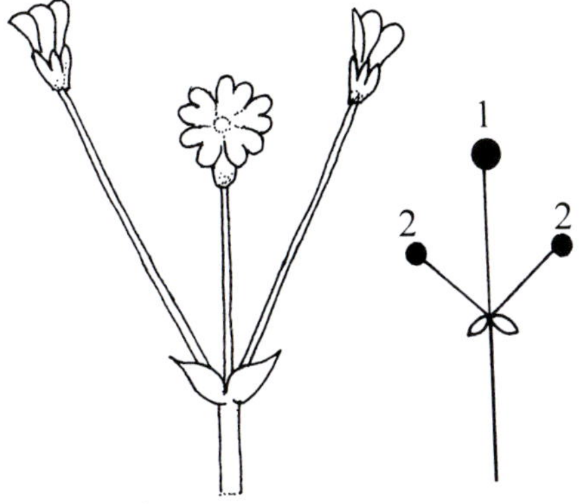

Simple cyme (*Cerastium* sp.)

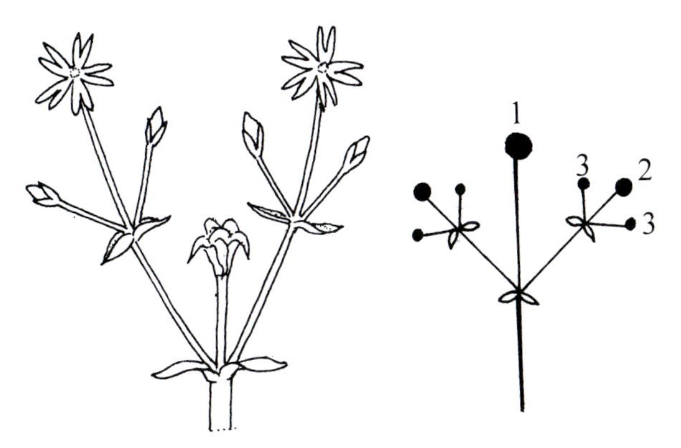

Compound cyme (*Stellaria* sp.)

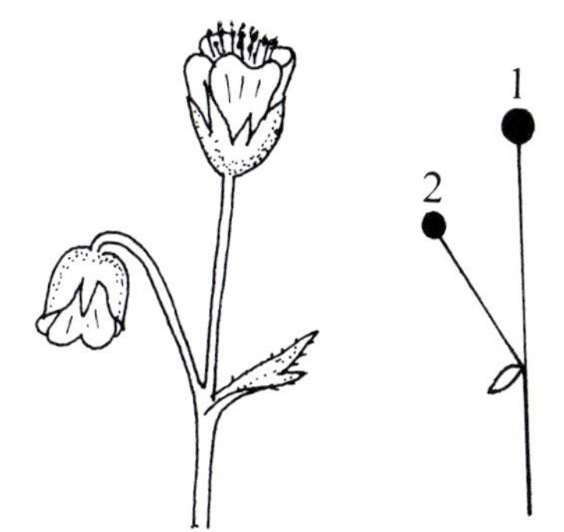

Simple cyme (*Geum* sp.)

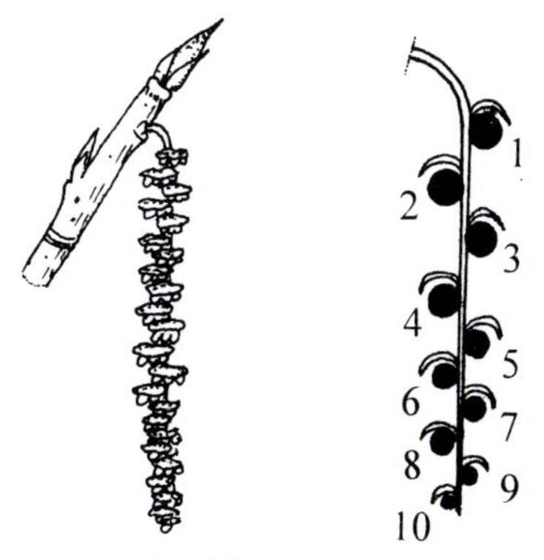

Catkin (*Populus* sp.)

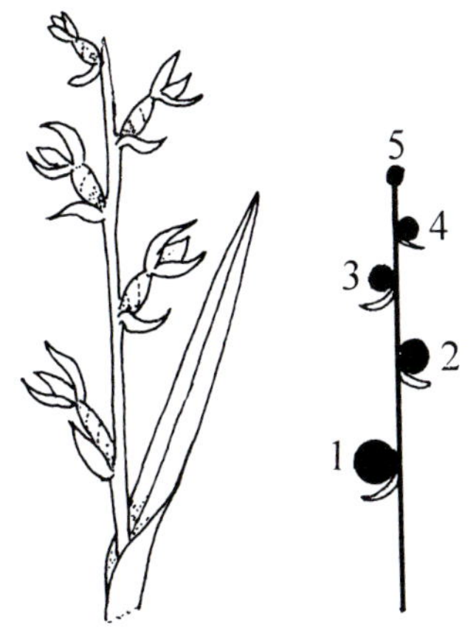

Spike (*Cephalanthera* sp.)

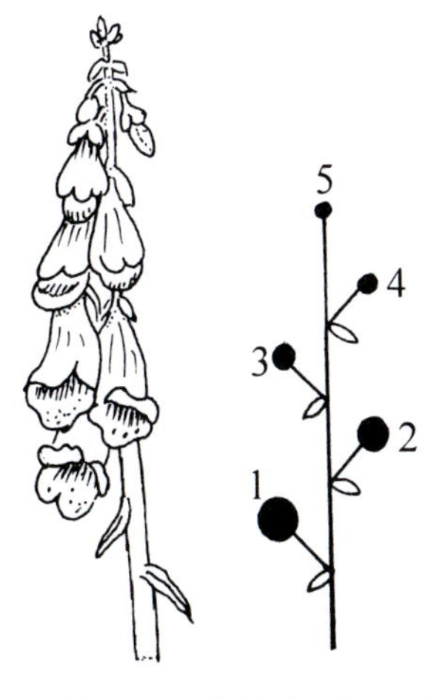

Raceme (*Digitalis* sp.)

Inflorescences 2

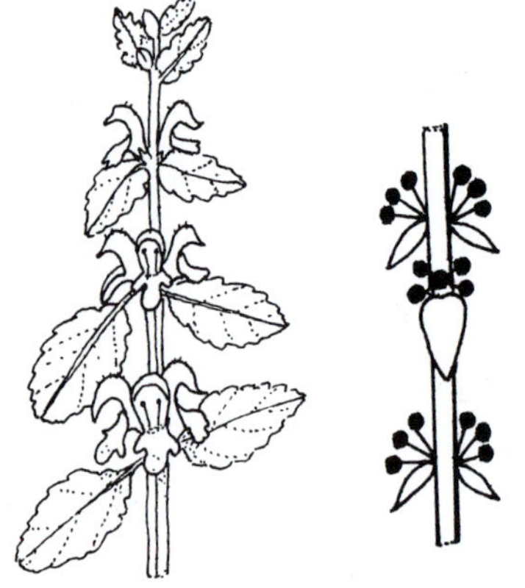

Verticillaster (*Lamium* sp.)

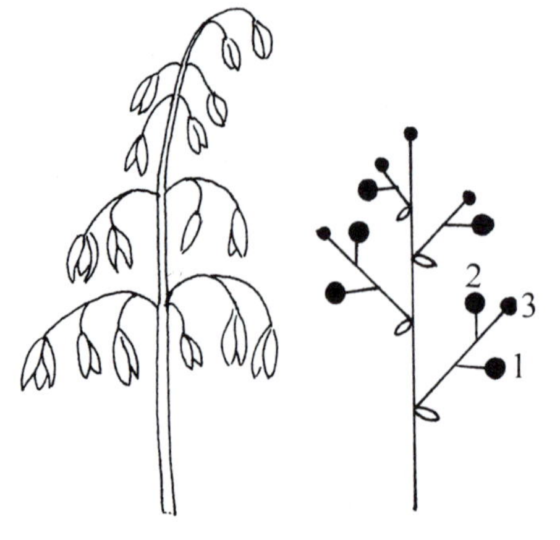

Panicle (*Avena* sp.)

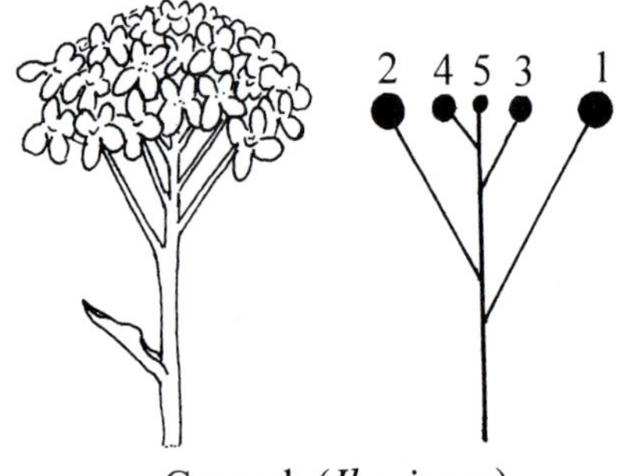

Corymb (*Iberis* sp.)

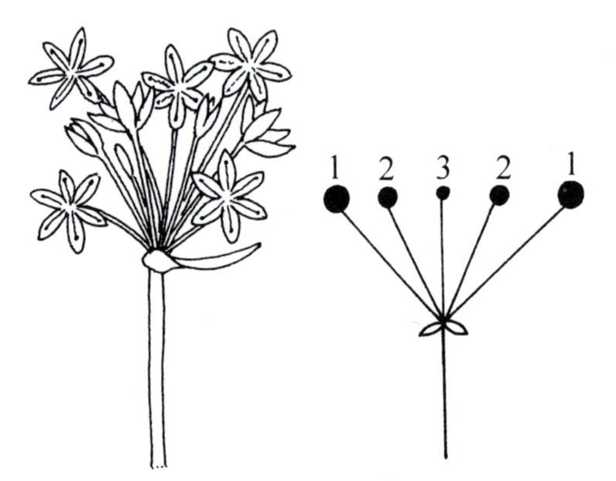

Simple umbel (*Allium* sp.)

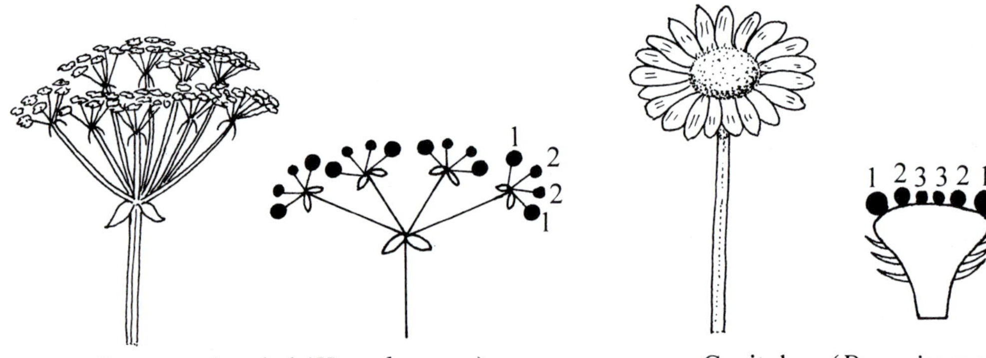

Compound umbel (*Heracleum* sp.)

Capitulum (*Doronicum* sp.)

Inflorescences 3

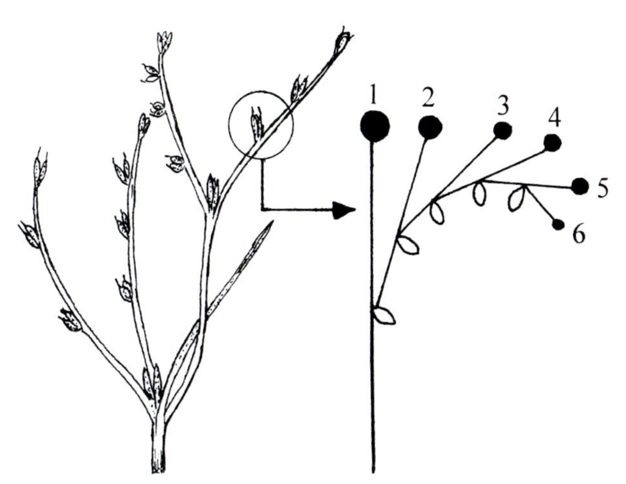

Drepanium (*Juncus bufonis*)
(also arrangement of drepania on plant)

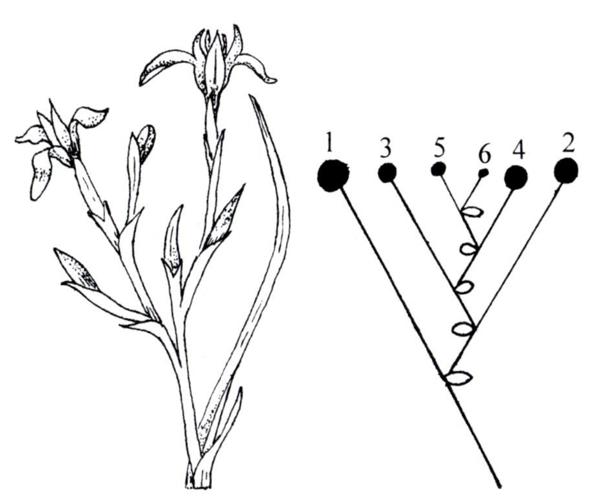

Rhipidium (*Moraea macgregorii*)

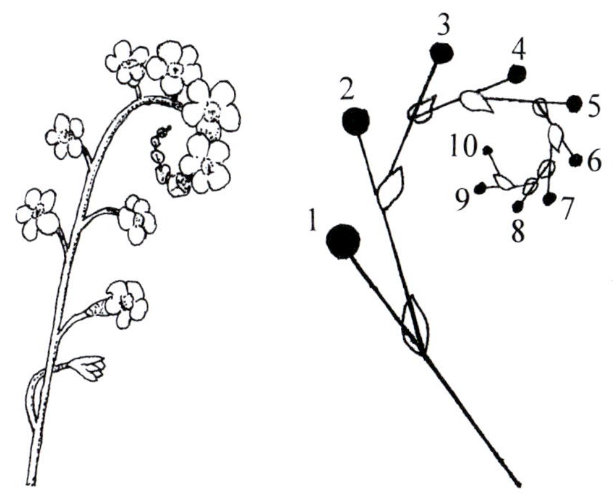

Cincinnus (*Myosotis scorpioides*)

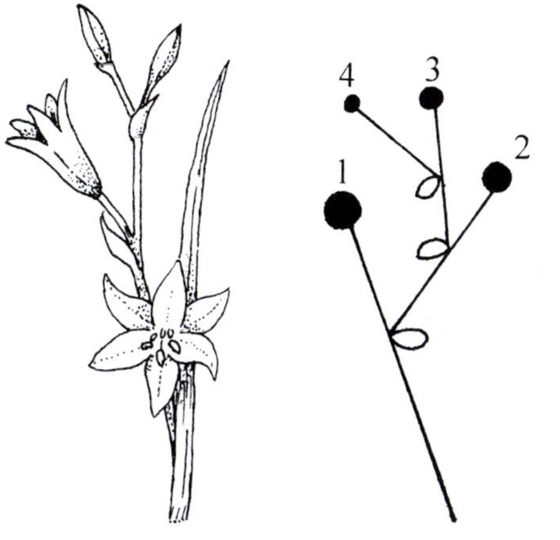

Bostryx (*Hemerocallis minor*)

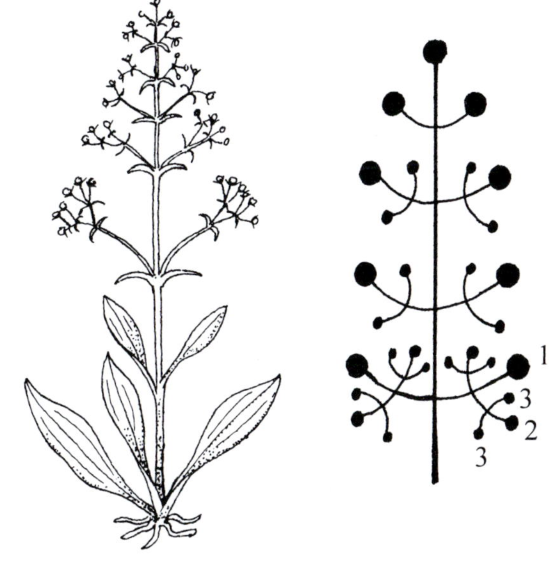

Thyrse (*Valeriana saxatilis*)

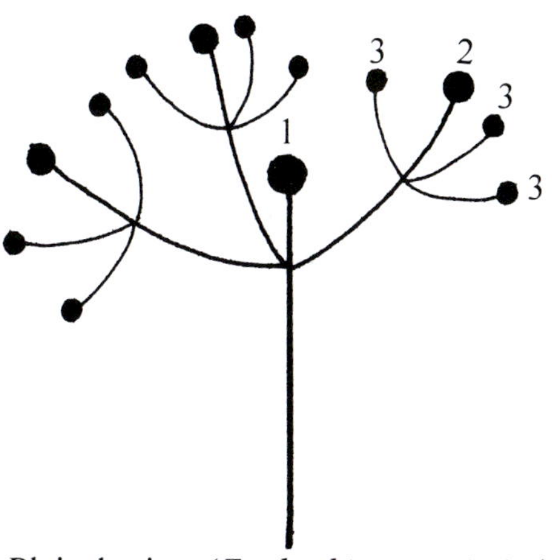

Pleiochasium (*Euphorbia cyparissias*)

Inflorescences 4

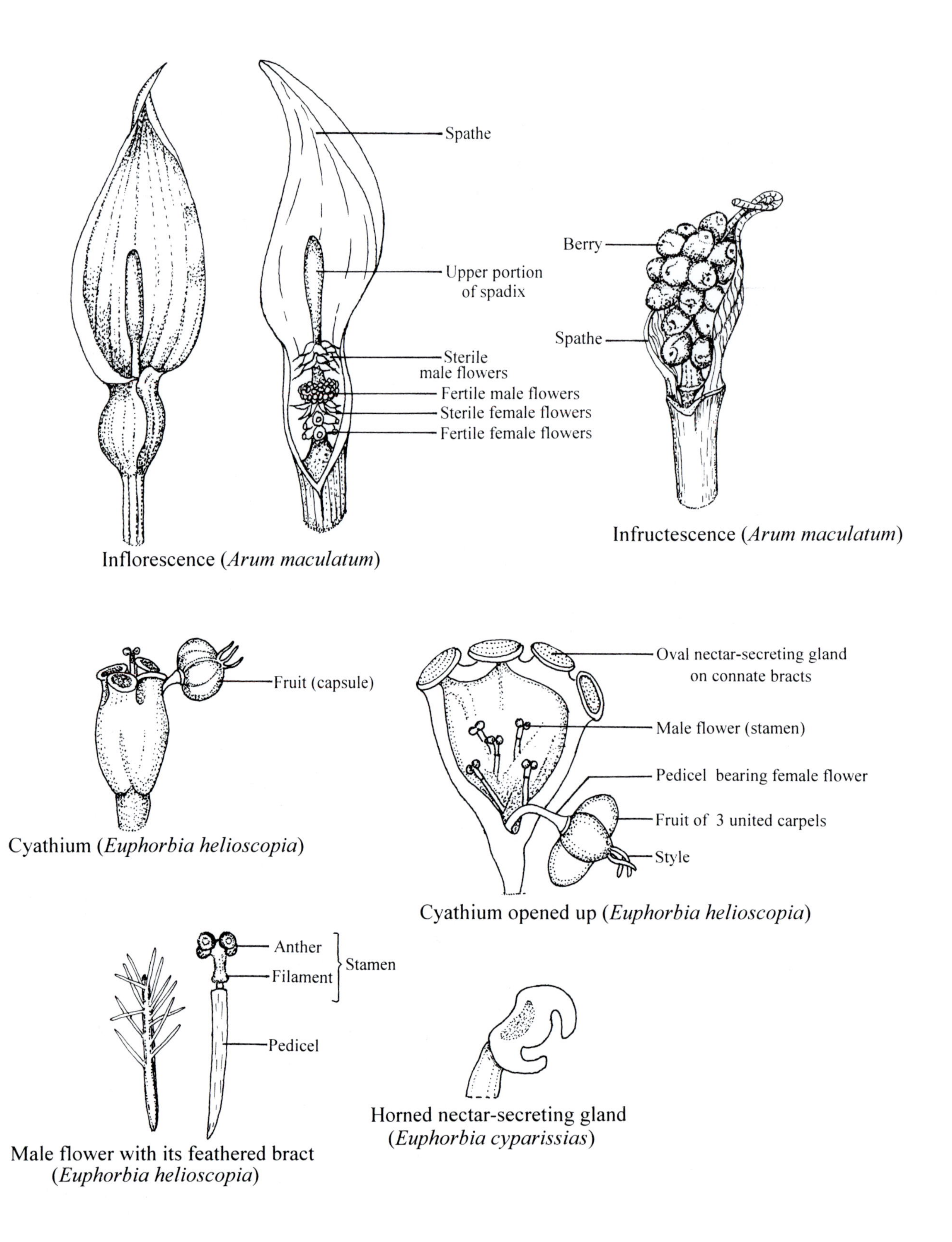

Inflorescence (*Arum maculatum*)

Infructescence (*Arum maculatum*)

Cyathium (*Euphorbia helioscopia*)

Cyathium opened up (*Euphorbia helioscopia*)

Male flower with its feathered bract (*Euphorbia helioscopia*)

Horned nectar-secreting gland (*Euphorbia cyparissias*)

Other Inflorescence and Flower Features 1

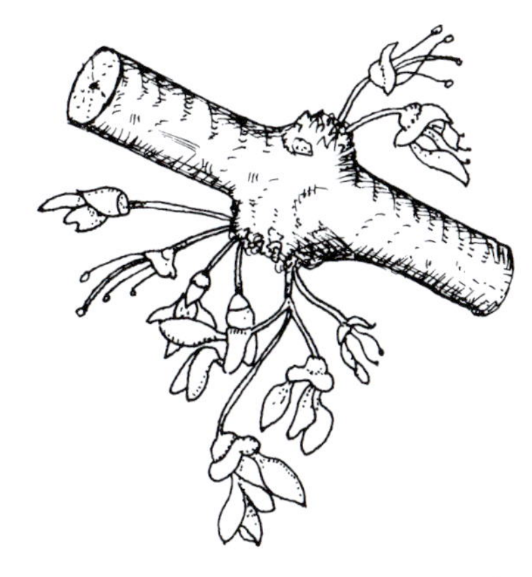

Cauliflory, showing flowers on main stem (*Cercis siliquastrum*)

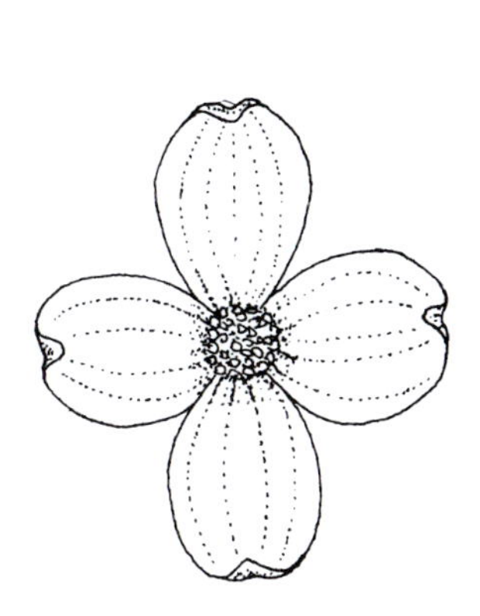

Pseudanthium subtended by bracts (*Cornus florida*)

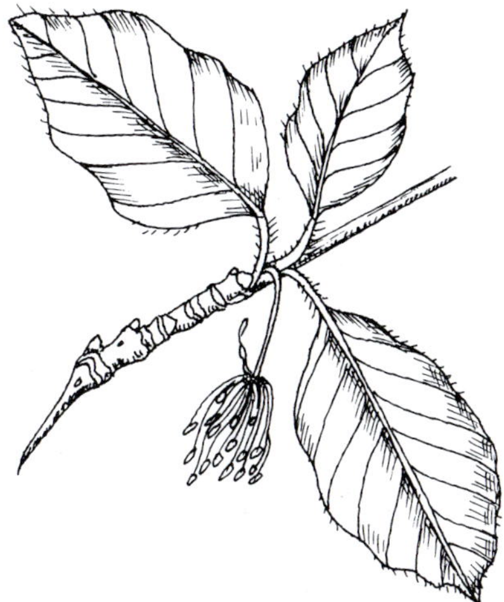

Cluster of male flowers (*Fagus sylvatica*)

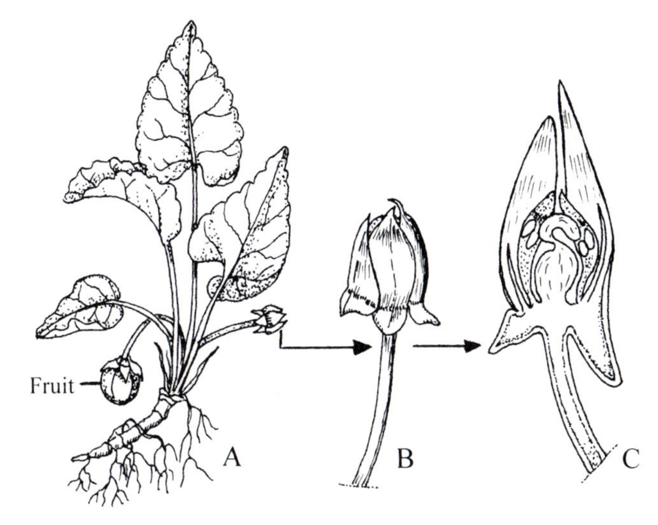

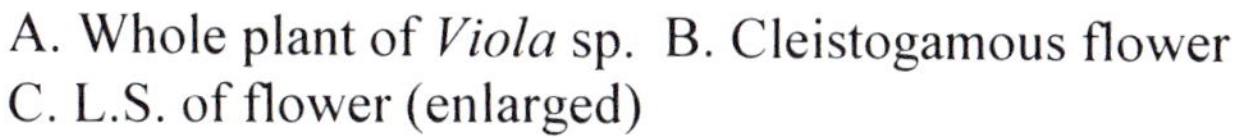

A. Whole plant of *Viola* sp. B. Cleistogamous flower
C. L.S. of flower (enlarged)

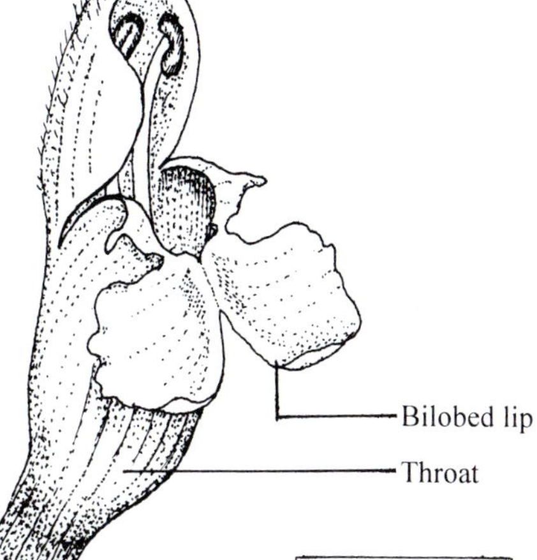

Upper portion of corolla (*Lamium purpureum*)

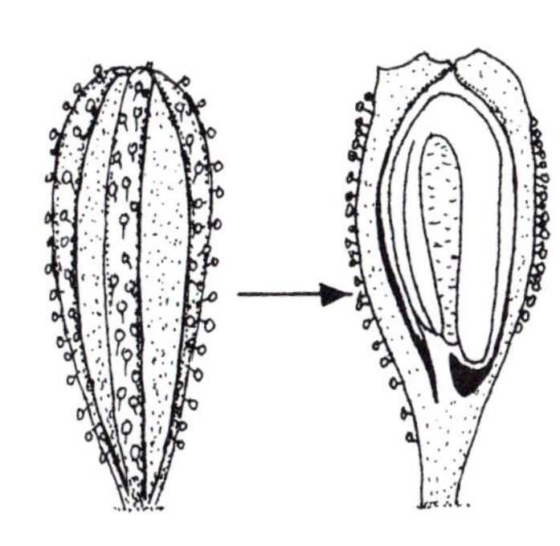

Anthocarp whole and in section (*Boerhavia diffusa*)

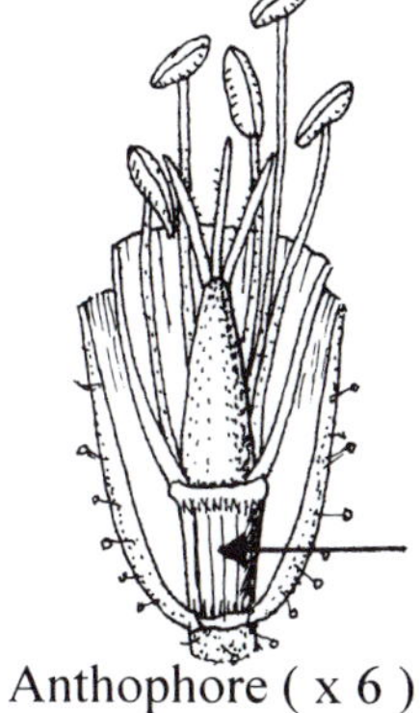

Anthophore (x 6) (*Silene gallica*)

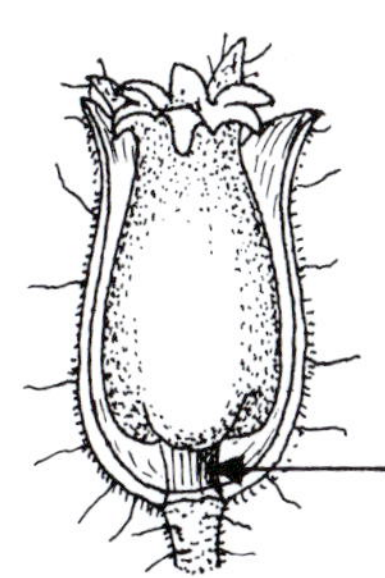

Carpophore (x 3) (*Silene gallica*)

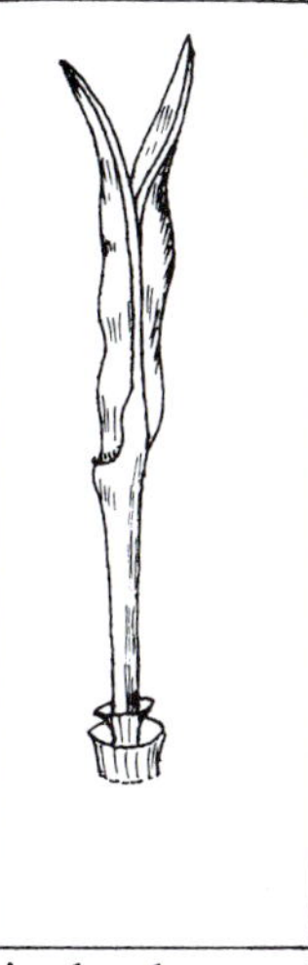

Bivalved capsule (*Gentiana* sp.)

Other Inflorescence and Flower Features 2

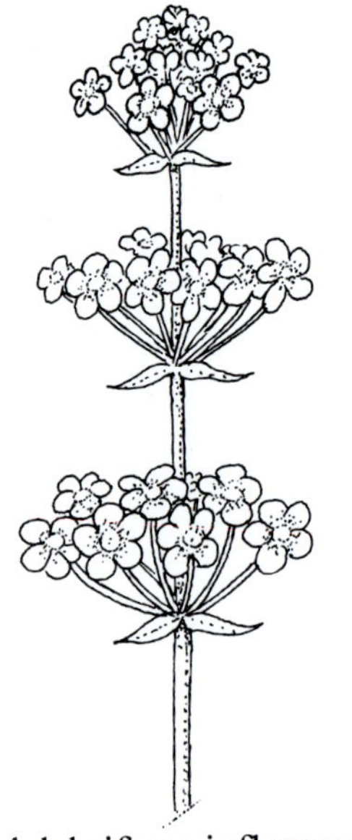

Candelabriform inflorescence
(*Primula japonica*)

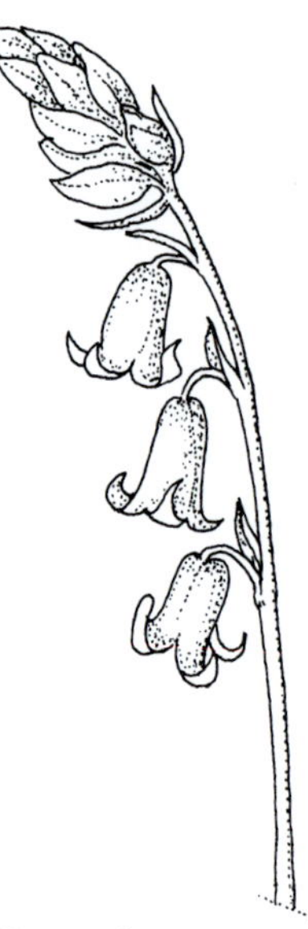

Secund racemose inflorescence
(*Hyacinthoides non-scripta*)

Diagram of an actinomorphic flower
showing radial symmetry

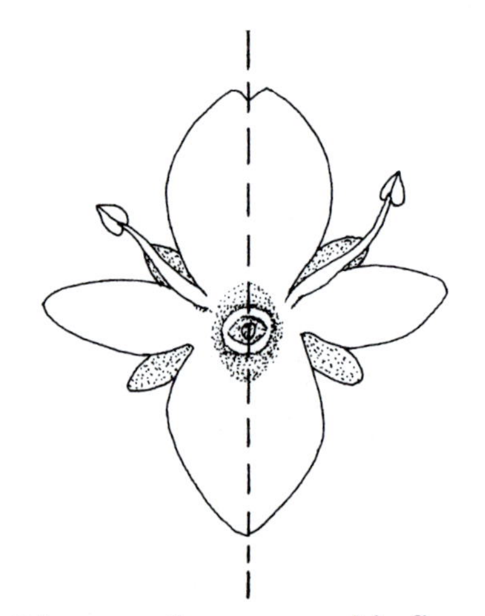

Diagram of a zygomorphic flower
showing bilateral symmetry

Calyces and Epicalyces

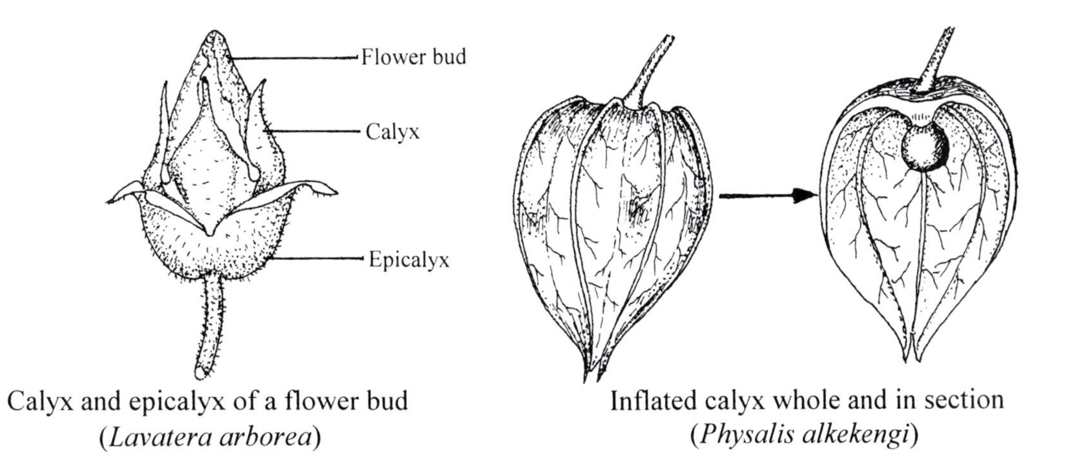

Calyx and epicalyx of a flower bud (*Lavatera arborea*)

Inflated calyx whole and in section (*Physalis alkekengi*)

Flower Forms 1

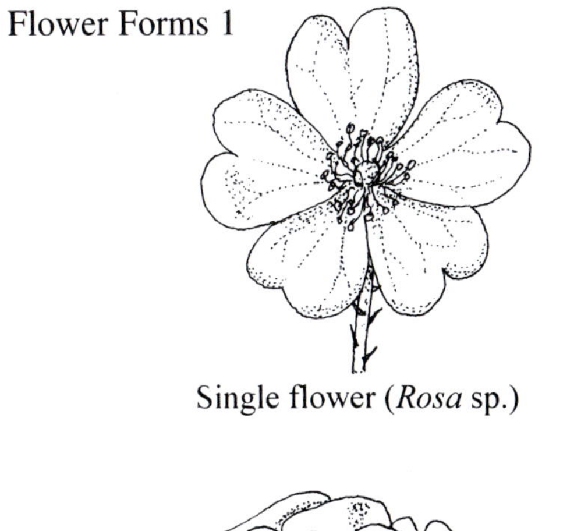

Single flower (*Rosa* sp.)

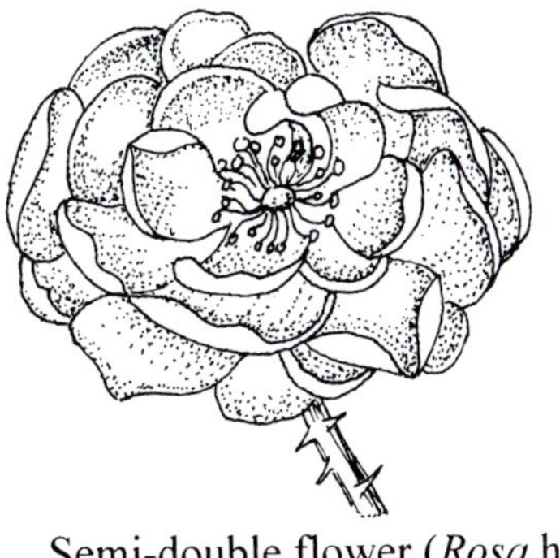

Semi-double flower (*Rosa* hybrid)

Double flower (*Rosa* hybrid)

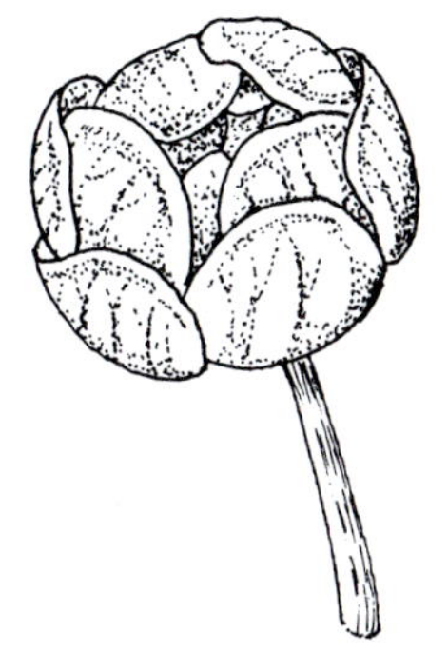

Globular flower (*Trollius europaeus*)

Flower Forms 2

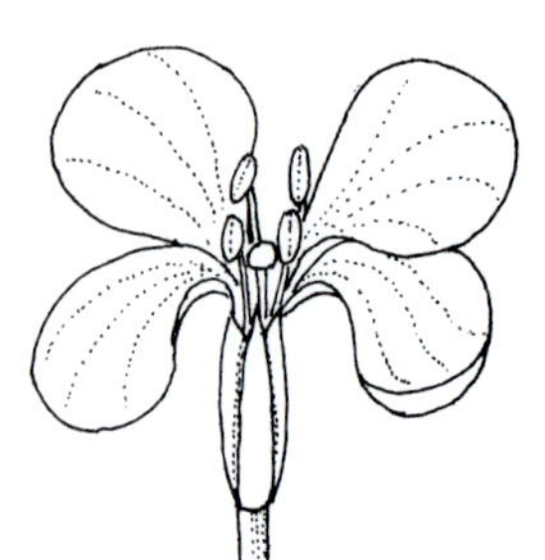
Cruciform

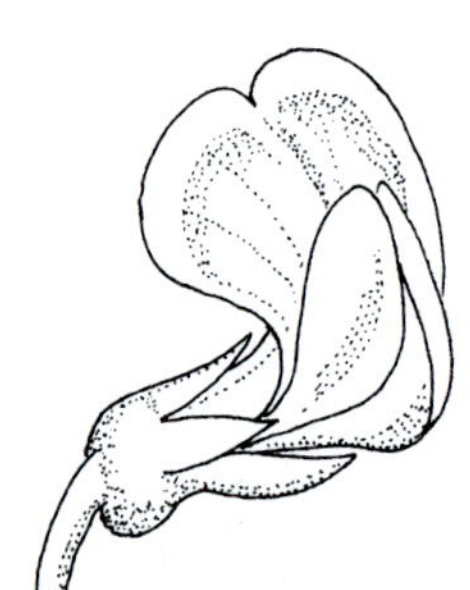
Papilionaceous

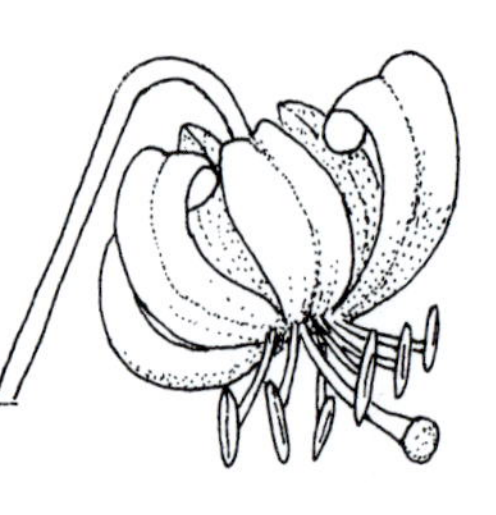
Liliaceous

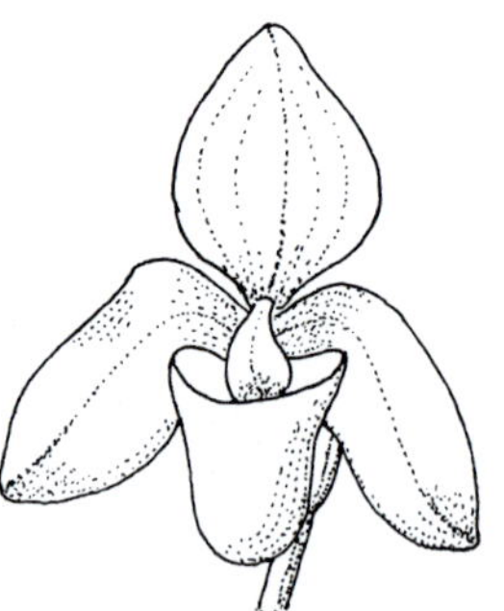
Orchidaceous

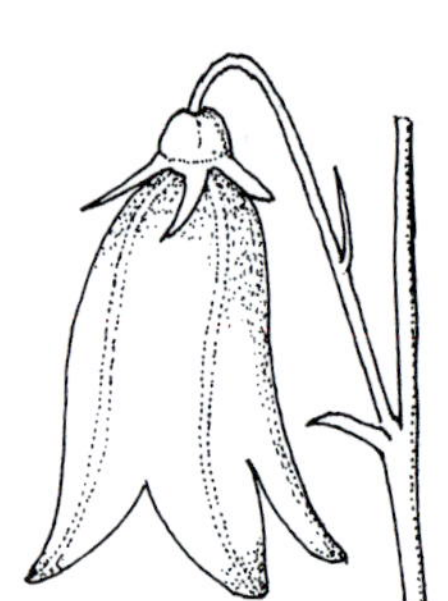
Campanulate

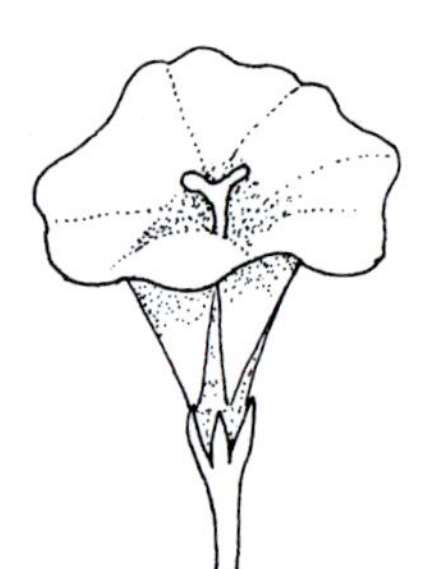
Funnel-shaped

Rotate

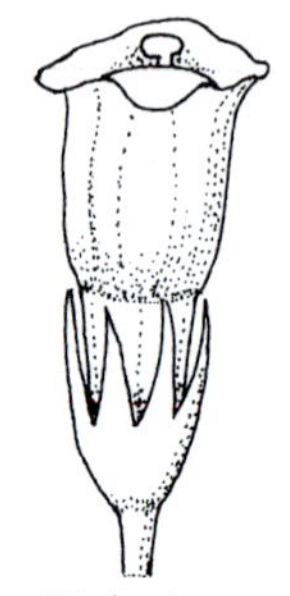
Tubular

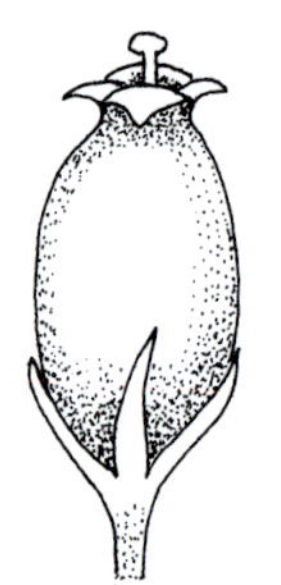
Urceolate

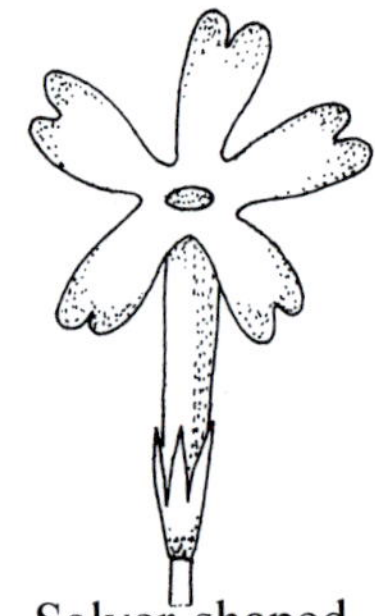
Salver-shaped

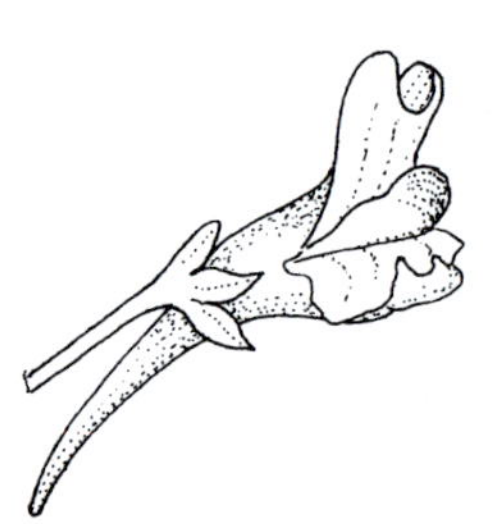
Personate with spur

Saccate

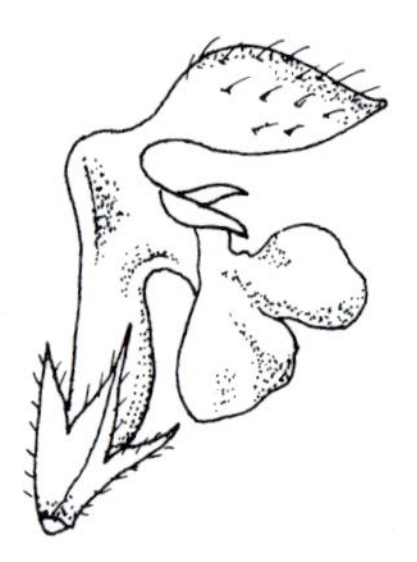
Labiate

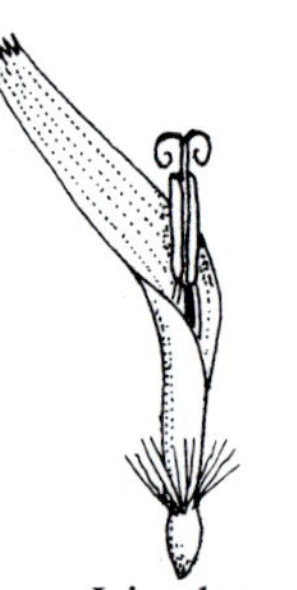
Ligulate

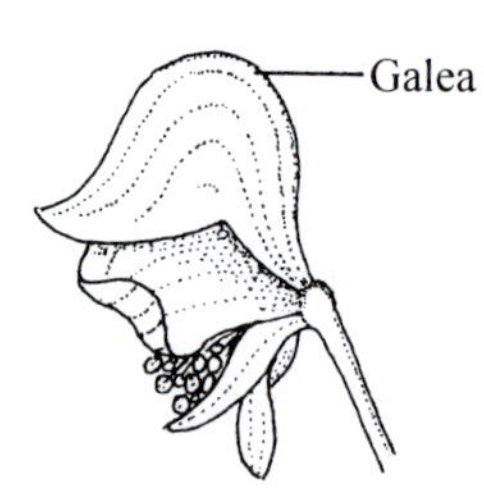

Galeate (*Consolida* sp.)

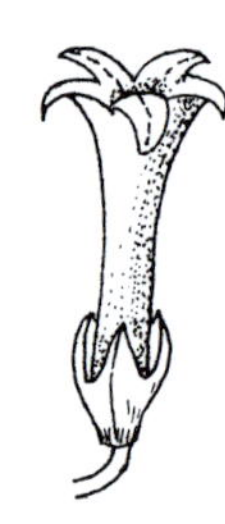
Trumpet-shaped

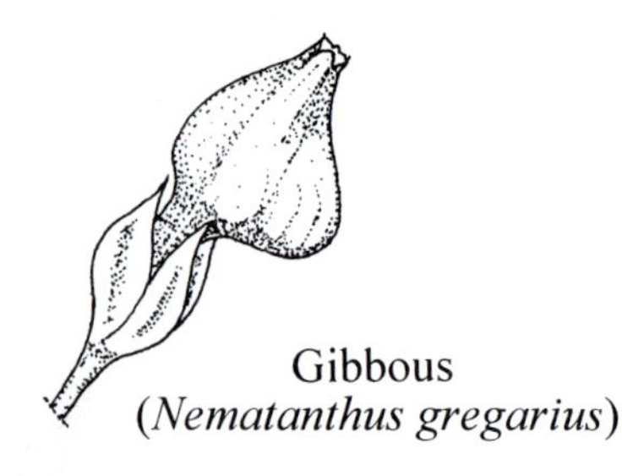

Gibbous (*Nematanthus gregarius*)

Dissimilar segments (*Iris* sp.)

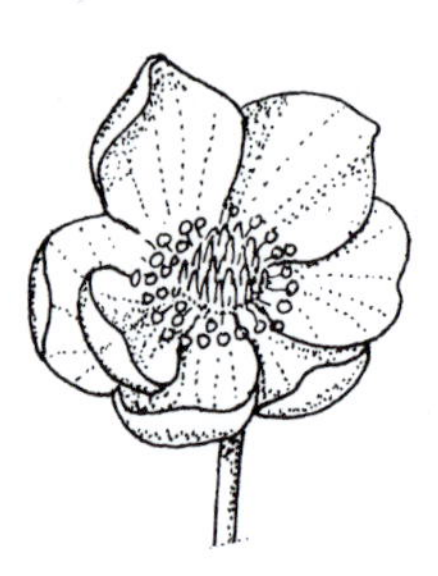
Sepals petaloid (*Anemone* sp.)

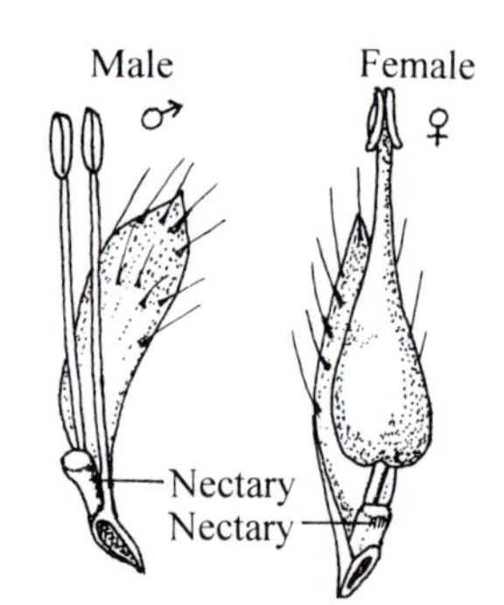

Achlamydeous (*Salix* sp.)

Pollen

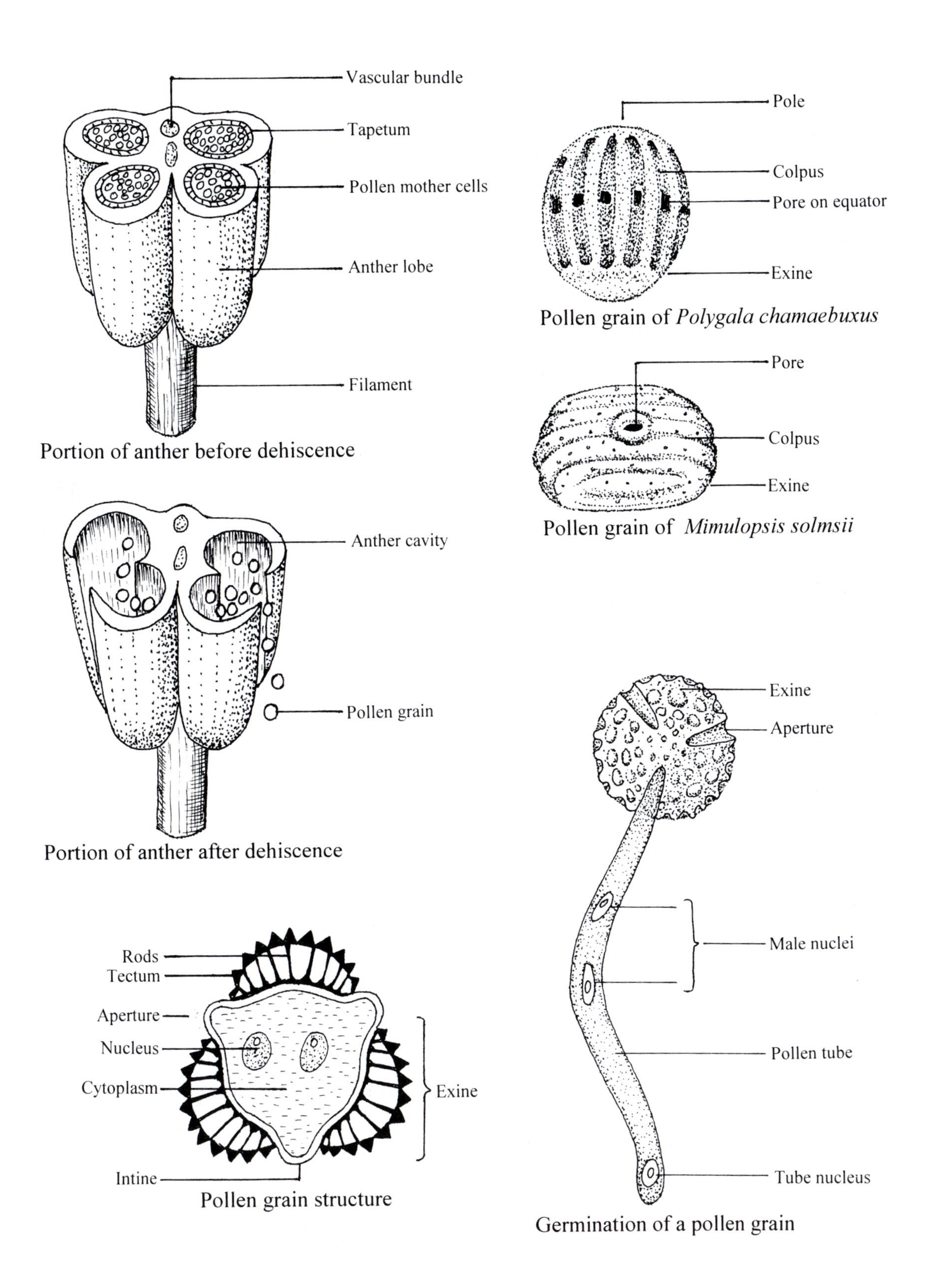

Portion of anther before dehiscence

Portion of anther after dehiscence

Pollen grain structure

Pollen grain of *Polygala chamaebuxus*

Pollen grain of *Mimulopsis solmsii*

Germination of a pollen grain

Angiosperm Fertilisation

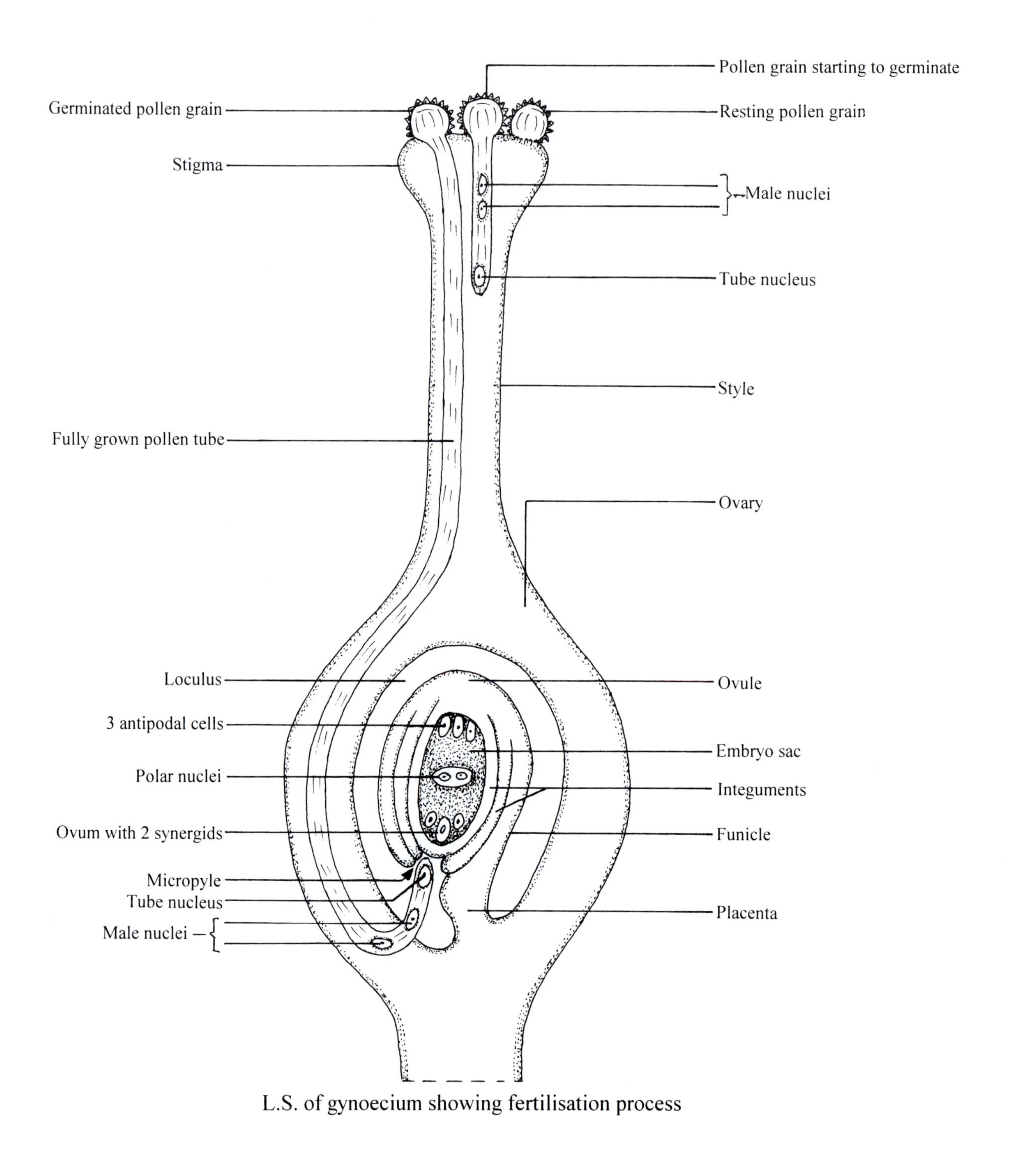

L.S. of gynoecium showing fertilisation process

Stamen Arrangement 1

Antisepalous

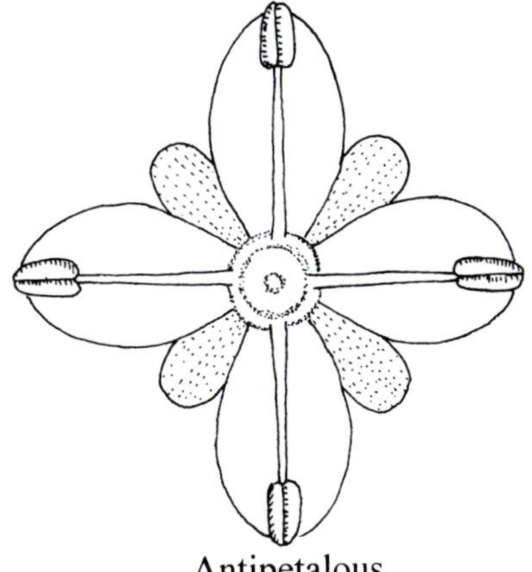
Antipetalous

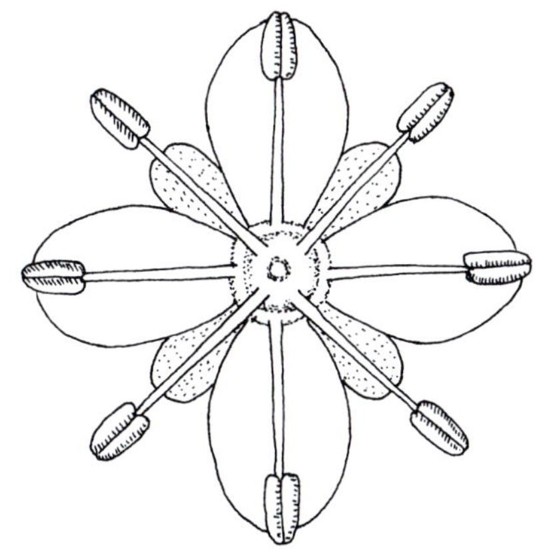
Obdiplostemonous

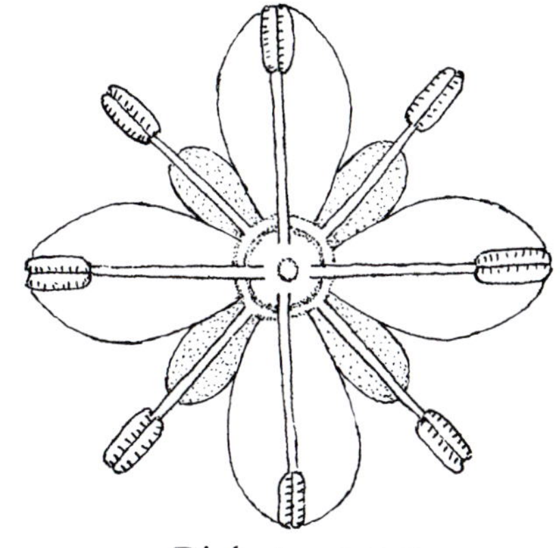
Diplostemonous

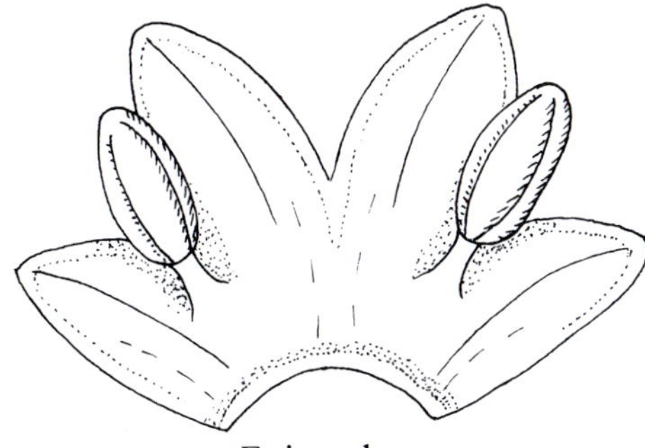
Epipetalous

Stamen Arrangement 2

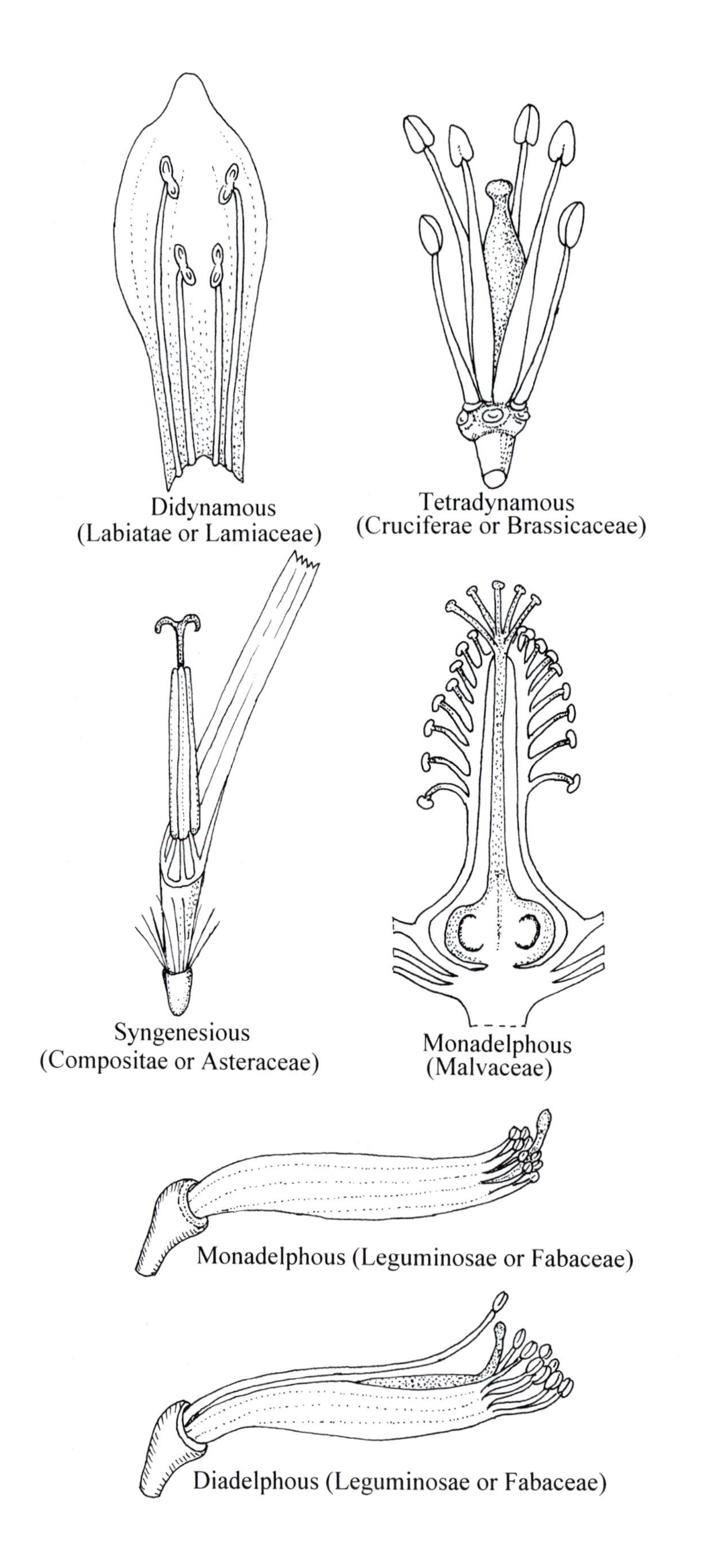

Stamen Types 1

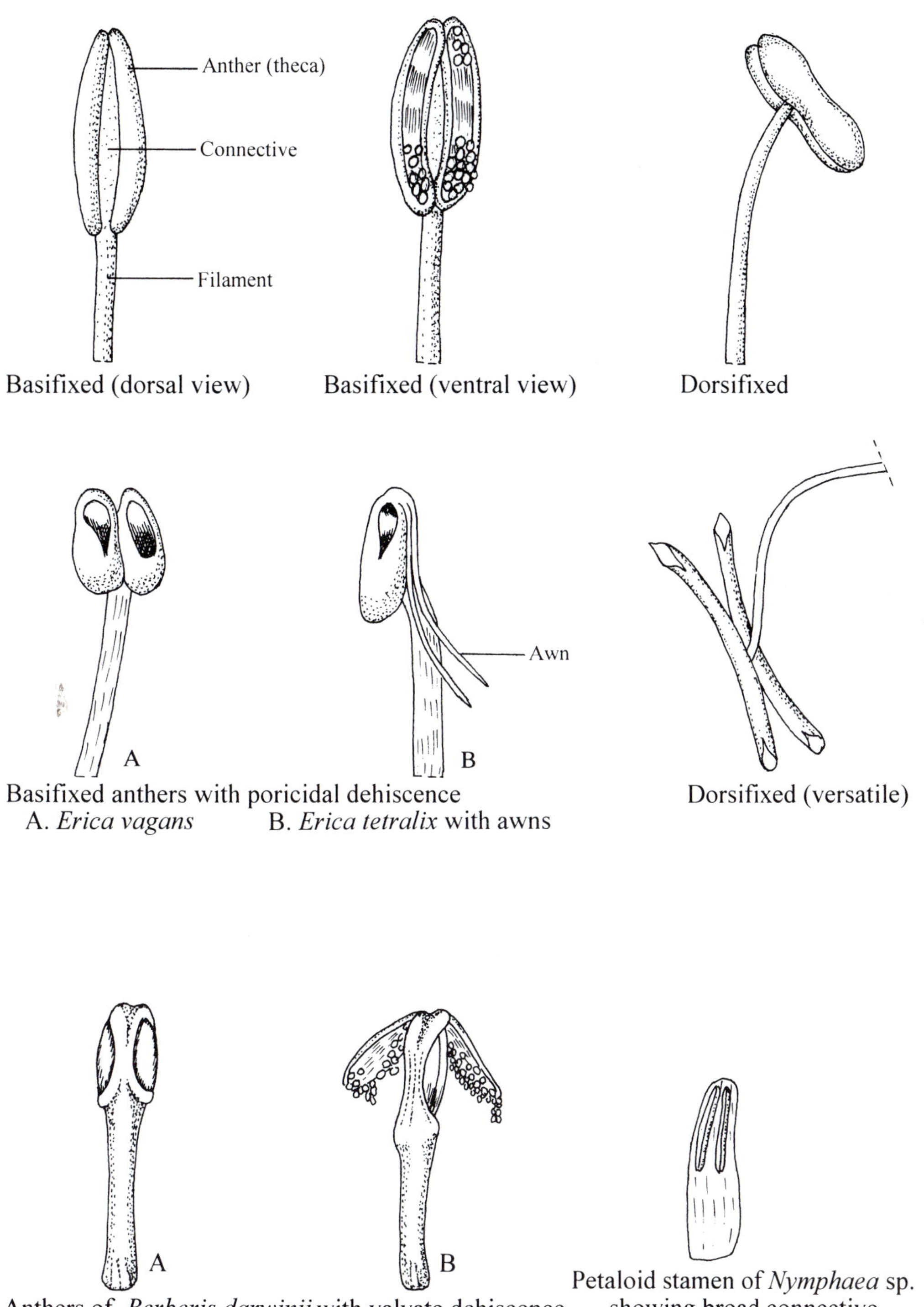

Basifixed (dorsal view)

Basifixed (ventral view)

Dorsifixed

Basifixed anthers with poricidal dehiscence
A. *Erica vagans* B. *Erica tetralix* with awns

Dorsifixed (versatile)

Anthers of *Berberis darwinii* with valvate dehiscence
A. Before dehiscence B. After dehiscence

Petaloid stamen of *Nymphaea* sp. showing broad connective

Stamen Types 2

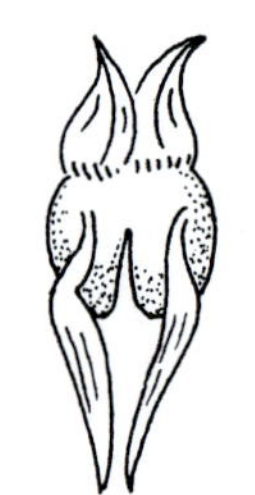

Anthers with nectar spurs
(*Viola* sp.)

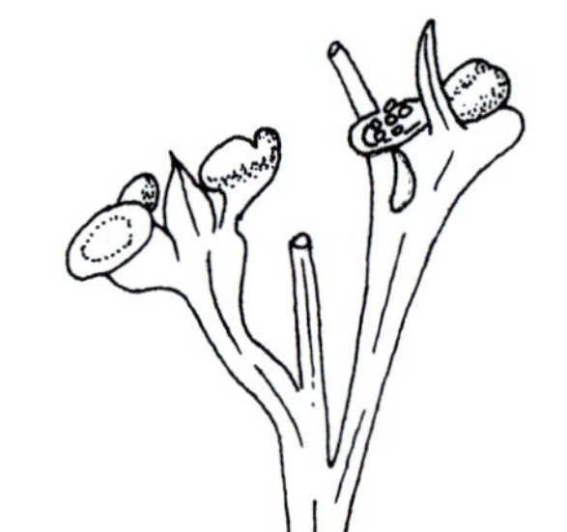

Several anthers on
a branched filament
(*Ricinus communis*)

Anthers with transverse attachment
(*Russelia* sp.)

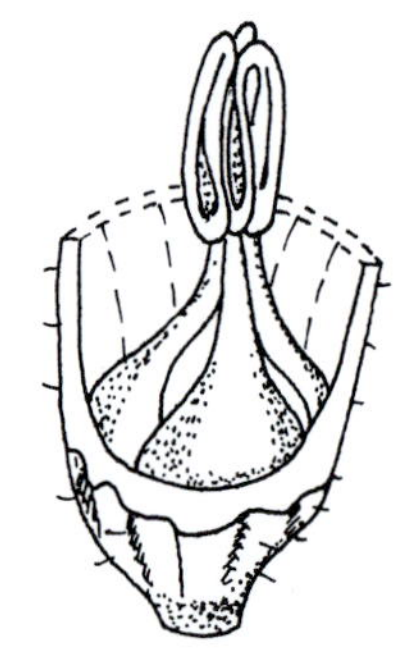

Anthers coherent, arching over nectary
(*Cucurbita* sp.)

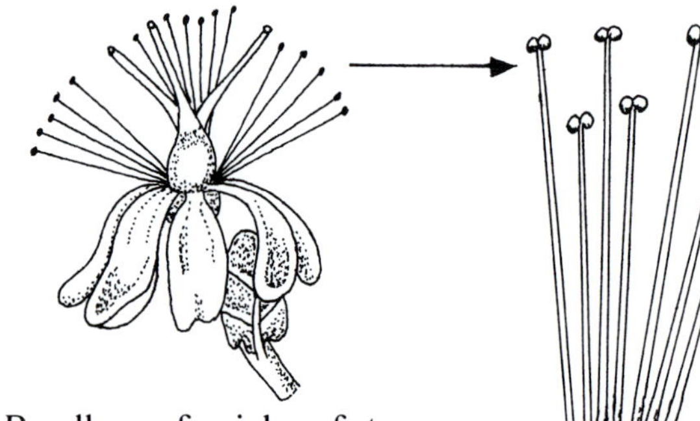

Bundles or fascicles of stamens
surrounding the gynoecium
of *Hypericum perforatum*

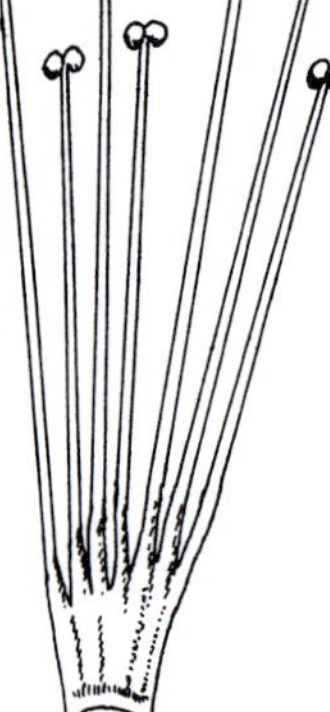

Fascicle of stamens
connate near the base

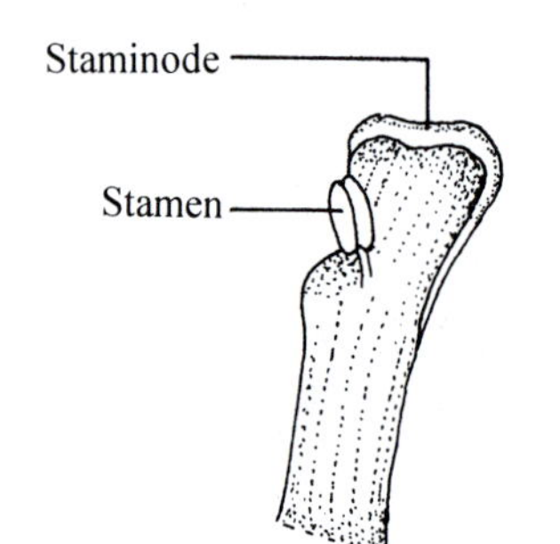

Fertile stamen with single cell
attached to a staminode (*Calathea* sp.)

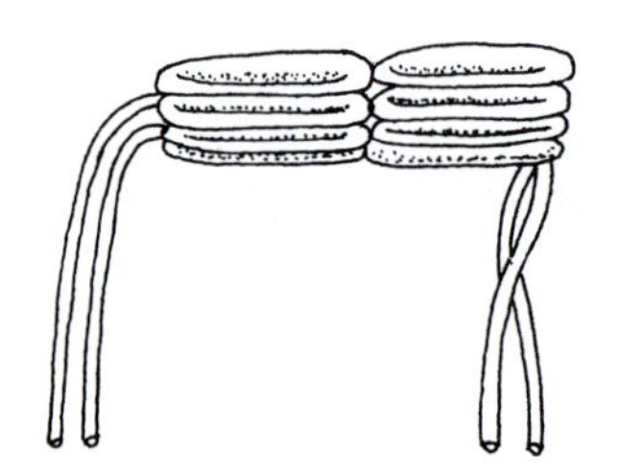

Connate anthers of *Columnea* x *banksii*

Dehiscence of Stamens

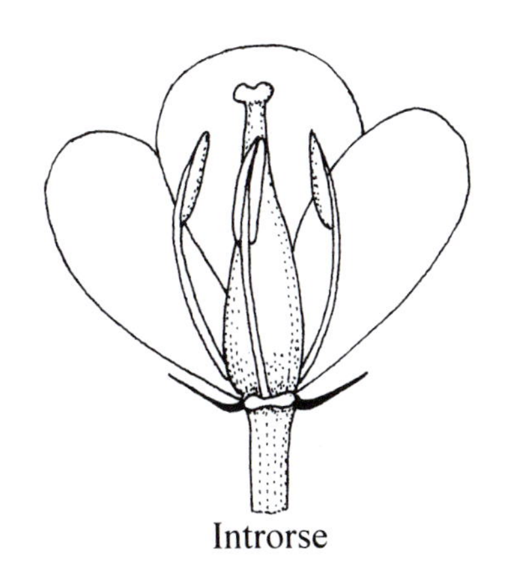

Introrse

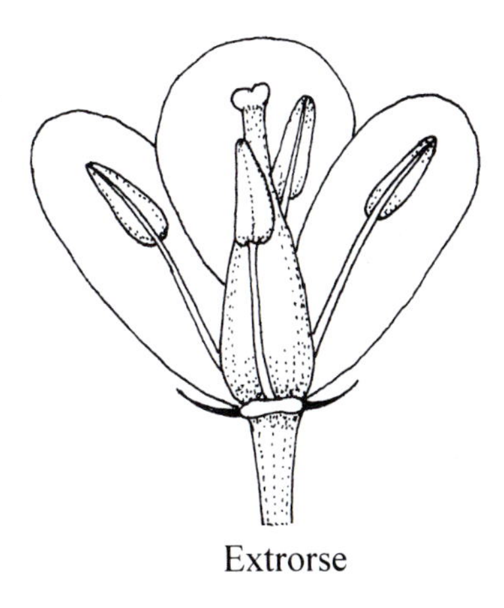

Extrorse

Ovary Position

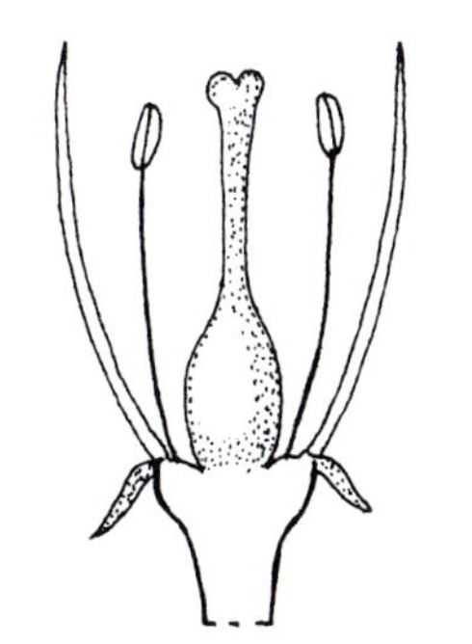

Ovary superior, flower hypogynous

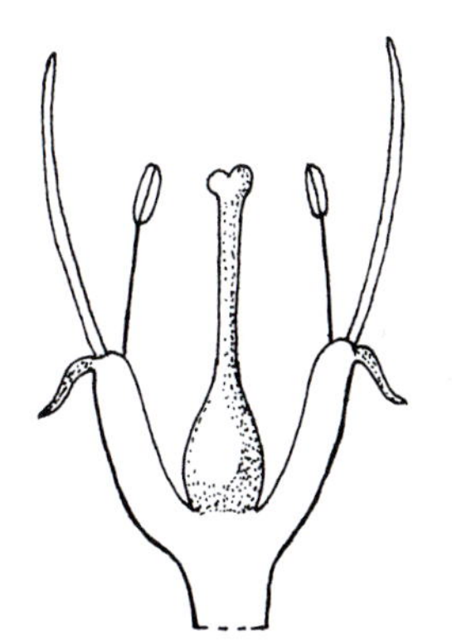

Ovary superior, flower perigynous

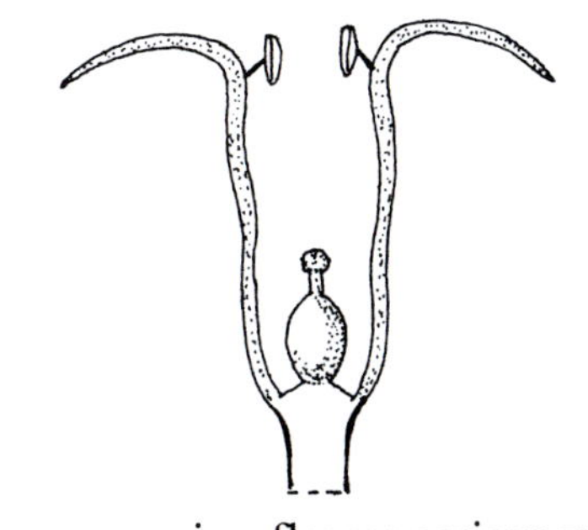

Ovary superior, flower perigynous

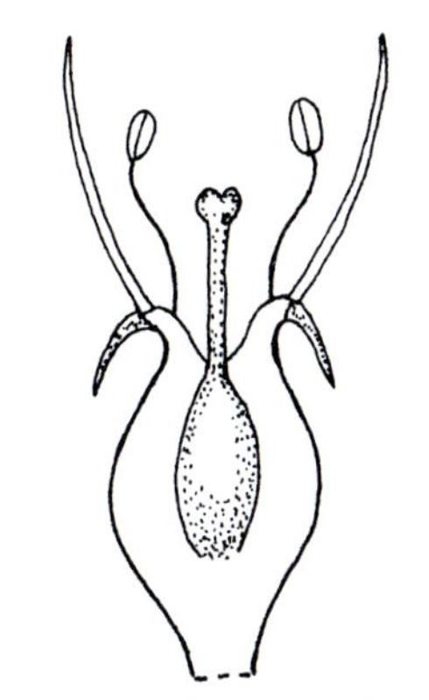

Ovary inferior, flower epigynous

Ovaries and Carpels

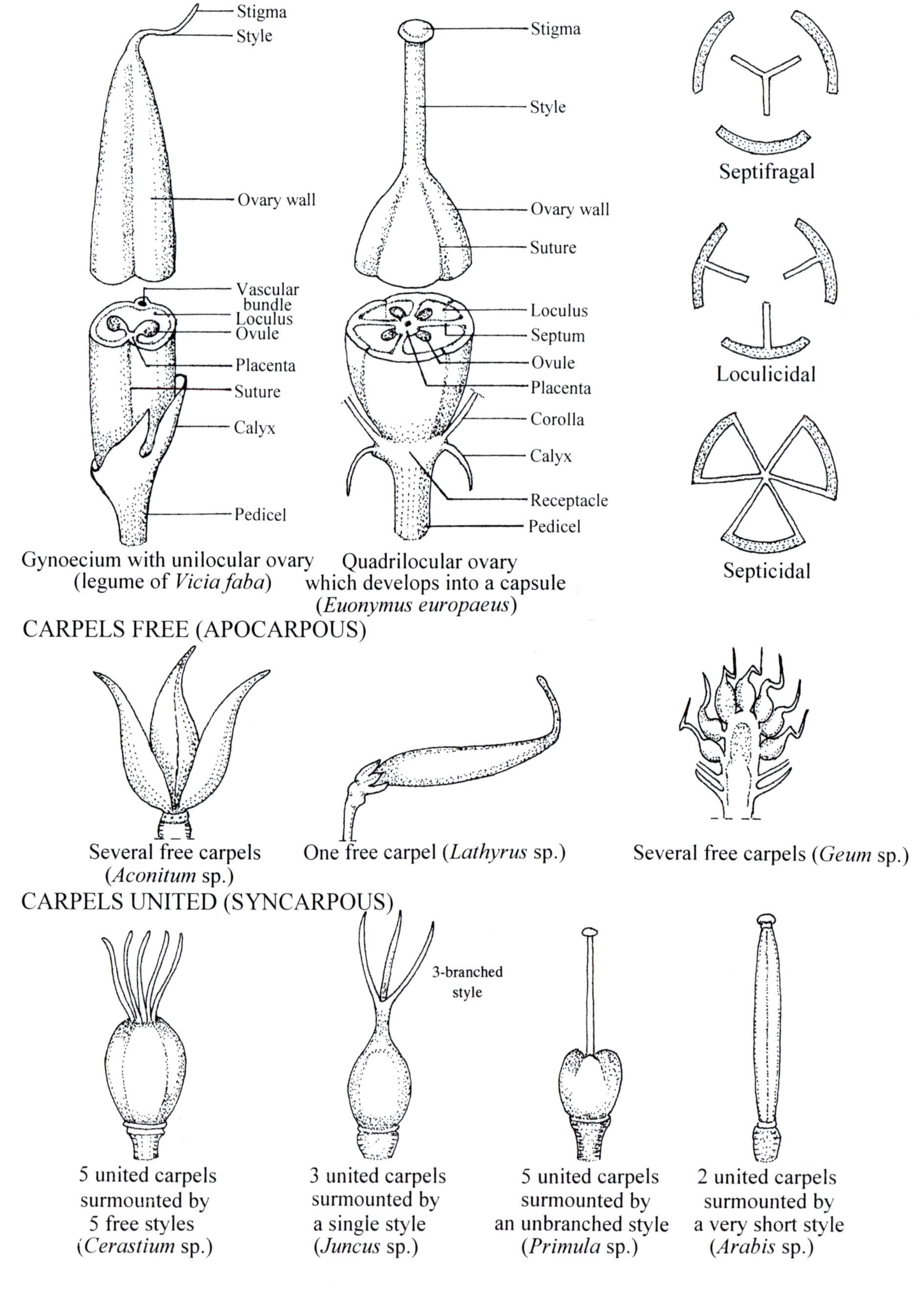

Gynoecium with unilocular ovary (legume of *Vicia faba*)

Quadrilocular ovary which develops into a capsule (*Euonymus europaeus*)

Several free carpels (*Aconitum* sp.)

One free carpel (*Lathyrus* sp.)

Several free carpels (*Geum* sp.)

5 united carpels surmounted by 5 free styles (*Cerastium* sp.)

3 united carpels surmounted by a single style (*Juncus* sp.)

5 united carpels surmounted by an unbranched style (*Primula* sp.)

2 united carpels surmounted by a very short style (*Arabis* sp.)

Styles and Stigmas 1

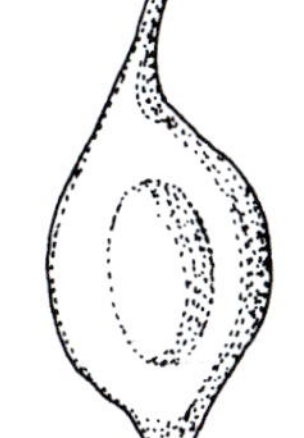

Style eccentric (*Ranunculus* sp.)

Style flabellate (*Viola* sp.)

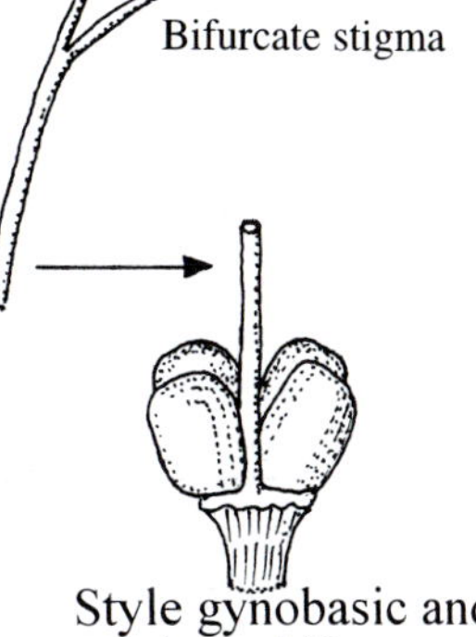

Style gynobasic and stigma bifurcate (*Lamium* sp.)

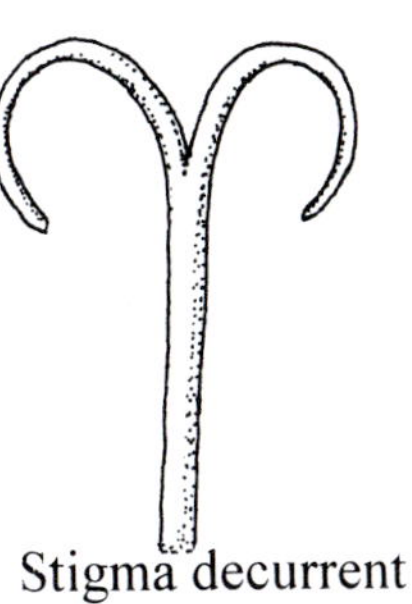

Stigma decurrent

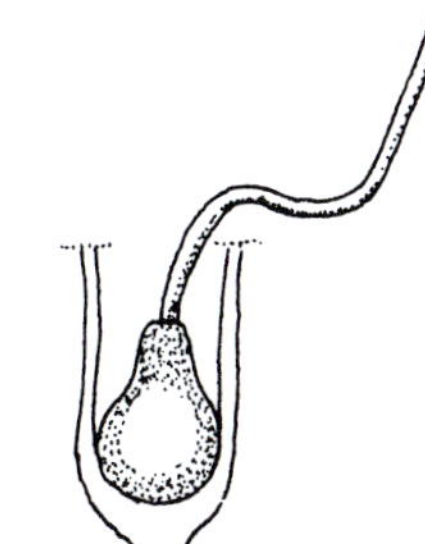

Style geniculate (*Rhexia* sp.)

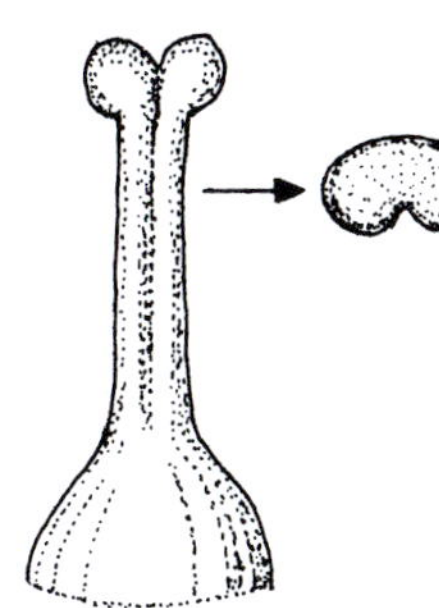

Style conduplicate

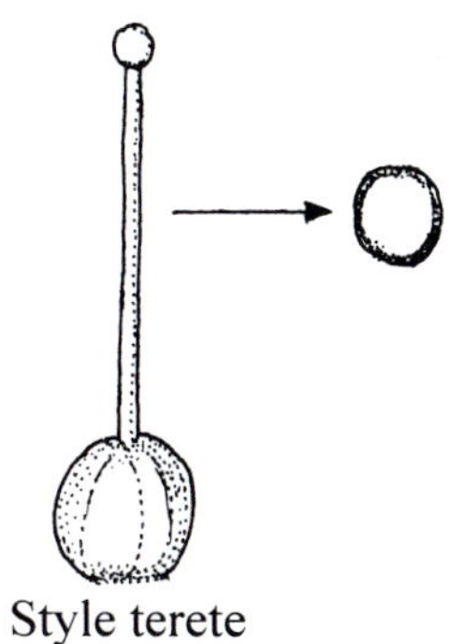

Style terete (*Primula* sp.)

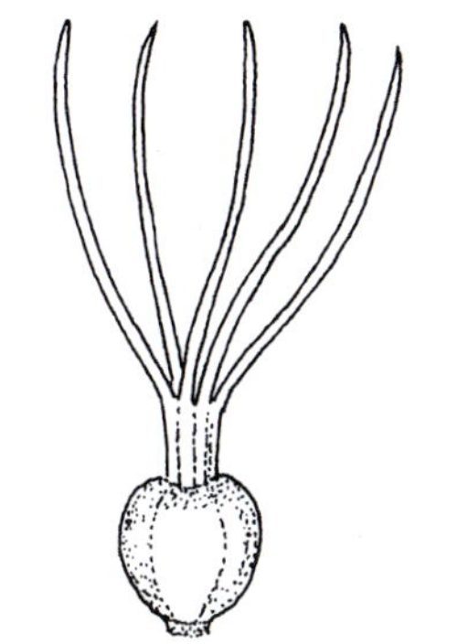

Style with filiform branches (*Armeria* sp.)

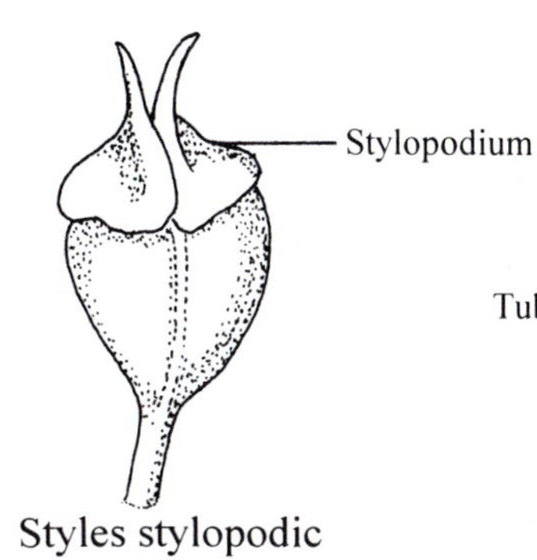

Styles stylopodic (*Heracleum* sp.)

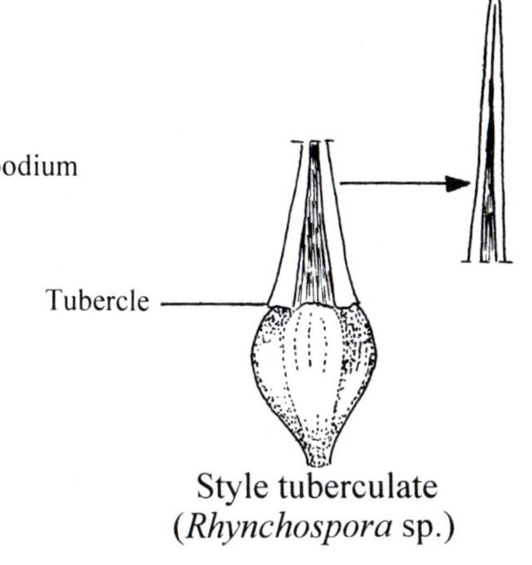

Style tuberculate (*Rhynchospora* sp.)

Styles and Stigmas 2

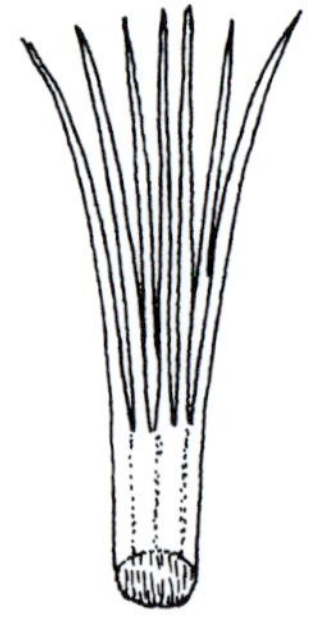

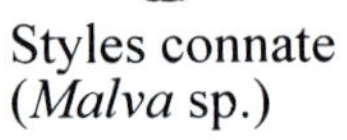
Styles connate
(*Malva* sp.)

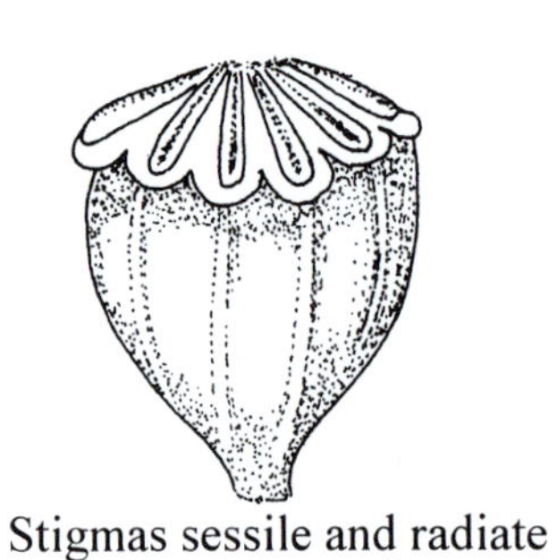

Stigmas sessile and radiate
(*Papaver* sp.)

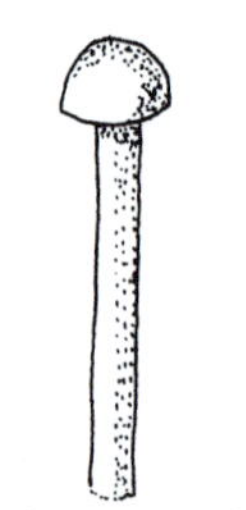

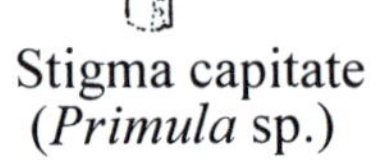
Stigma capitate
(*Primula* sp.)

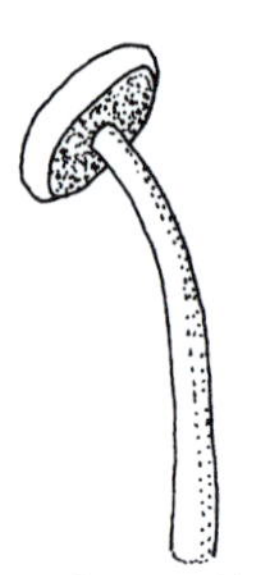

Stigma discoid
(*Hibiscus* sp.)

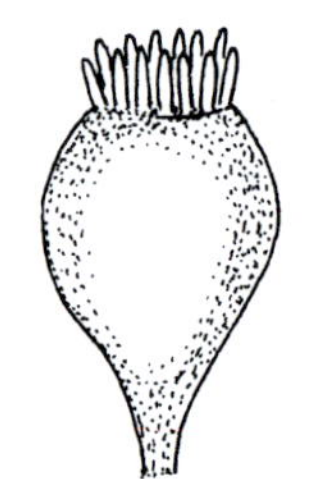

Stigmas forming a crest

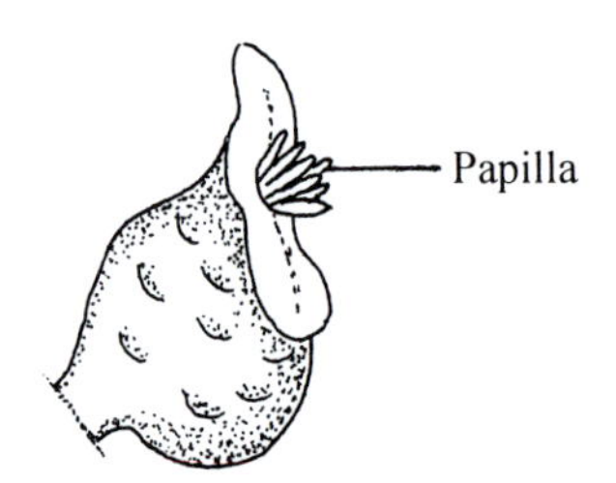

Stigma sessile with papillae
(*Peperomia* sp.)

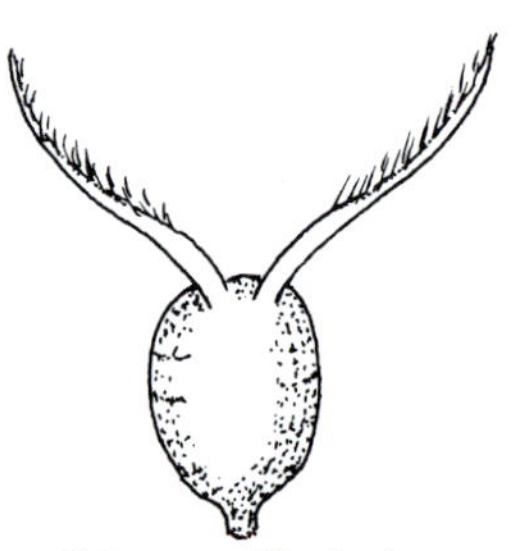

Stigmas fimbriate

Stigmas plumose
(Gramineae or Poaceae)

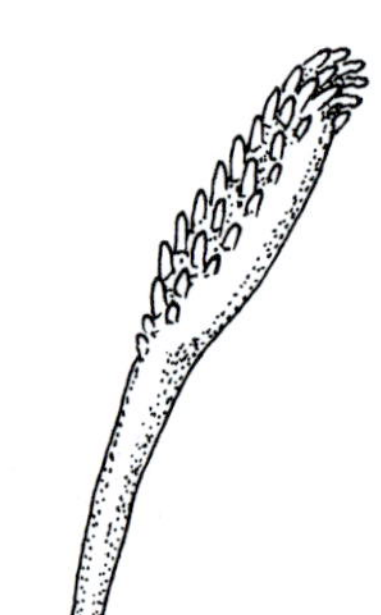

Stigma linear
(*Kitaibelia* sp.)

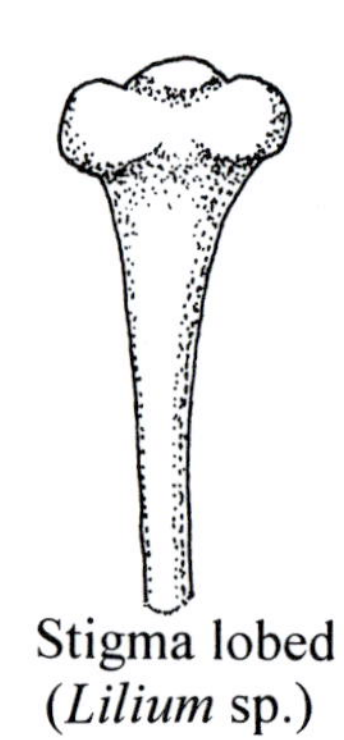

Stigma lobed
(*Lilium* sp.)

Placentation 1

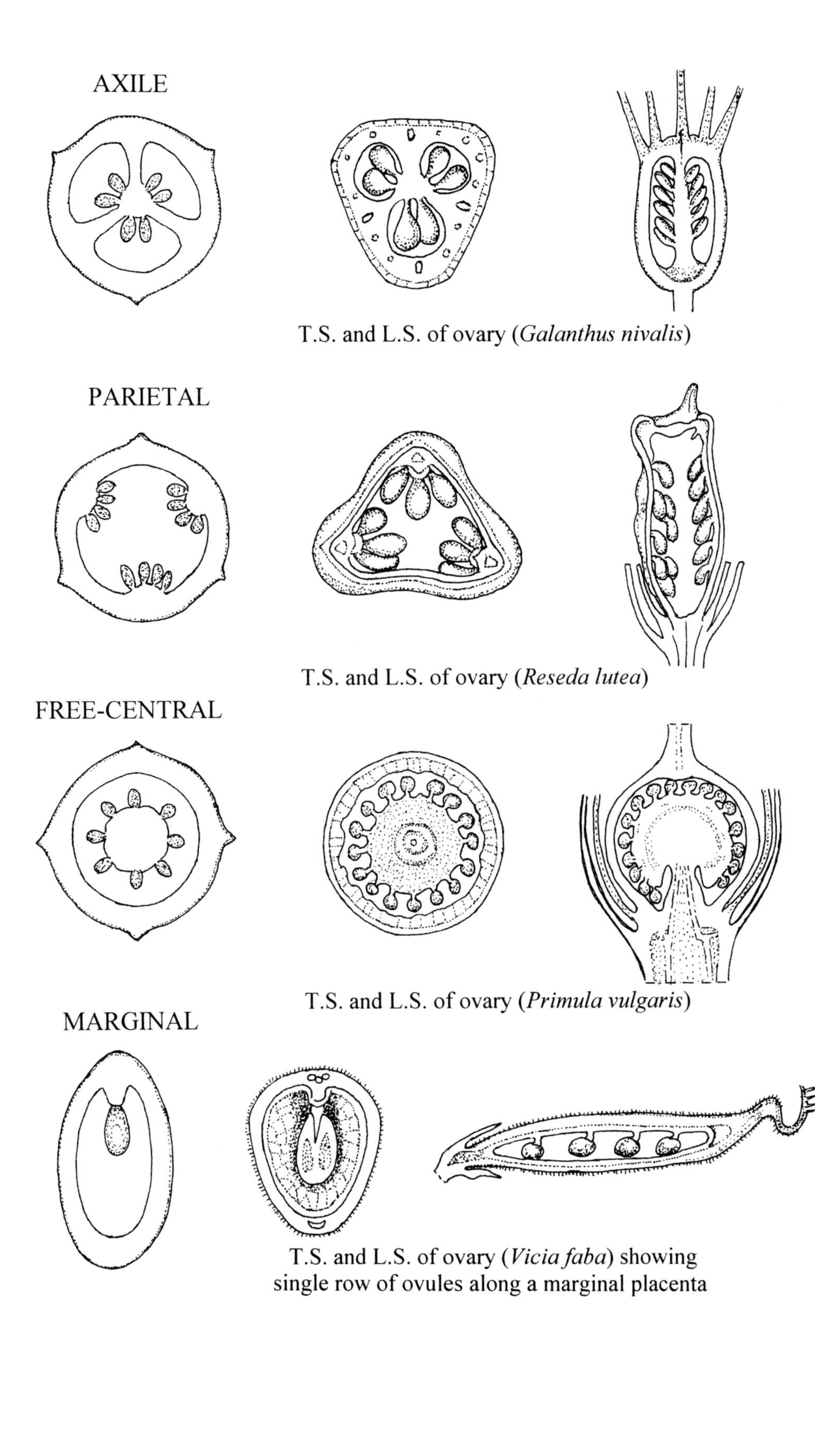

T.S. and L.S. of ovary (*Galanthus nivalis*)

T.S. and L.S. of ovary (*Reseda lutea*)

T.S. and L.S. of ovary (*Primula vulgaris*)

T.S. and L.S. of ovary (*Vicia faba*) showing single row of ovules along a marginal placenta

Placentation 2

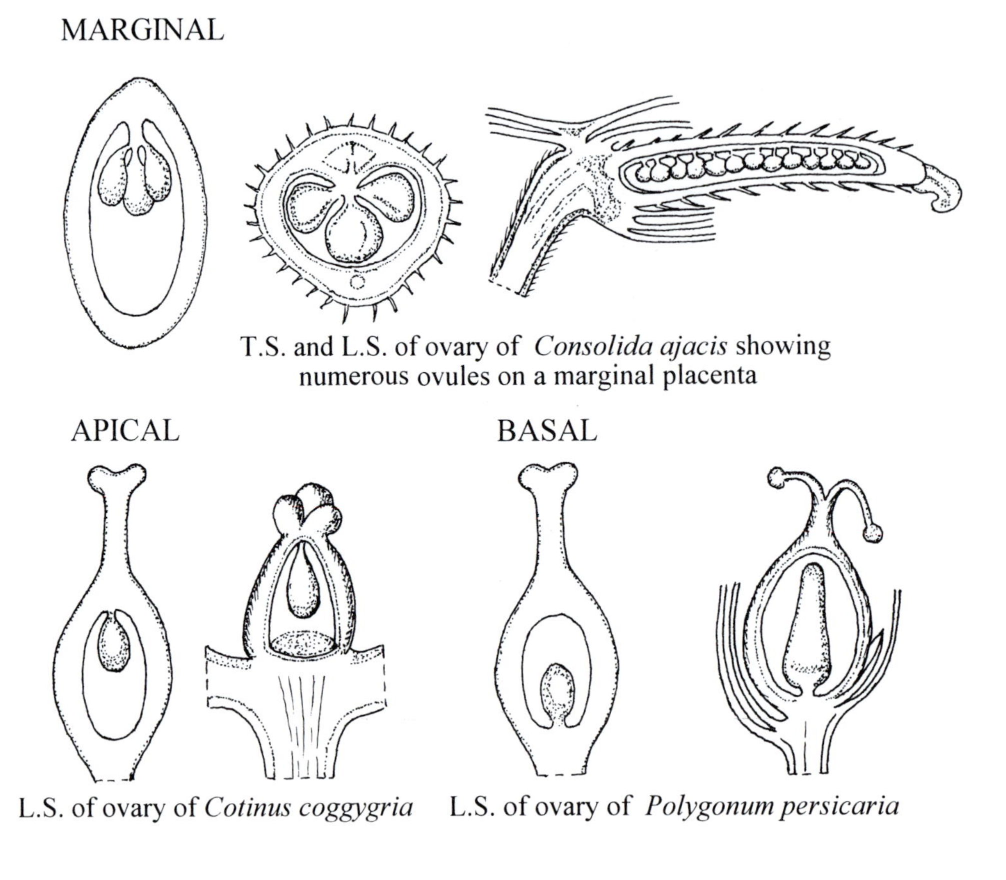

T.S. and L.S. of ovary of *Consolida ajacis* showing numerous ovules on a marginal placenta

L.S. of ovary of *Cotinus coggygria*

L.S. of ovary of *Polygonum persicaria*

Ovules

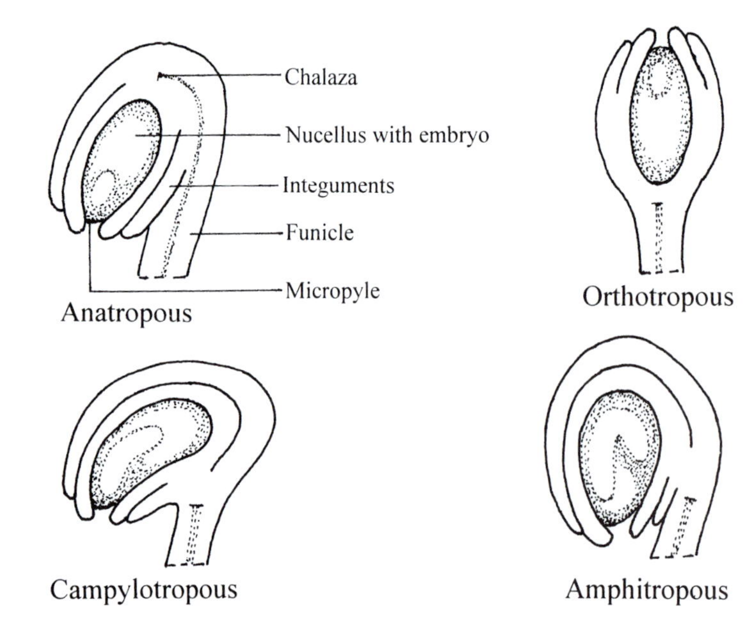

10. Flower Structure

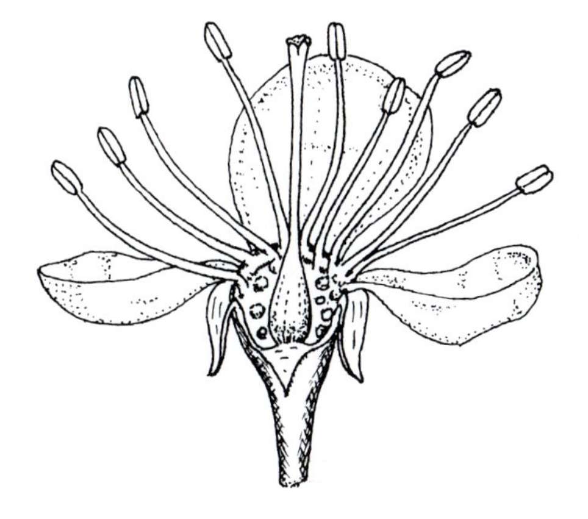

Flower Structure 1 *Aristolochia* and *Sarracenia*

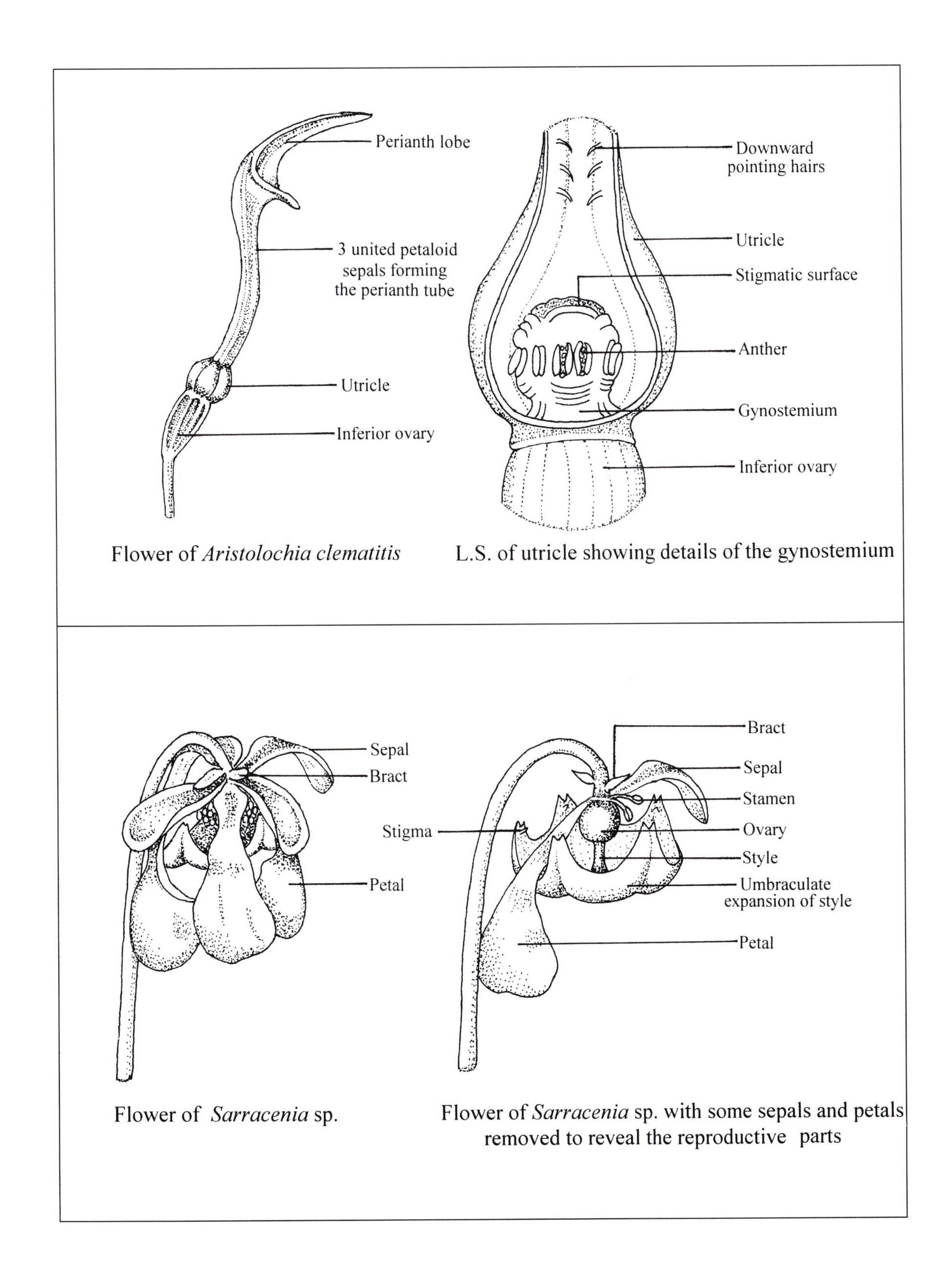

Flower of *Aristolochia clematitis*

L.S. of utricle showing details of the gynostemium

Flower of *Sarracenia* sp.

Flower of *Sarracenia* sp. with some sepals and petals removed to reveal the reproductive parts

Flower Structure 2 *Consolida* and *Salvia*

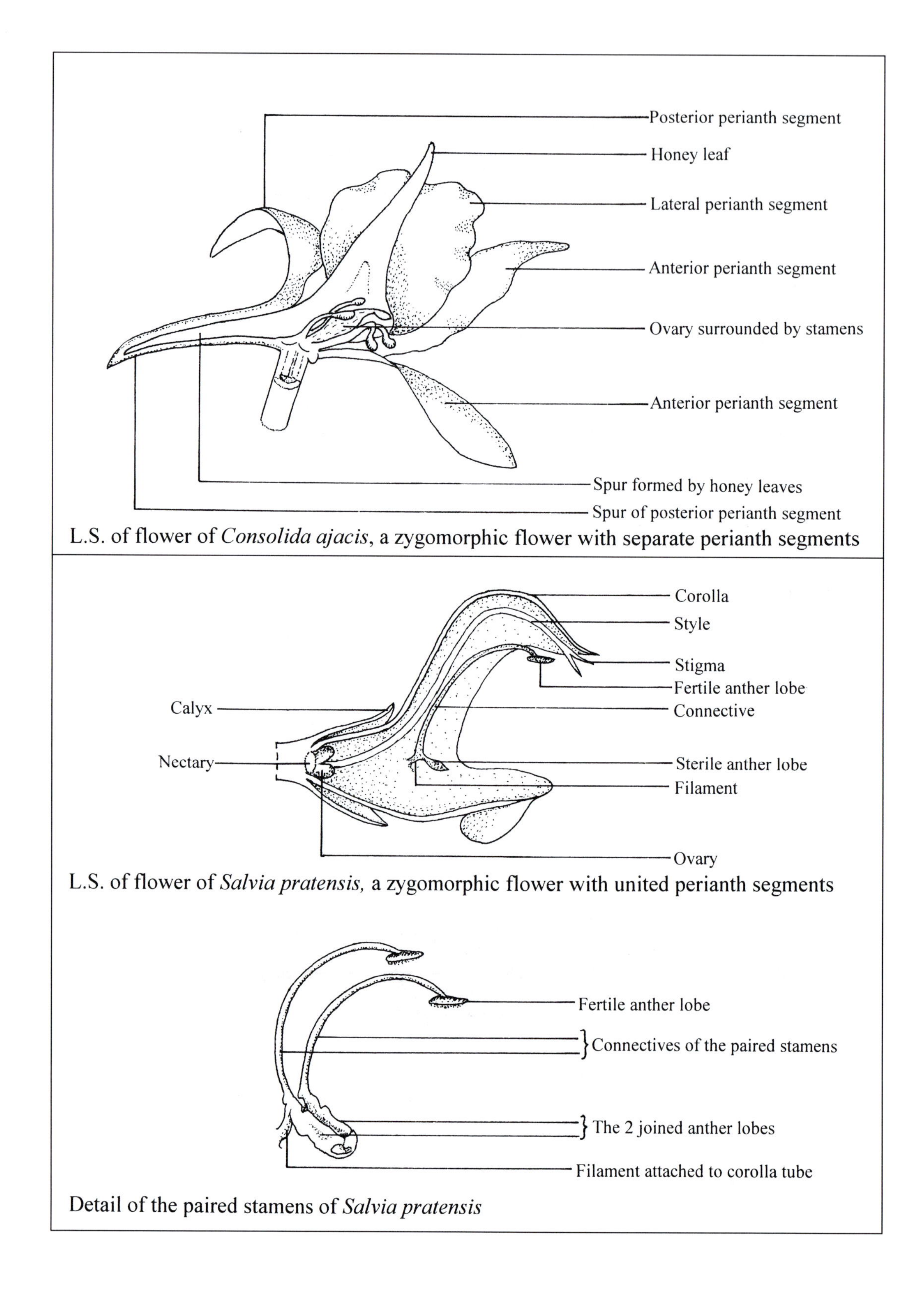

L.S. of flower of *Consolida ajacis*, a zygomorphic flower with separate perianth segments

L.S. of flower of *Salvia pratensis,* a zygomorphic flower with united perianth segments

Detail of the paired stamens of *Salvia pratensis*

Flower Structure 3 *Silene dioica*

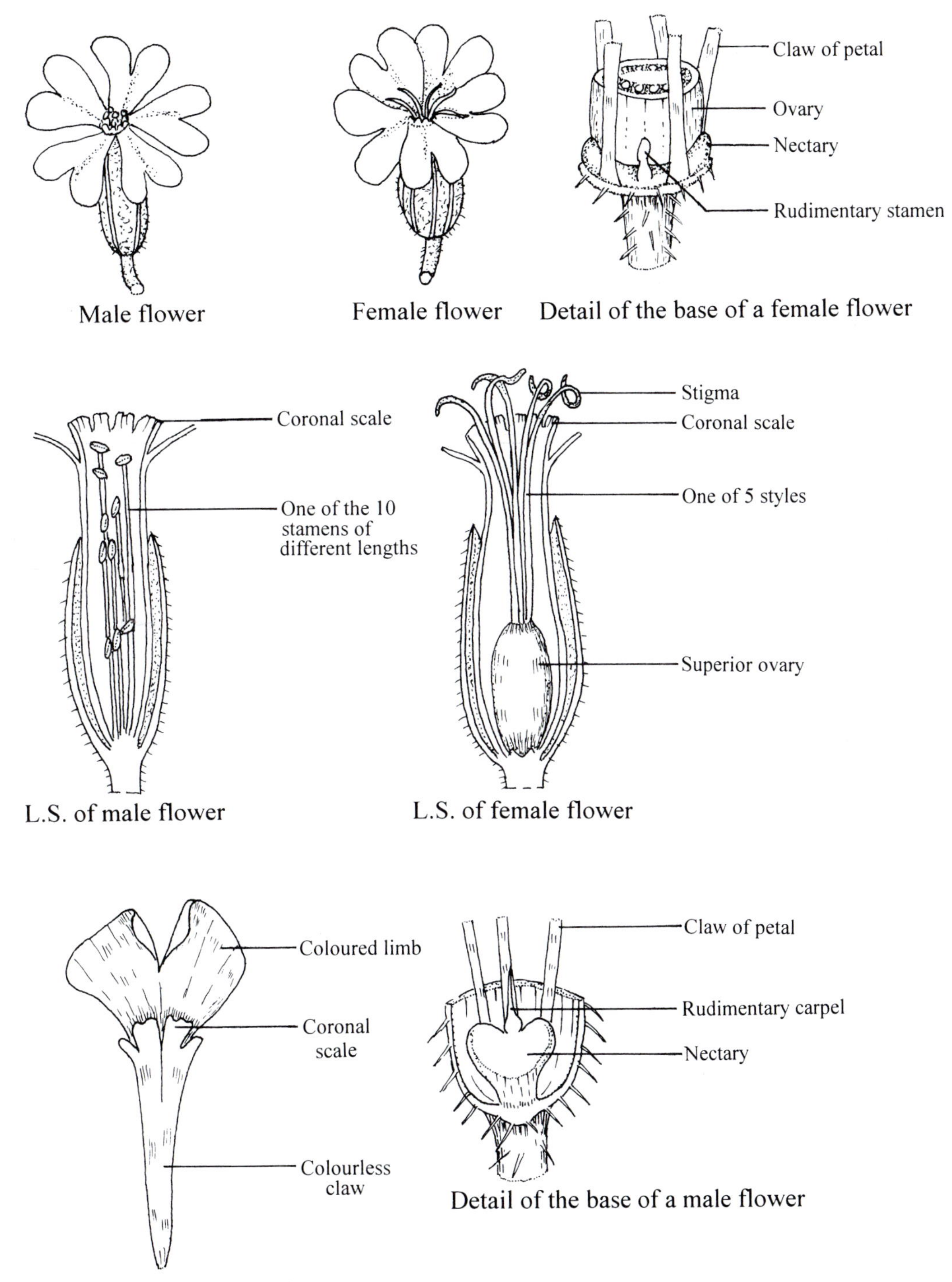

Flower Structure 4 *Passiflora*

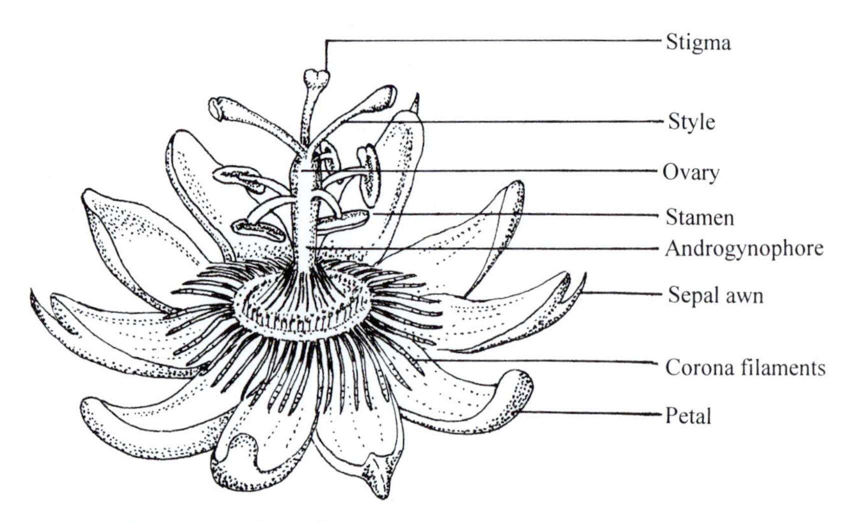

Flower of *Passiflora caerulea*

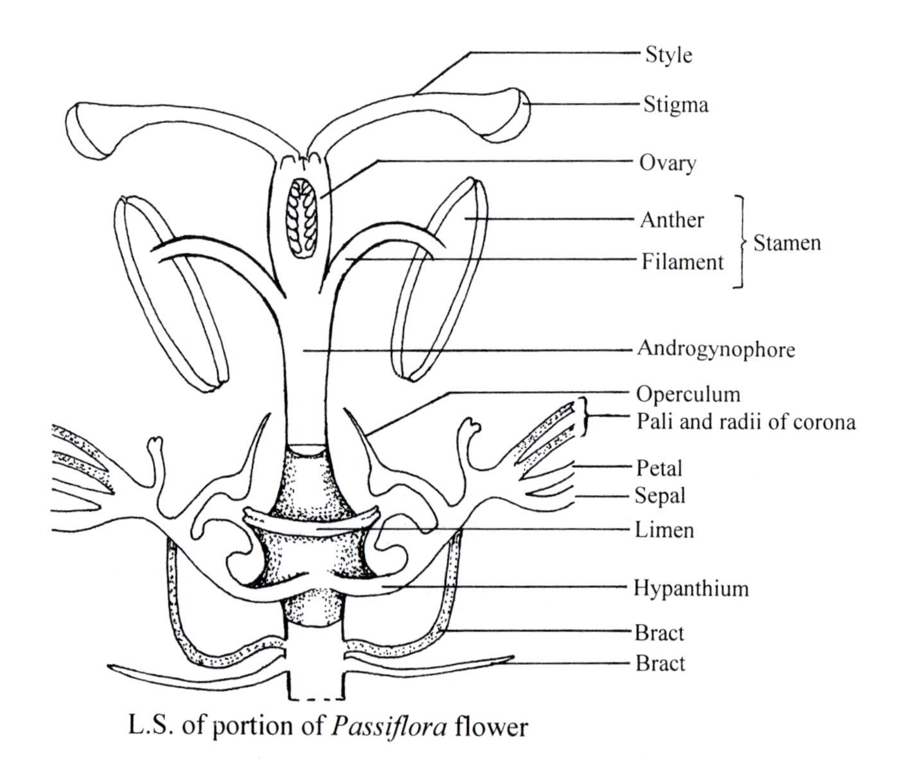

L.S. of portion of *Passiflora* flower

Flower Structure 5 *Primula, Cyclamen* and *Viola*

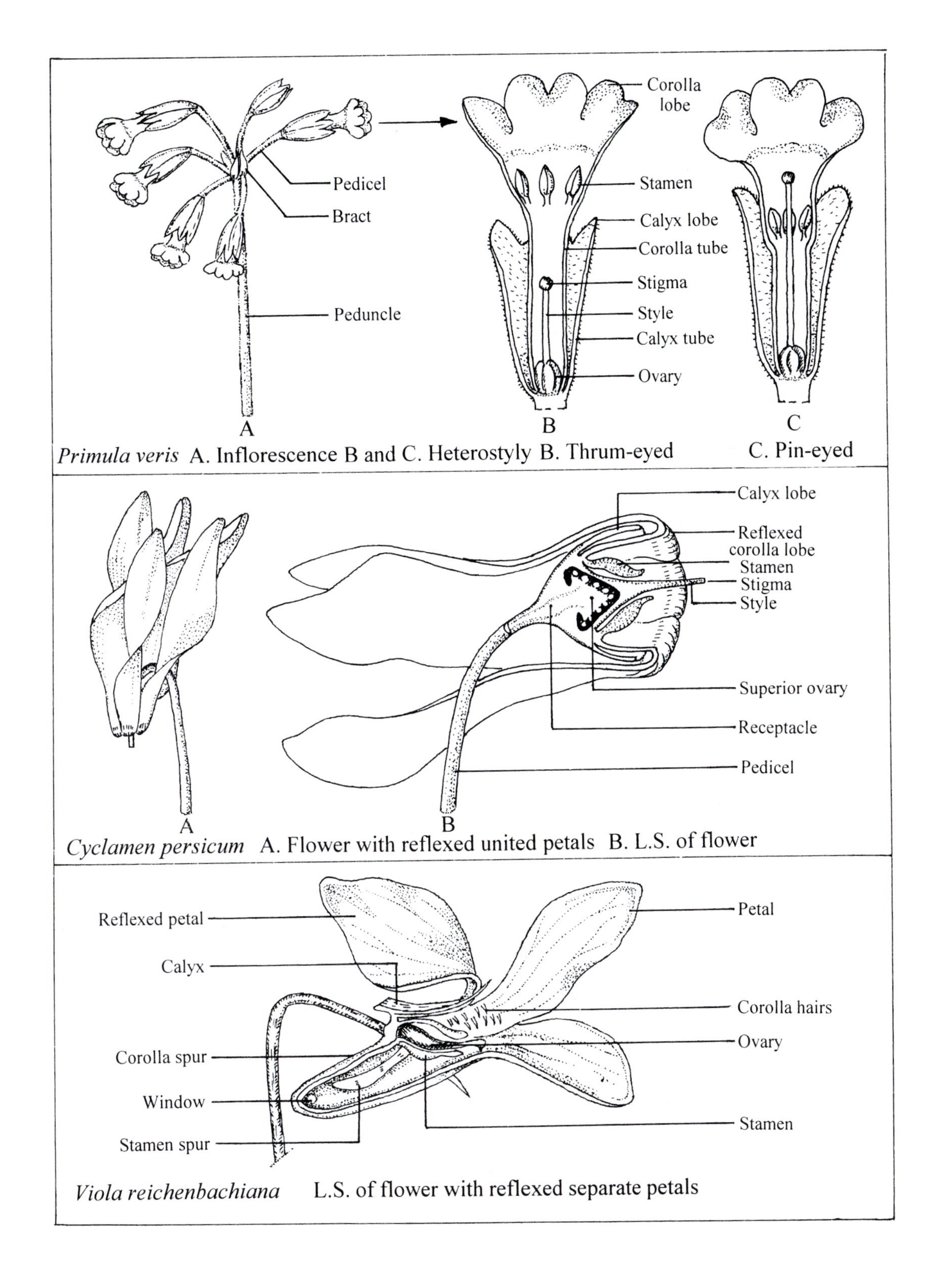

Primula veris A. Inflorescence B and C. Heterostyly B. Thrum-eyed C. Pin-eyed

Cyclamen persicum A. Flower with reflexed united petals B. L.S. of flower

Viola reichenbachiana L.S. of flower with reflexed separate petals

Flower Structure 6 *Lathyrus*

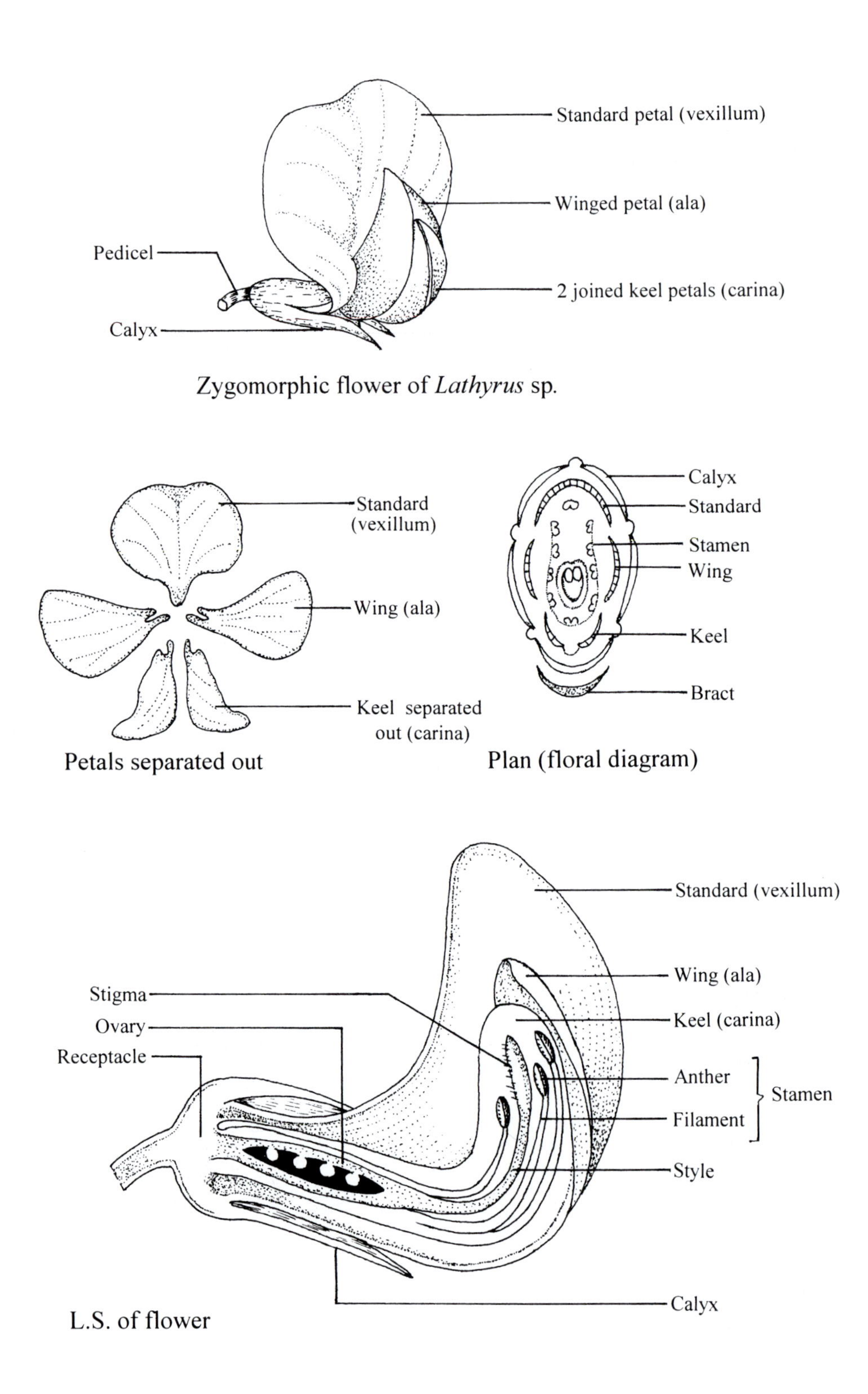

Flower Structure 7 *Eucalyptus,* Nectar Guides and Nectaries

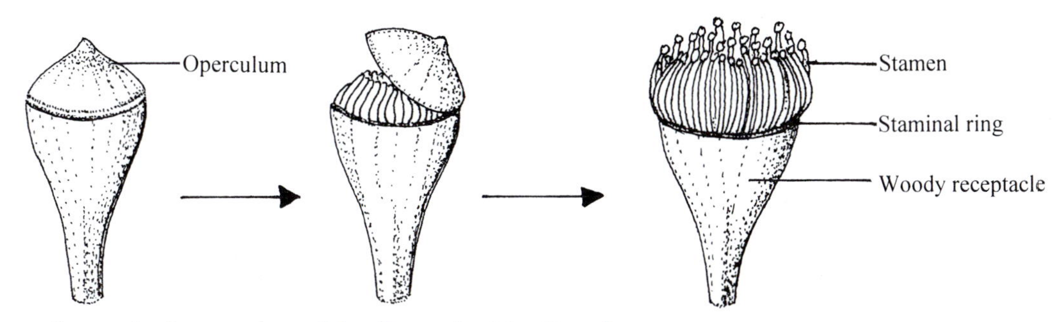

Stages in the opening of the flower bud in *Eucalyptus* sp.

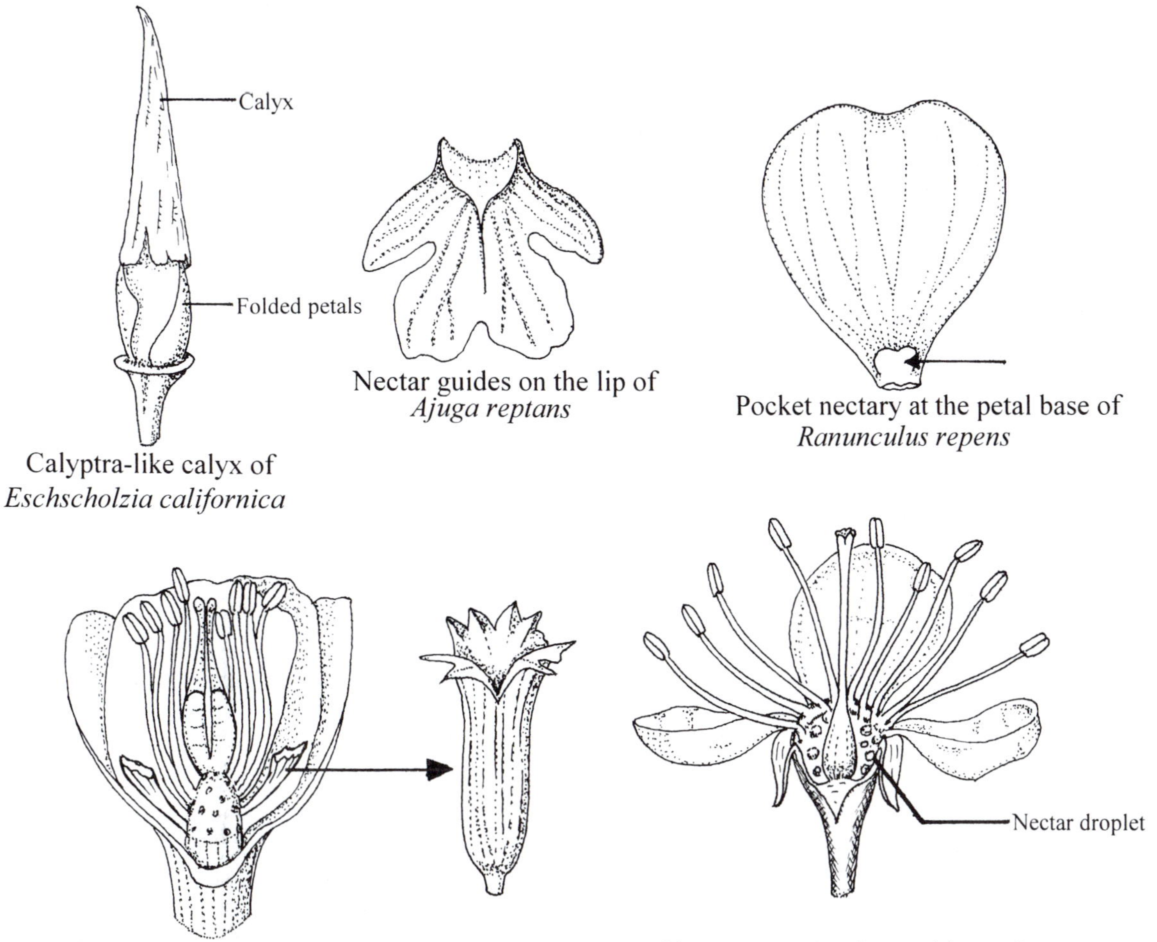

Calyptra-like calyx of *Eschscholzia californica*

Nectar guides on the lip of *Ajuga reptans*

Pocket nectary at the petal base of *Ranunculus repens*

Petal nectary *in situ* and separated out from the flower of *Helleborus foetidus*

Nectar-secreting hypanthium of the flower of *Prunus spinosa*

Flower Structure 8 *Geranium*

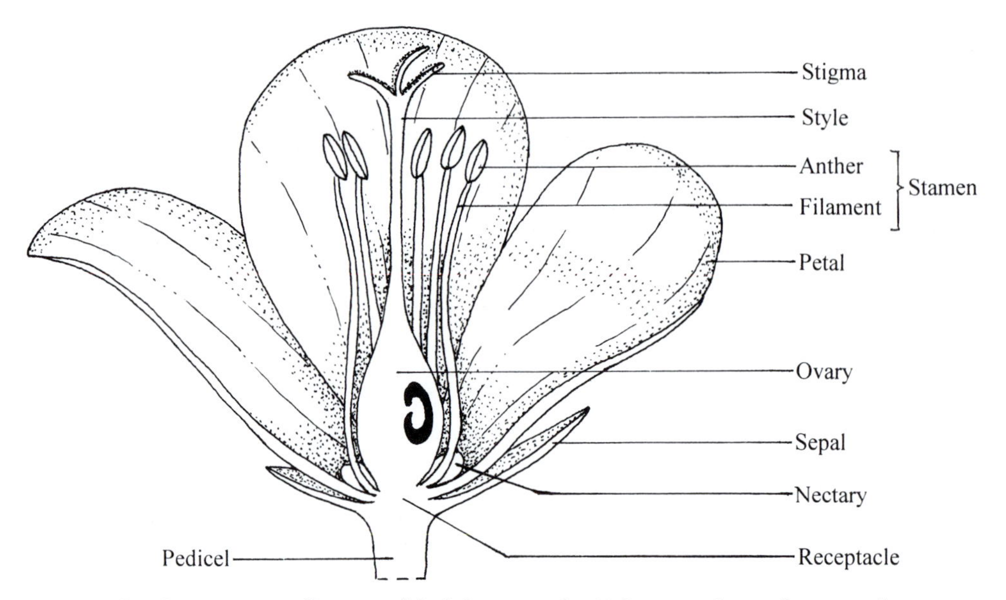

L.S. of a *Geranium,* a flower with 5 free sepals, 5 free petals, and a superior ovary

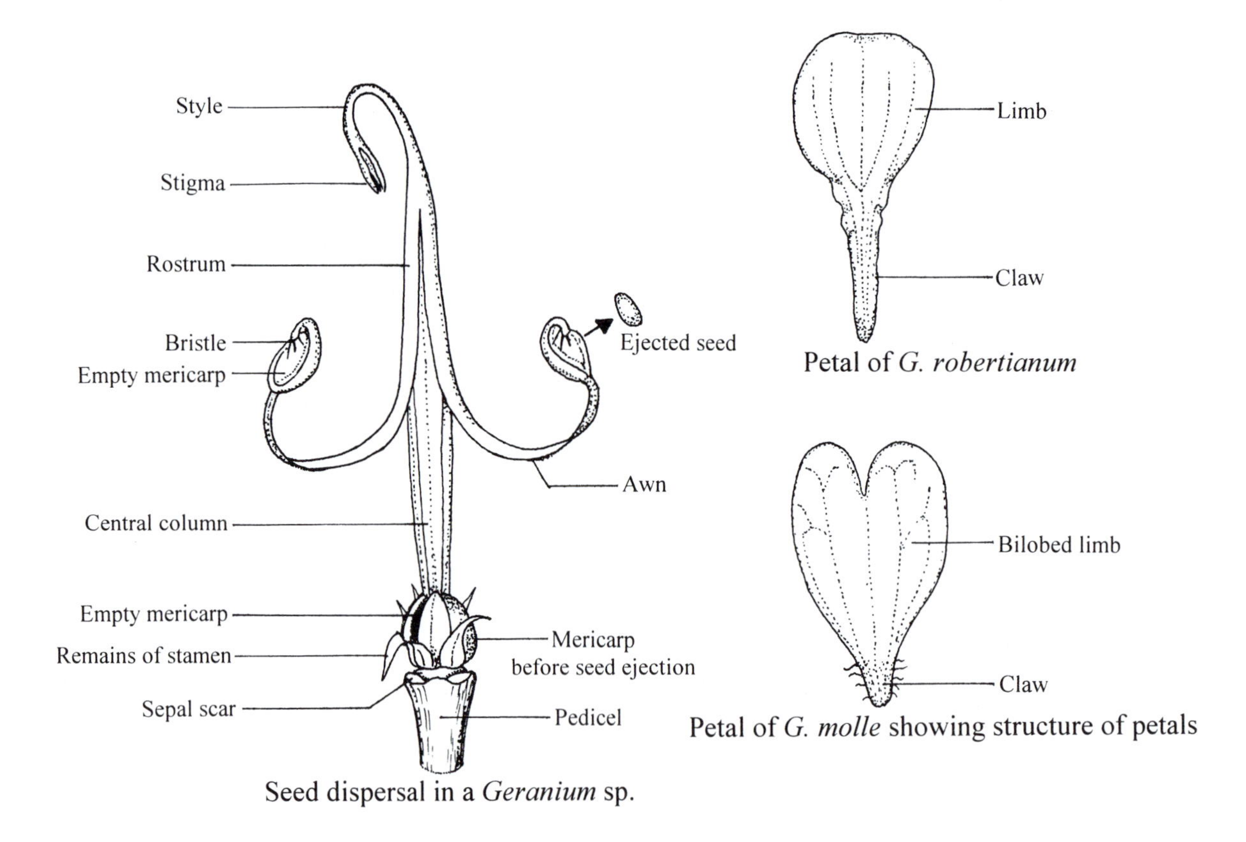

Seed dispersal in a *Geranium* sp.

Petal of *G. robertianum*

Petal of *G. molle* showing structure of petals

Flower Structure 9 *Asclepias* and *Polygala*

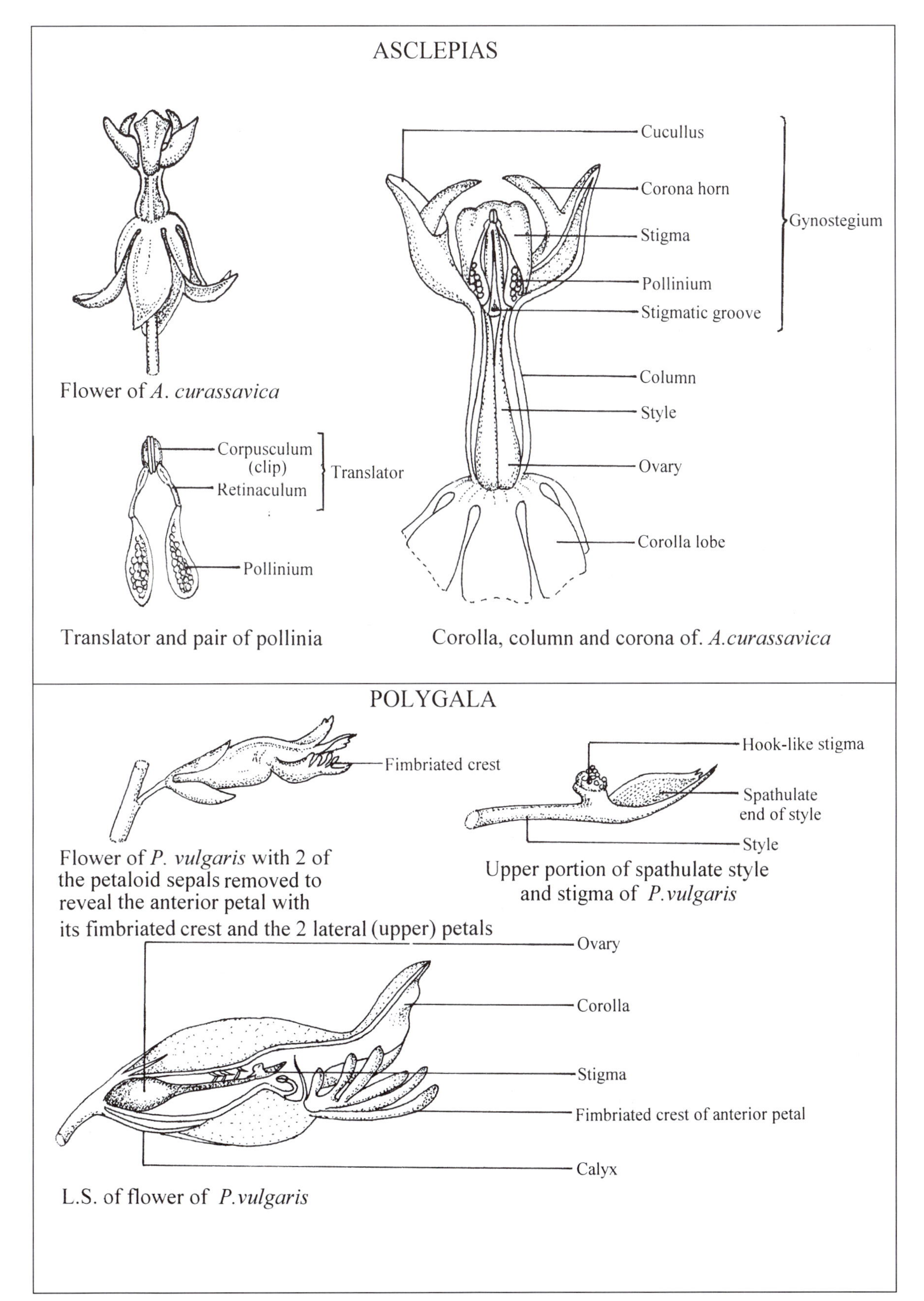

Flower Structure 10 Compositae (Asteraceae) 1

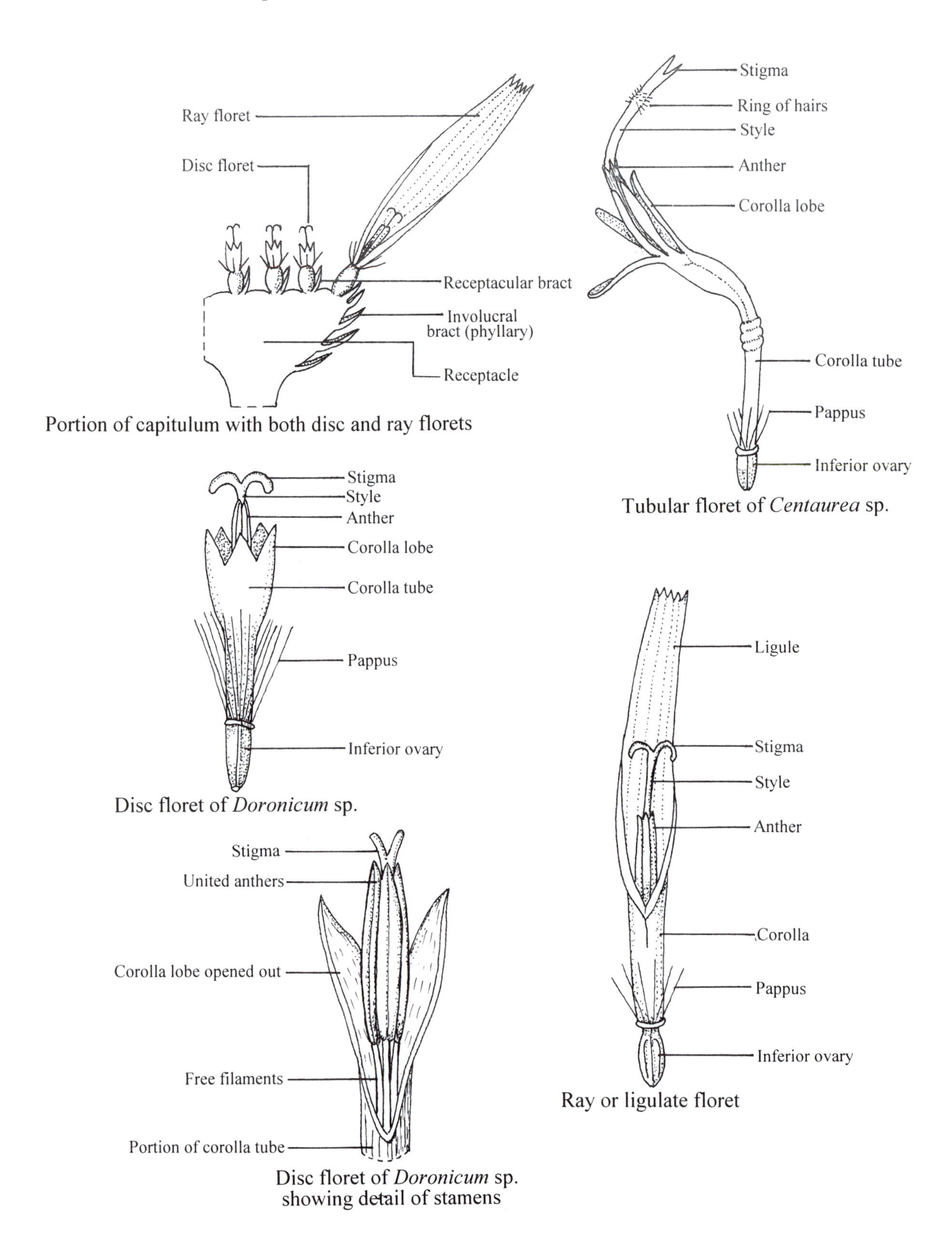

Portion of capitulum with both disc and ray florets

Tubular floret of *Centaurea* sp.

Disc floret of *Doronicum* sp.

Ray or ligulate floret

Disc floret of *Doronicum* sp. showing detail of stamens

Flower Structure 11 Compositae (Asteraceae) 2

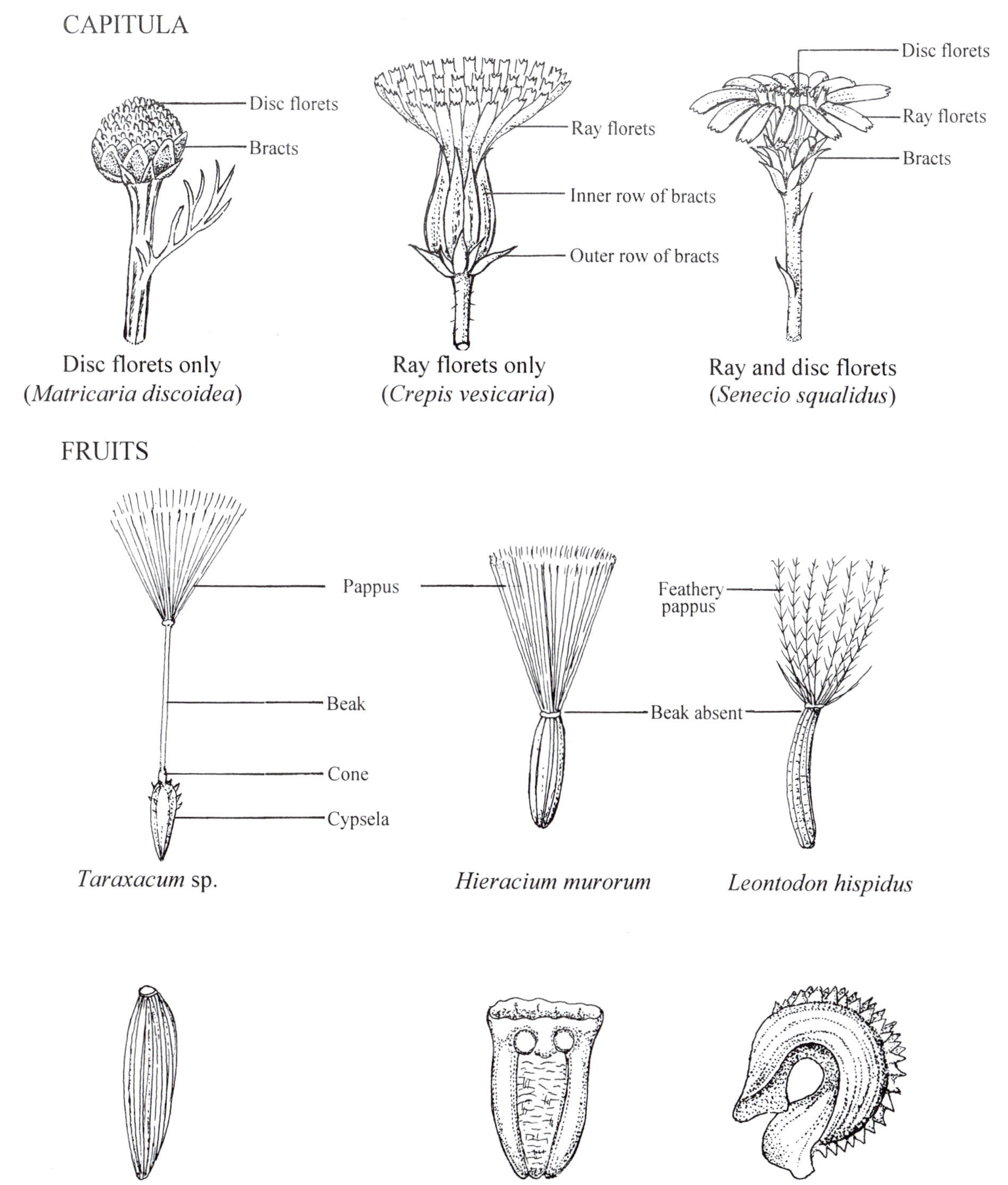

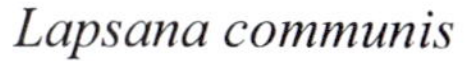

Lapsana communis

Matricaria recutita

Calendula arvensis

Flower Structure 12 *Strelitzia* and *Canna*

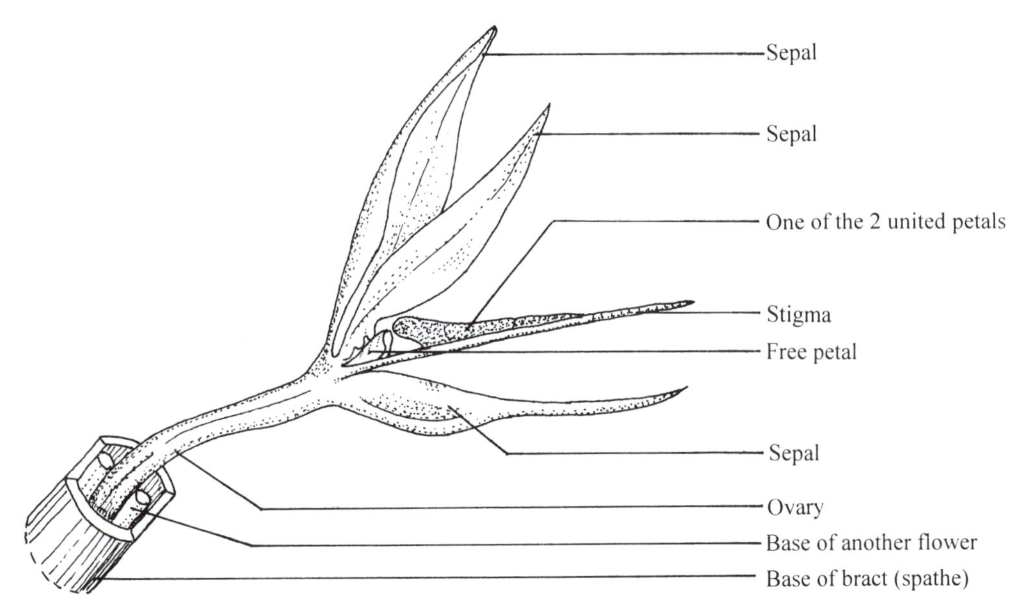

Single flower of *Strelitzia reginae*

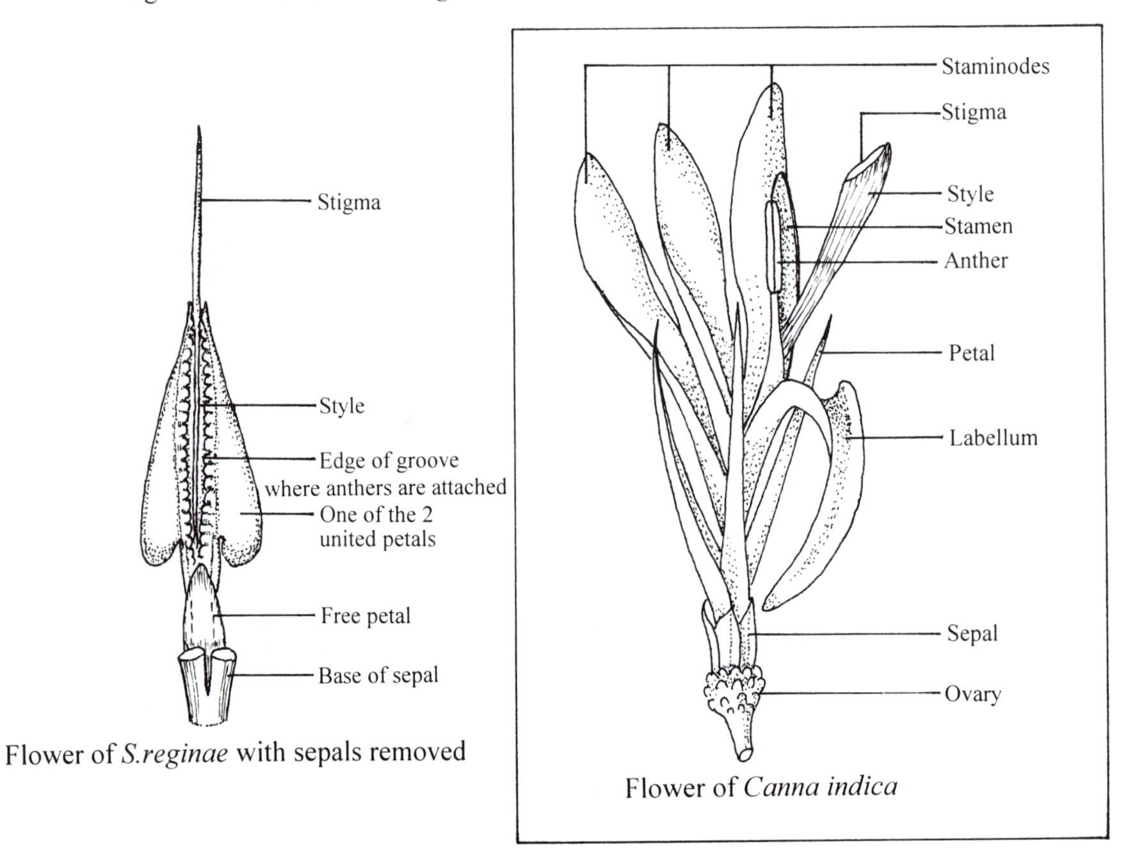

Flower of *S.reginae* with sepals removed

Flower of *Canna indica*

Flower Structure 13 *Tulipa* and *Narcissus*

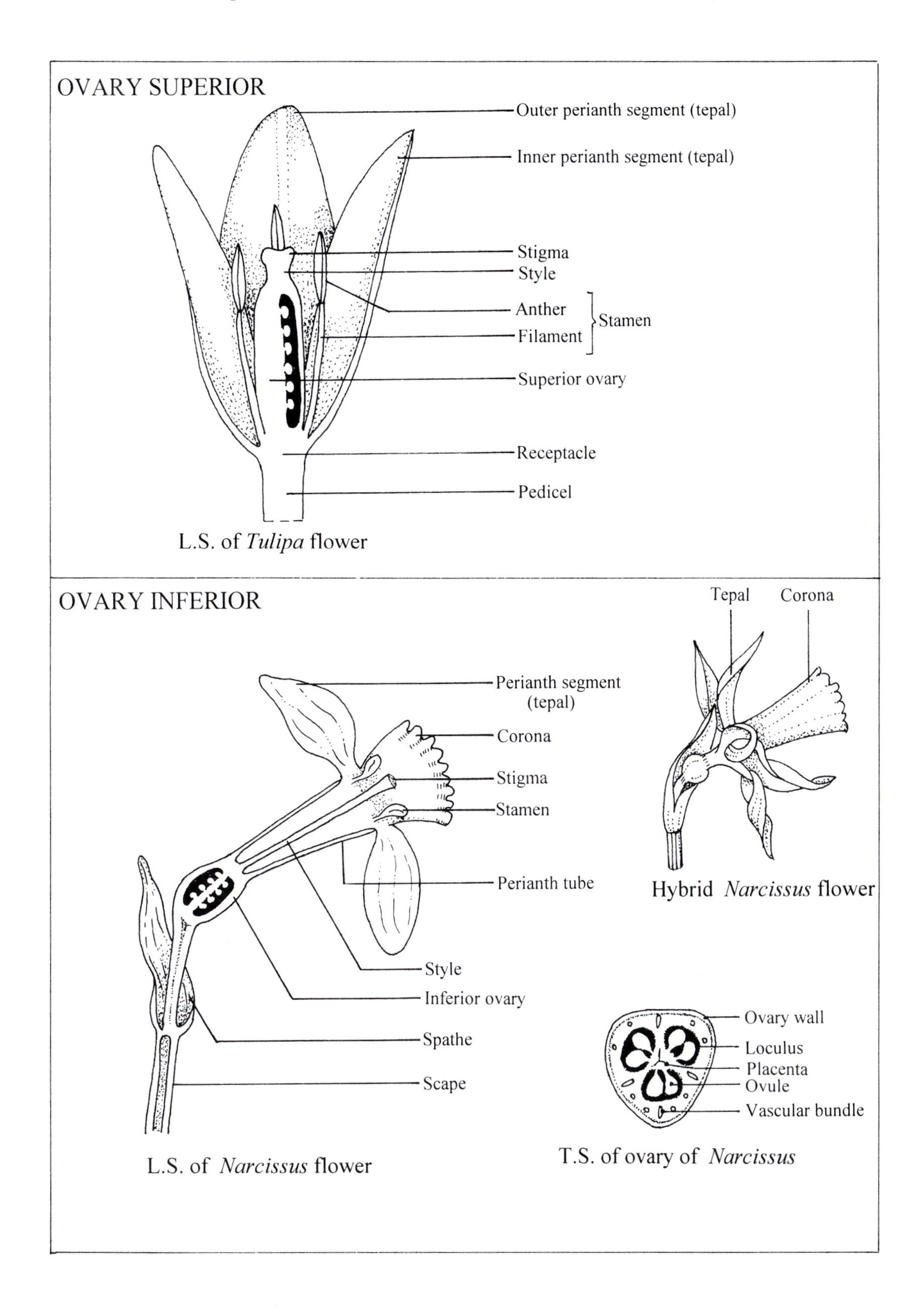

Flower Structure 14 *Iris*

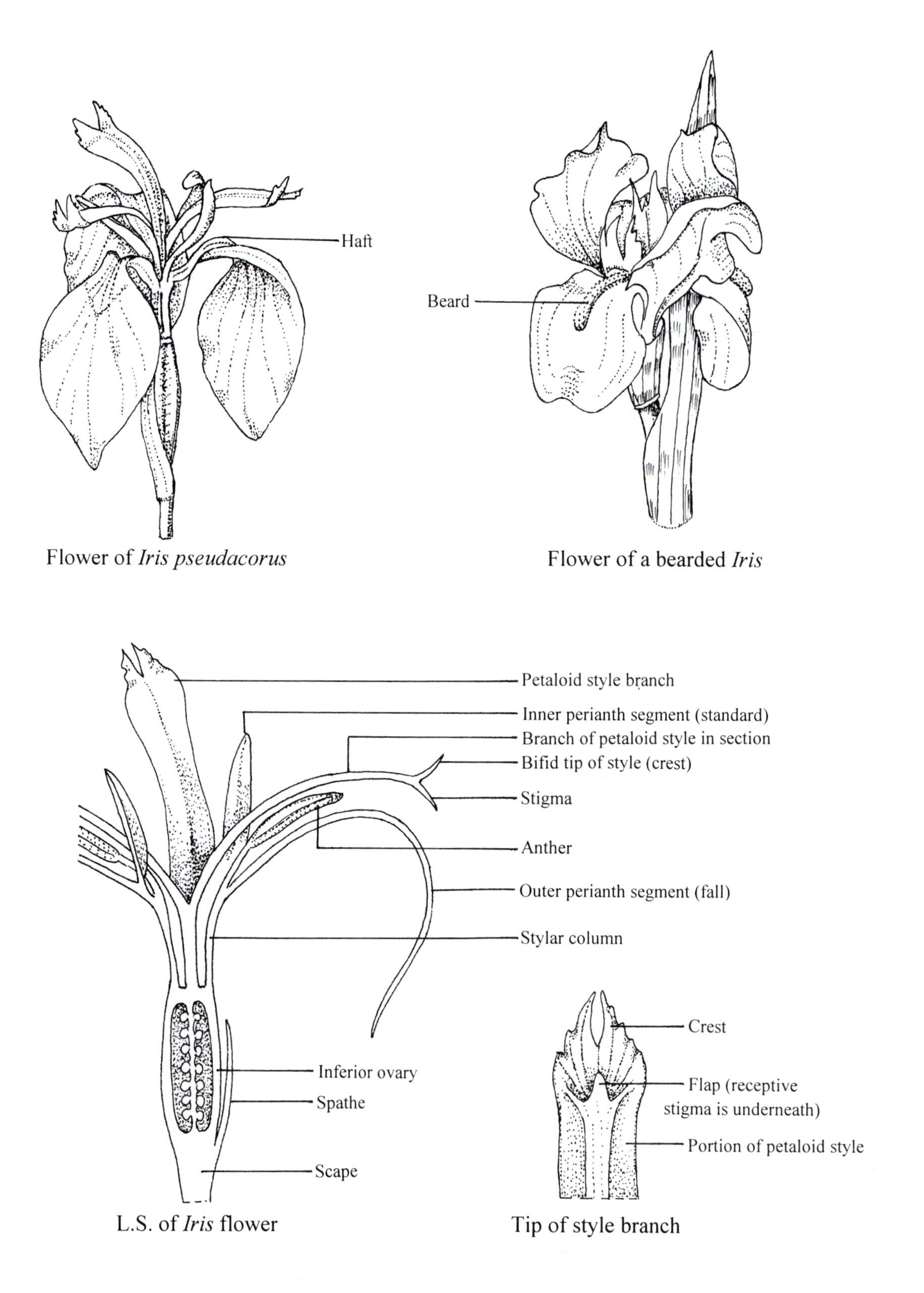

Flower of *Iris pseudacorus*

Flower of a bearded *Iris*

L.S. of *Iris* flower

Tip of style branch

11. Features of Certain Plant Families

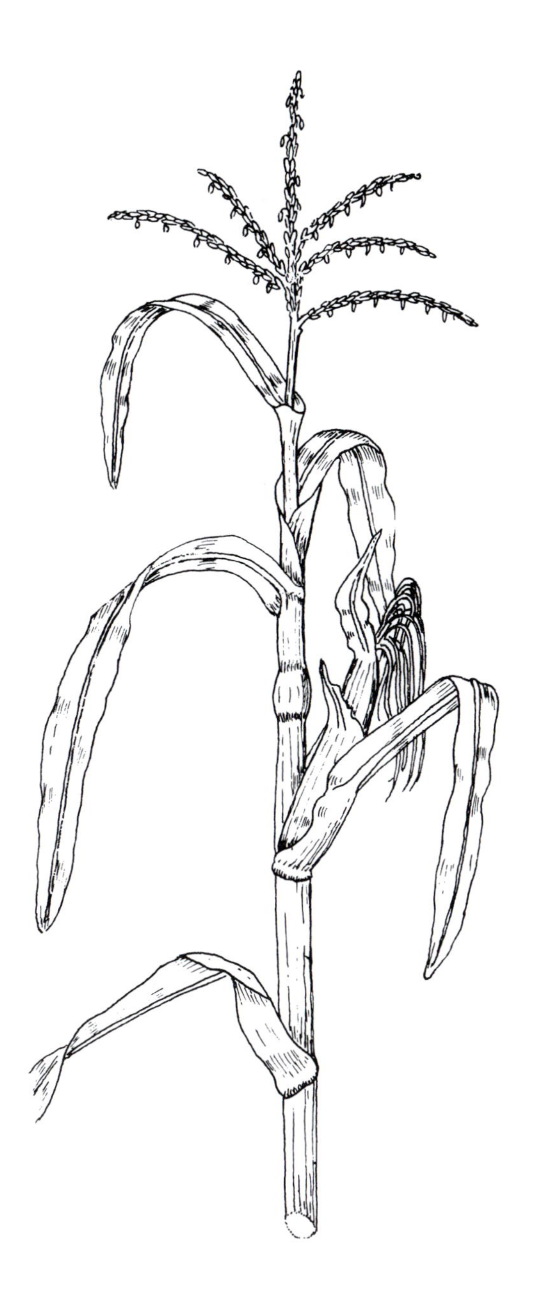

Structure of Cacti

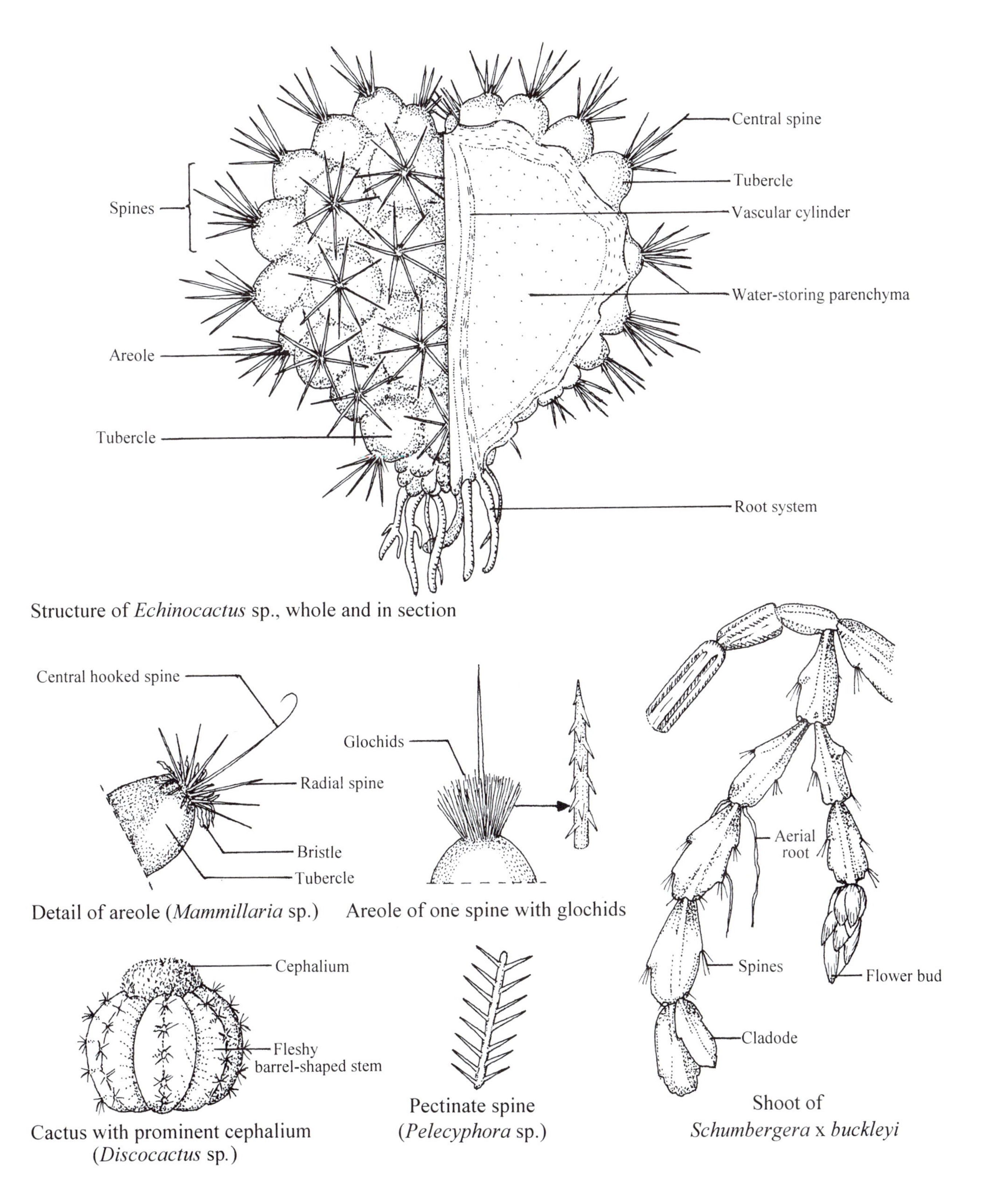

Structure of *Echinocactus* sp., whole and in section

Detail of areole (*Mammillaria* sp.)

Areole of one spine with glochids

Cactus with prominent cephalium (*Discocactus* sp.)

Pectinate spine (*Pelecyphora* sp.)

Shoot of *Schumbergera* x *buckleyi*

Structure of Grasses 1 Main Features

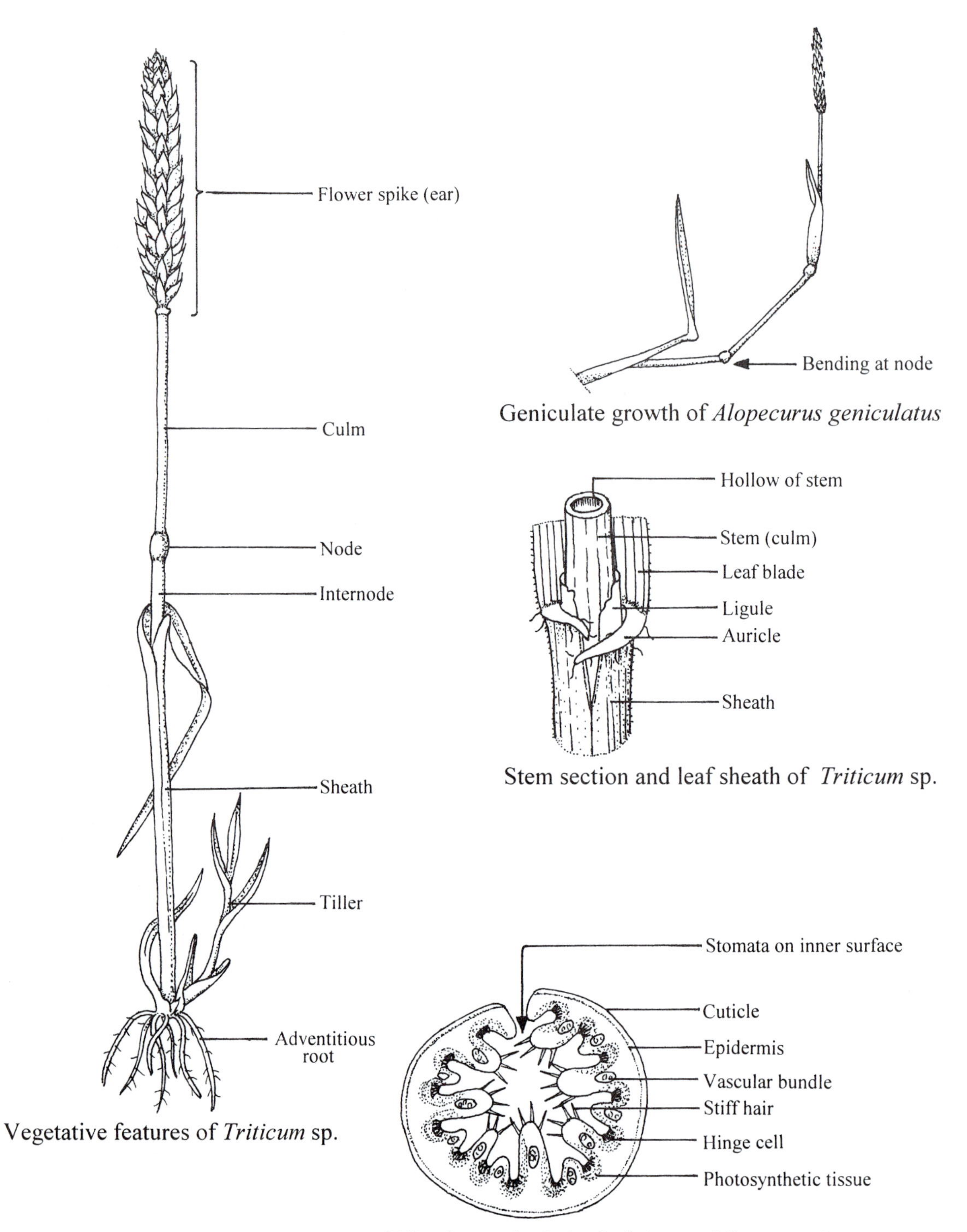

Vegetative features of *Triticum* sp.

Geniculate growth of *Alopecurus geniculatus*

Stem section and leaf sheath of *Triticum* sp.

T.S. of xerophytic leaf of *Ammophila arenaria*

Structure of Grasses 2 Inflorescence Forms

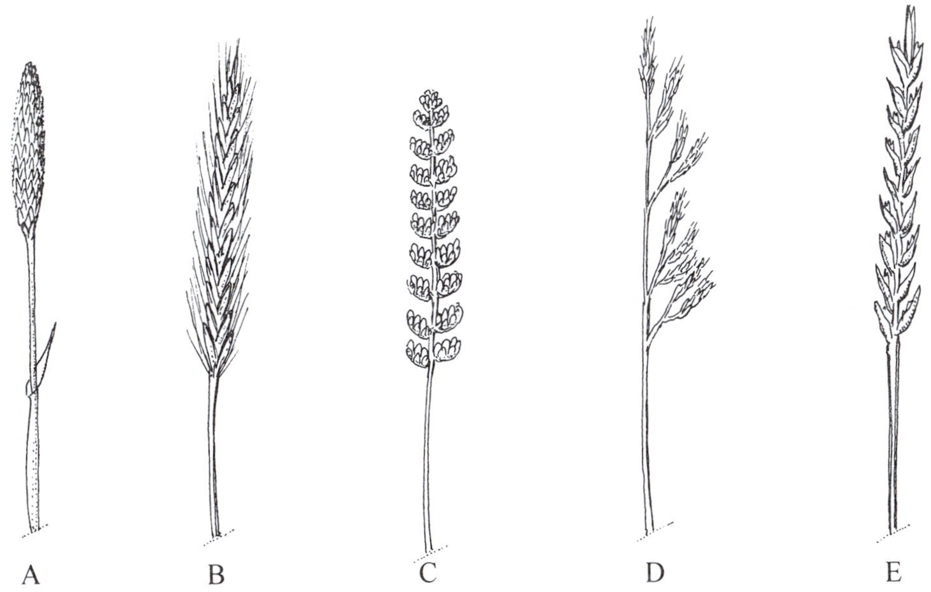

A B C D E

F G

A. *Phleum arenarium*

B. *Hordeum murinum*

C. *Cynosurus cristatus*

D. *Vulpia bromoides*

E. *Agropyron pungens*

F. *Poa annua*

G. *Dactylis glomerata*

Structure of Grasses 3

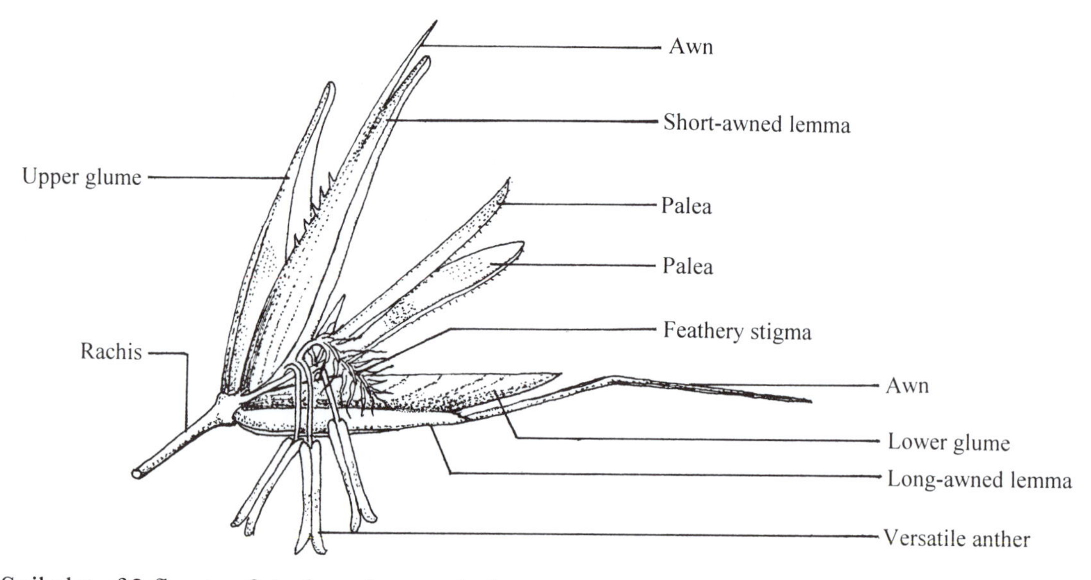

Spikelet of 2 florets of *Arrhenatherum elatius*

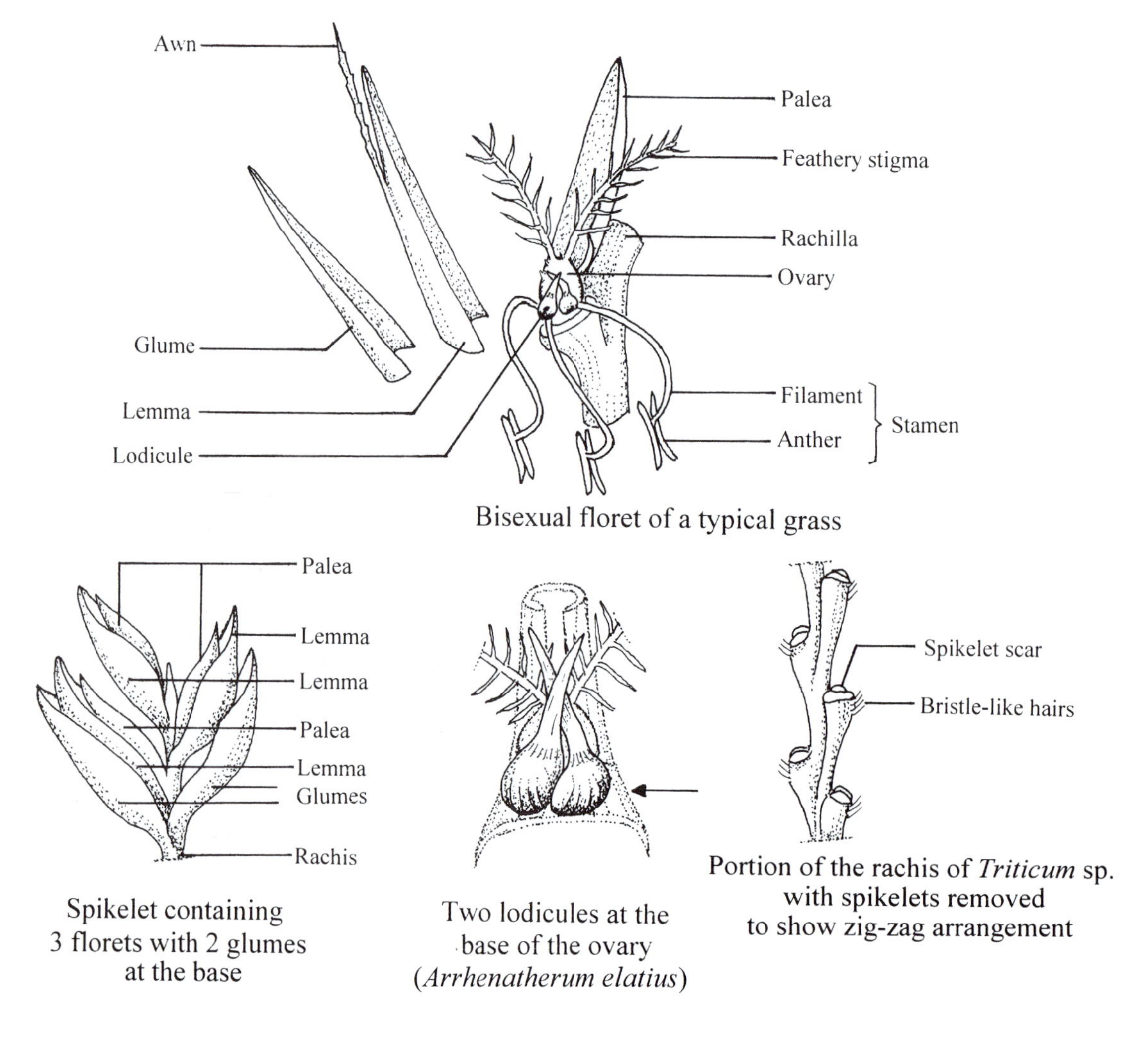

Bisexual floret of a typical grass

Spikelet containing 3 florets with 2 glumes at the base

Two lodicules at the base of the ovary (*Arrhenatherum elatius*)

Portion of the rachis of *Triticum* sp. with spikelets removed to show zig-zag arrangement

Structure of Grasses 4 *Zea mays*

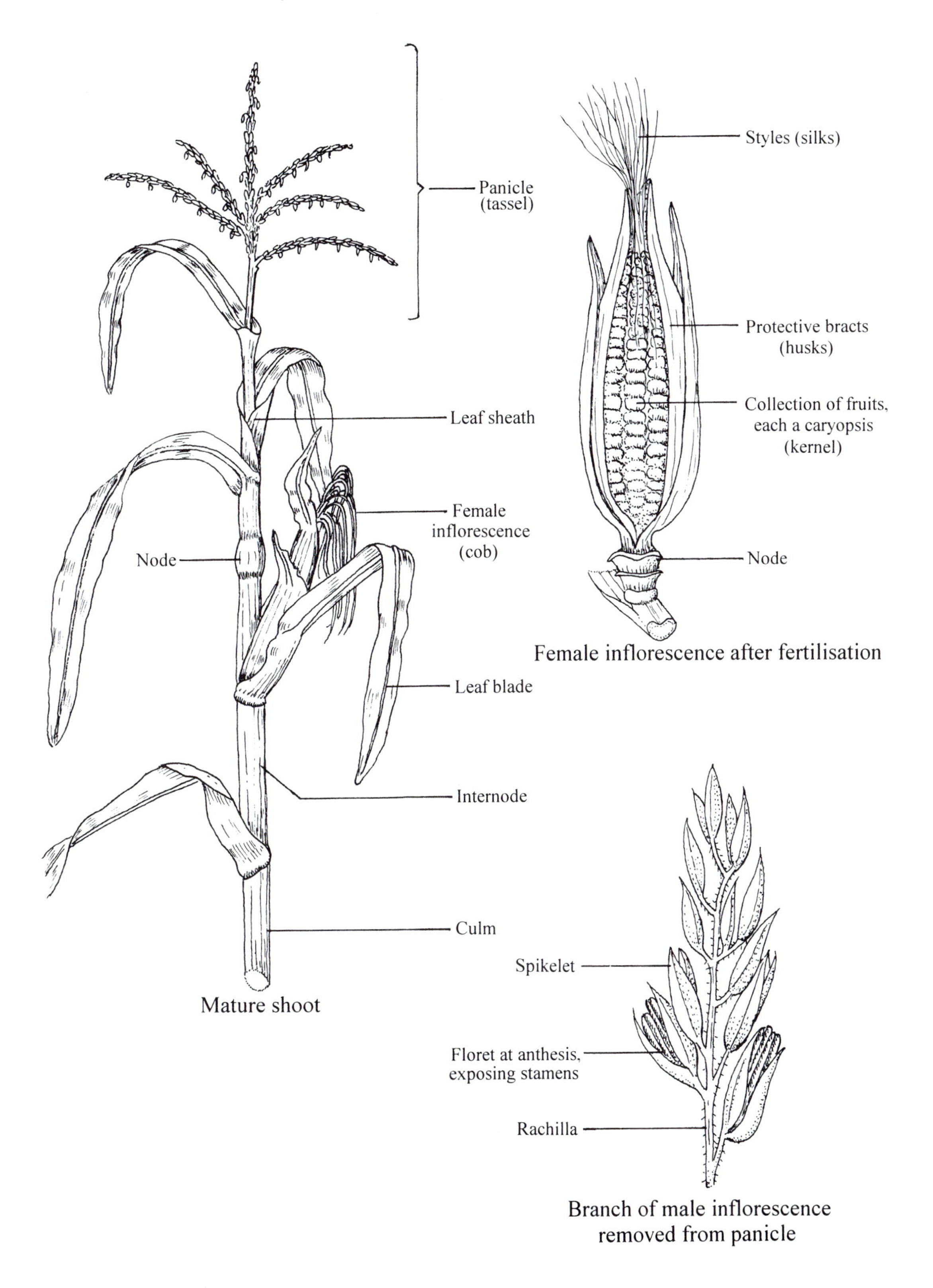

Mature shoot

Female inflorescence after fertilisation

Branch of male inflorescence
removed from panicle

Structure of Rushes *Juncus* and *Luzula*

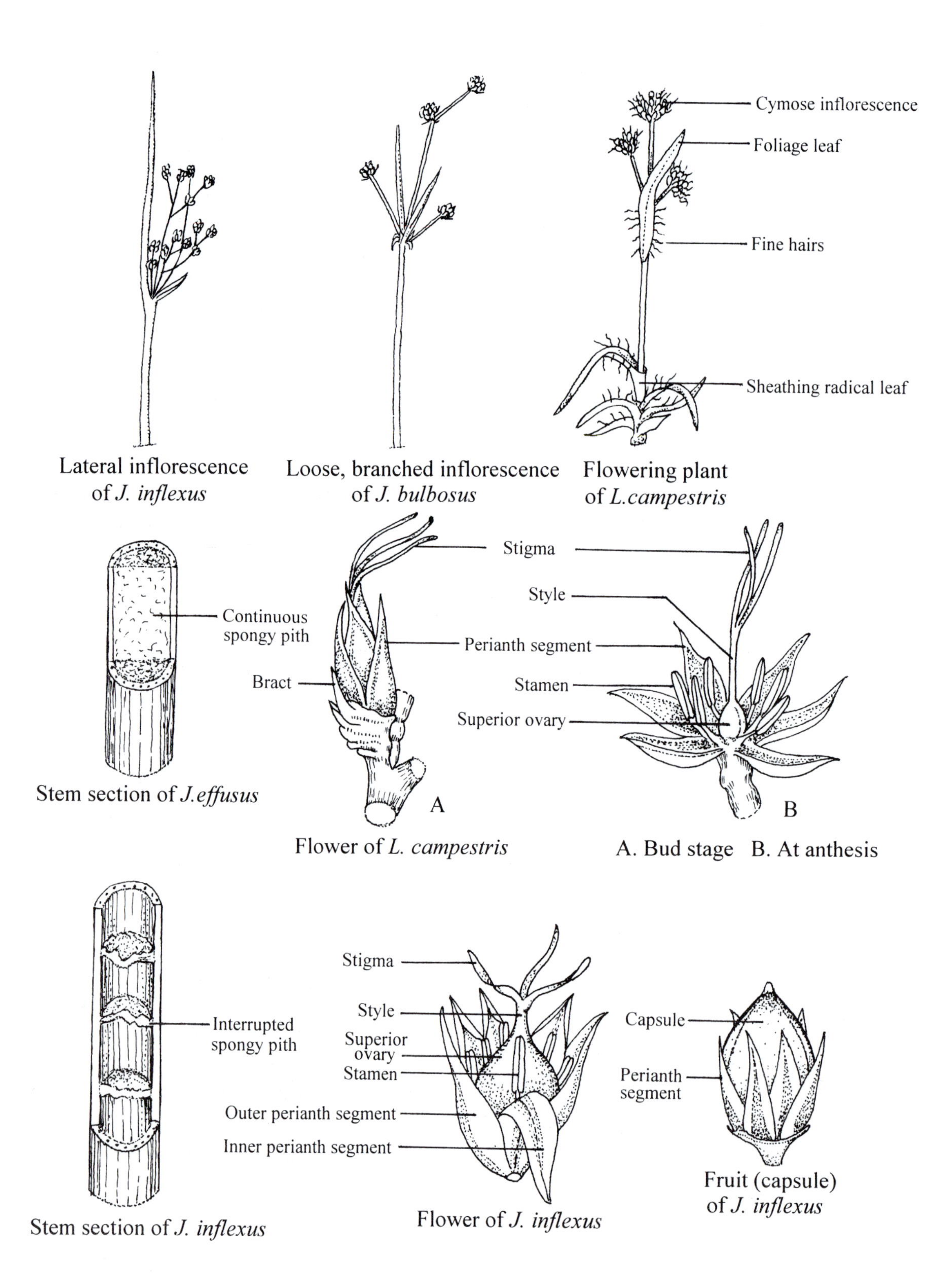

Structure of Sedges *Carex*

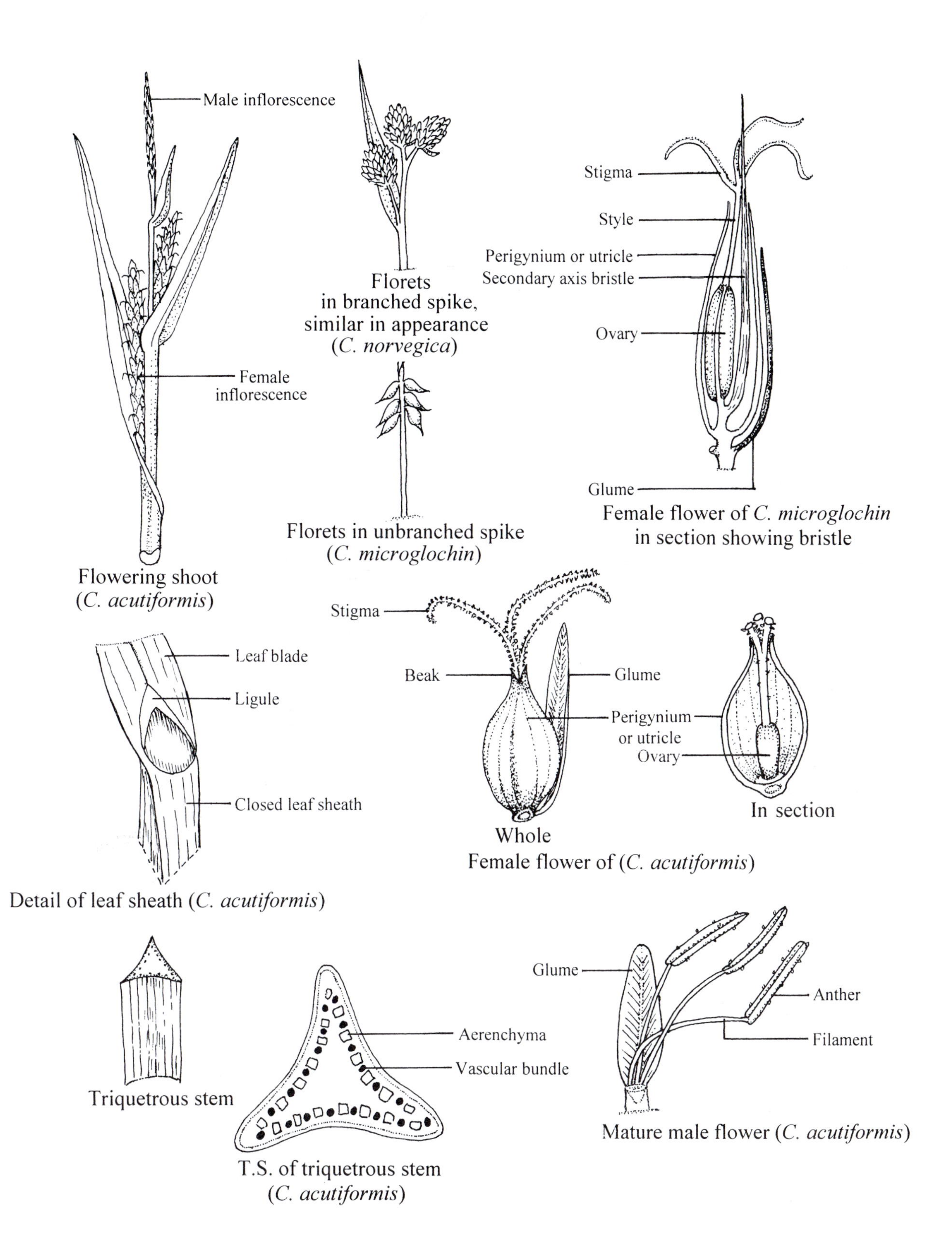

Structure of Palms 1

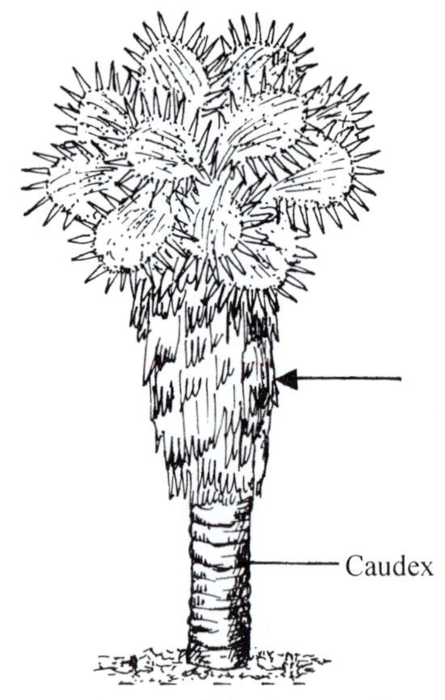

Marcescent leaves (*Washingtonia* sp.)

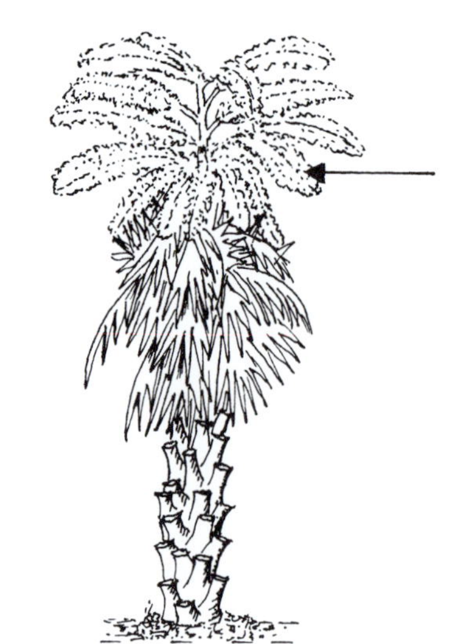

Suprafoliar inflorescence (*Corypha* sp.)

Dichotomously branched (*Hyphaene* sp.)

Acaulescent (*Allagoptera* sp.)

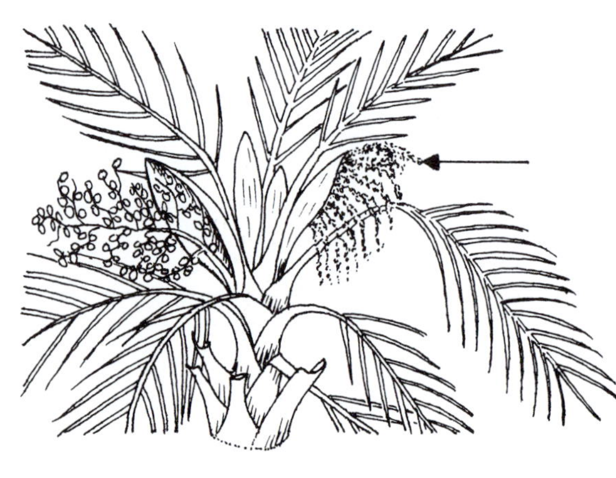

Interfoliar inflorescence (*Butia* sp.)

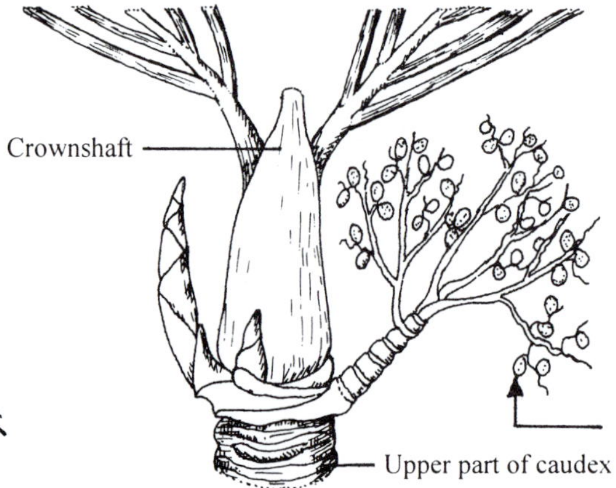

Infrafoliar inflorescence (*Hyophorbe* sp.)

Structure of Palms 2

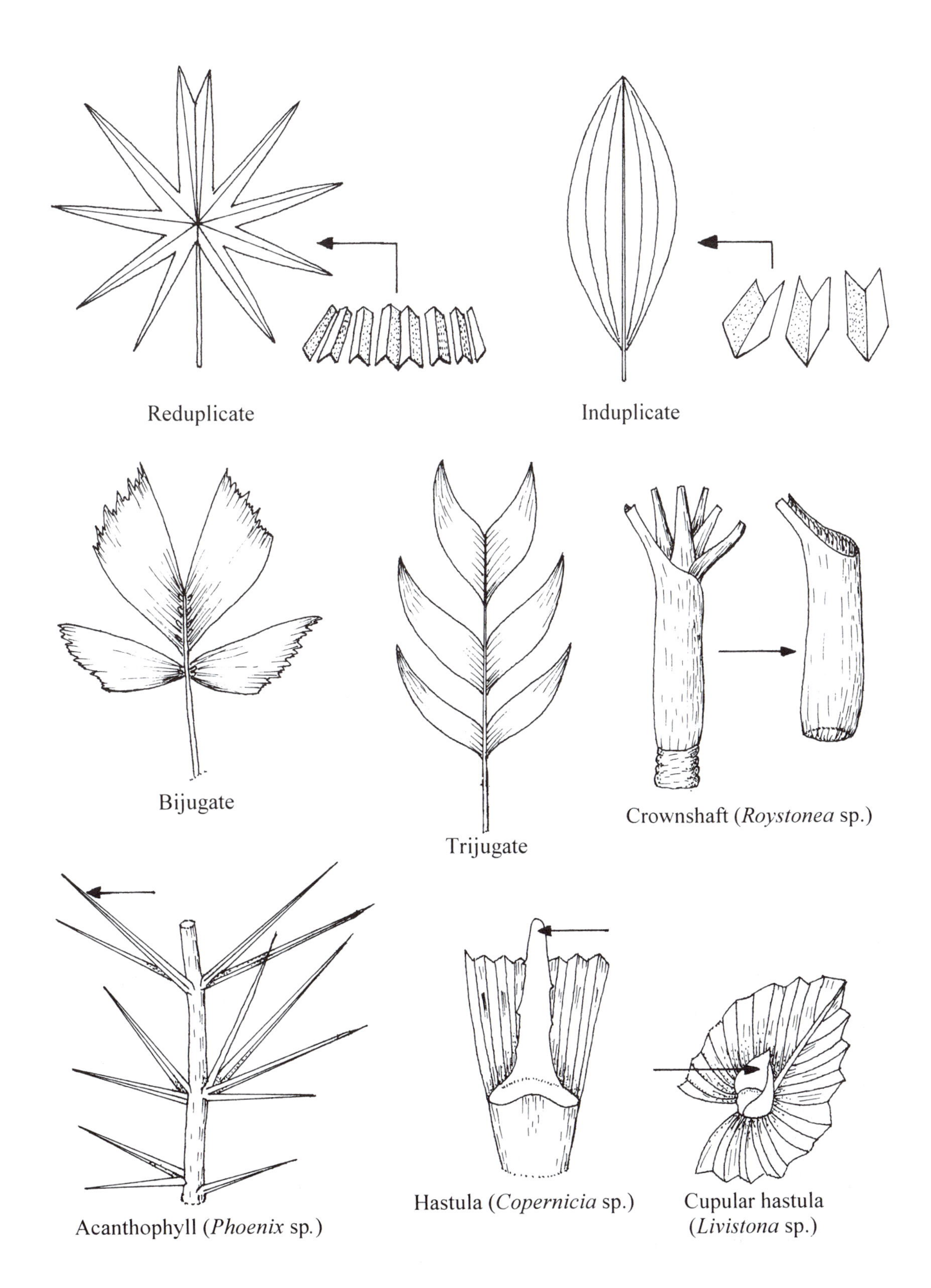
Reduplicate

Induplicate

Bijugate

Trijugate

Crownshaft (*Roystonea* sp.)

Acanthophyll (*Phoenix* sp.)

Hastula (*Copernicia* sp.)

Cupular hastula (*Livistona* sp.)

Structure of Palms 3

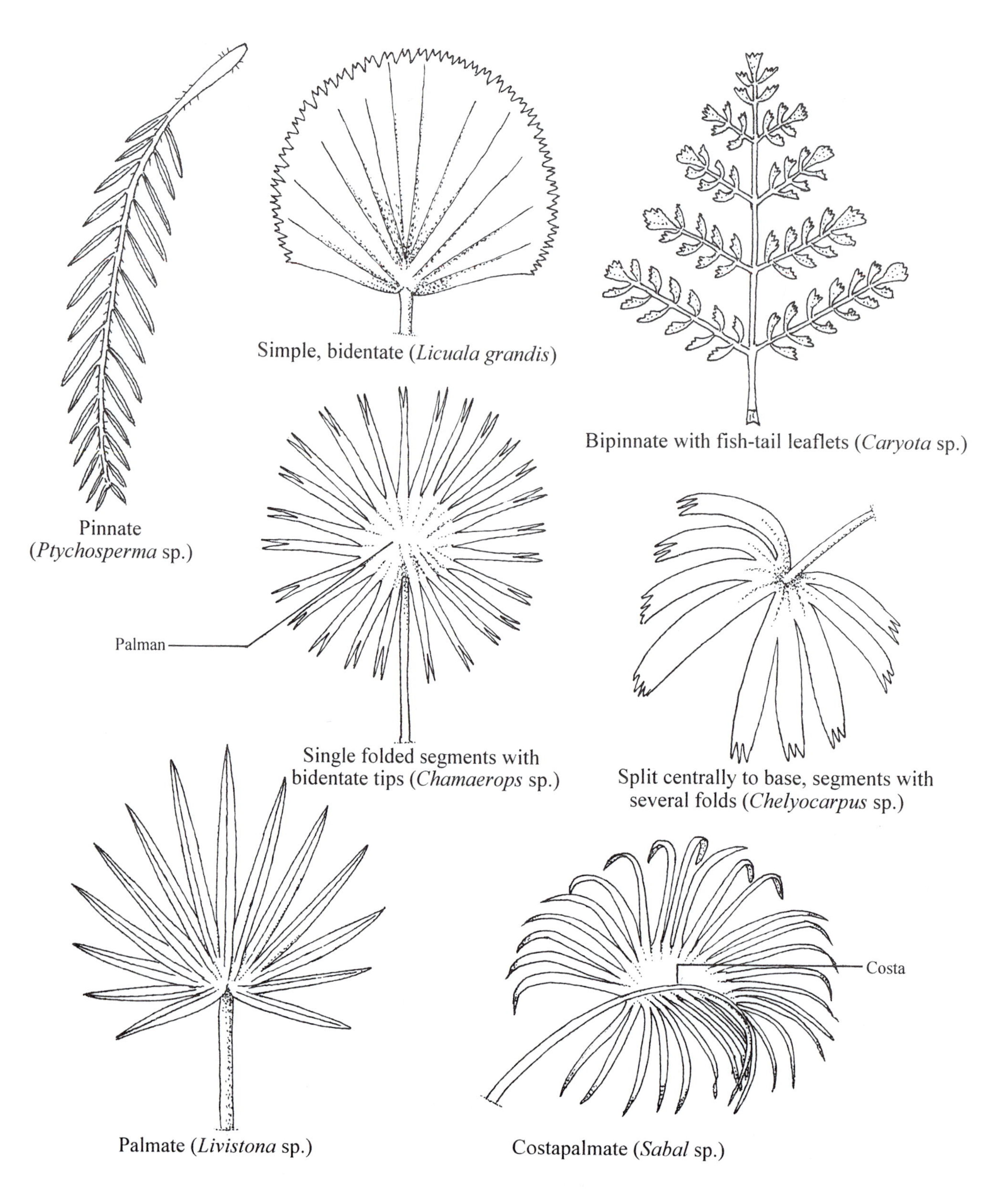

Structure of Orchids 1

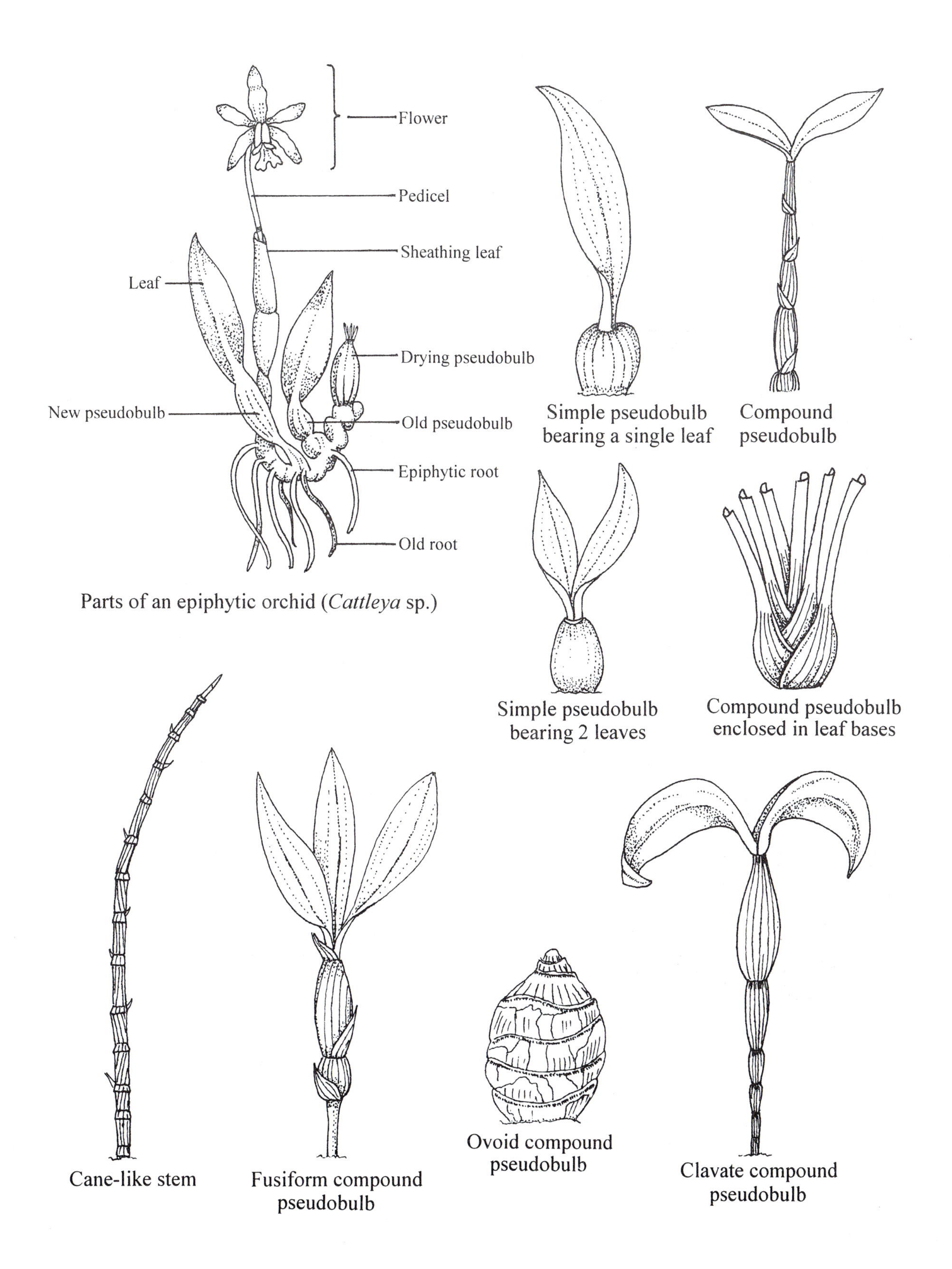

Parts of an epiphytic orchid (*Cattleya* sp.)

Structure of Orchids 2

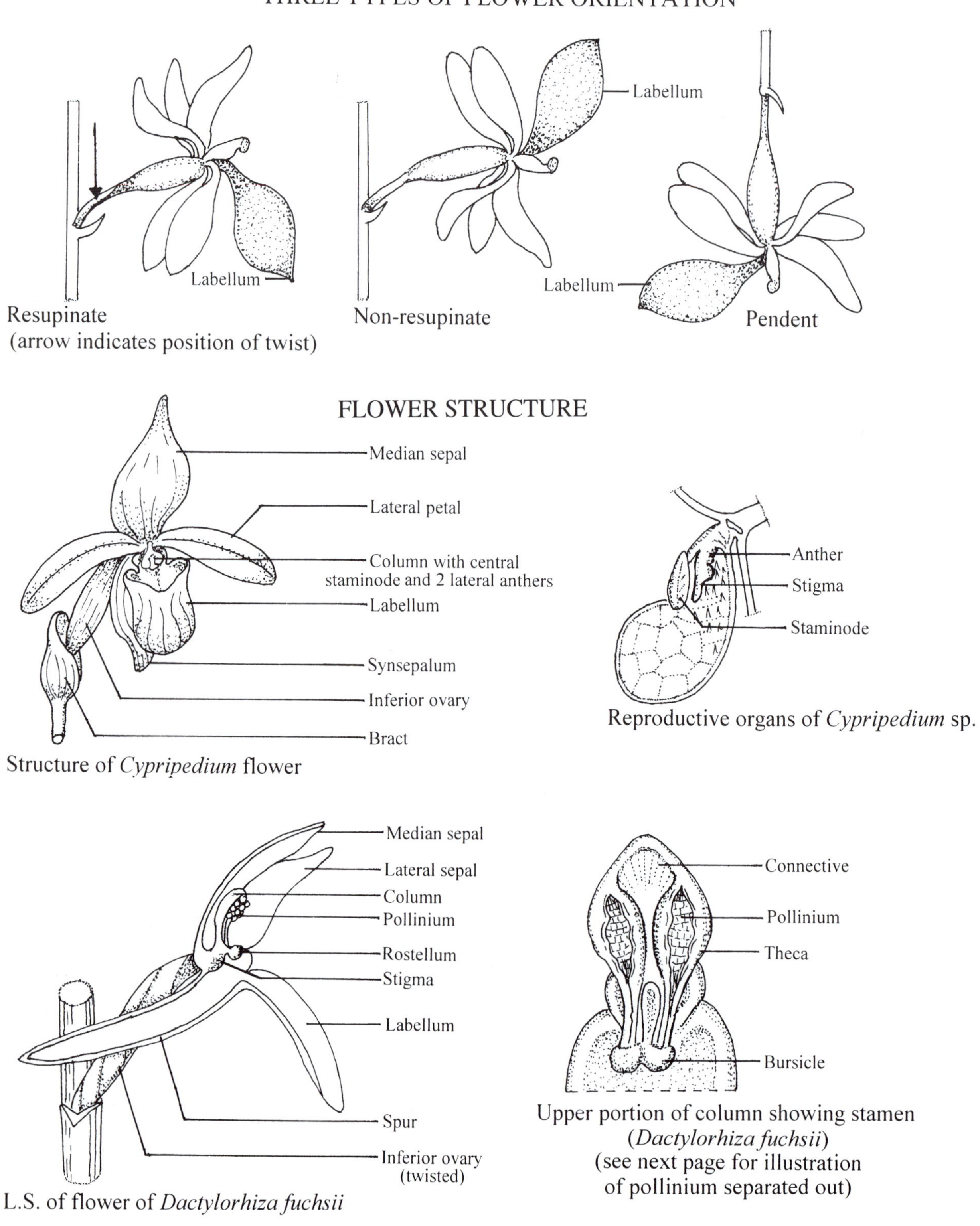

Structure of *Cypripedium* flower

Reproductive organs of *Cypripedium* sp.

L.S. of flower of *Dactylorhiza fuchsii*

Upper portion of column showing stamen (*Dactylorhiza fuchsii*) (see next page for illustration of pollinium separated out)

Structure of Orchids 3

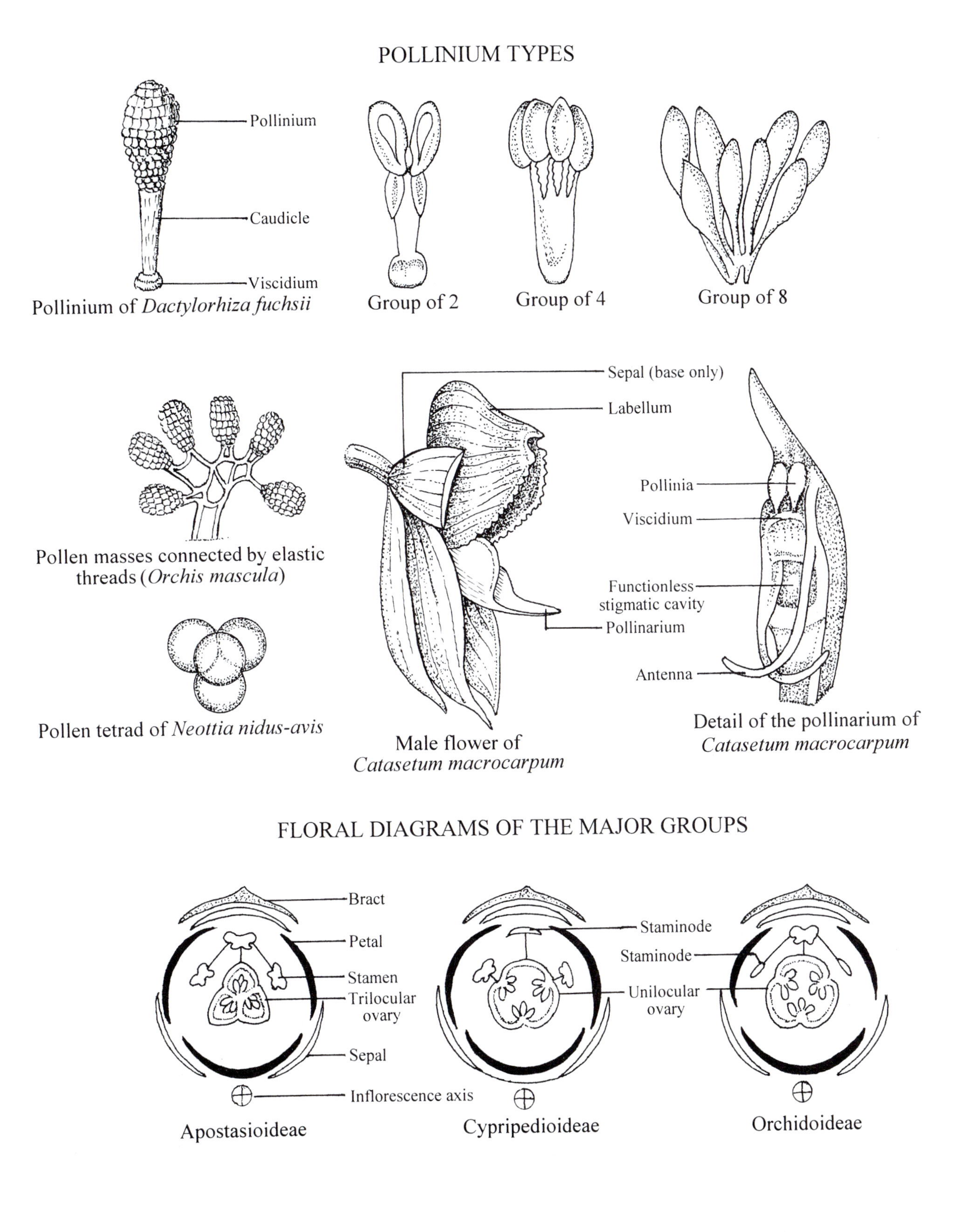

Structure of Orchids 4

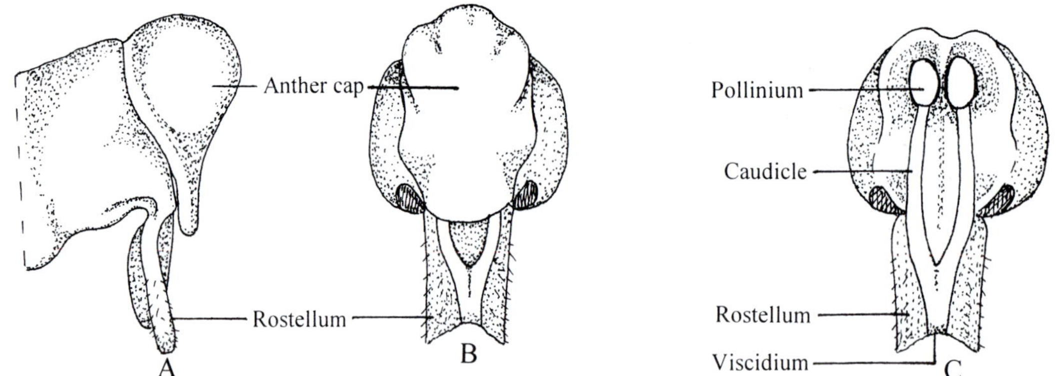

Mystacidium capense Details of column A. Side view B. Front view C. Front view (cap removed)

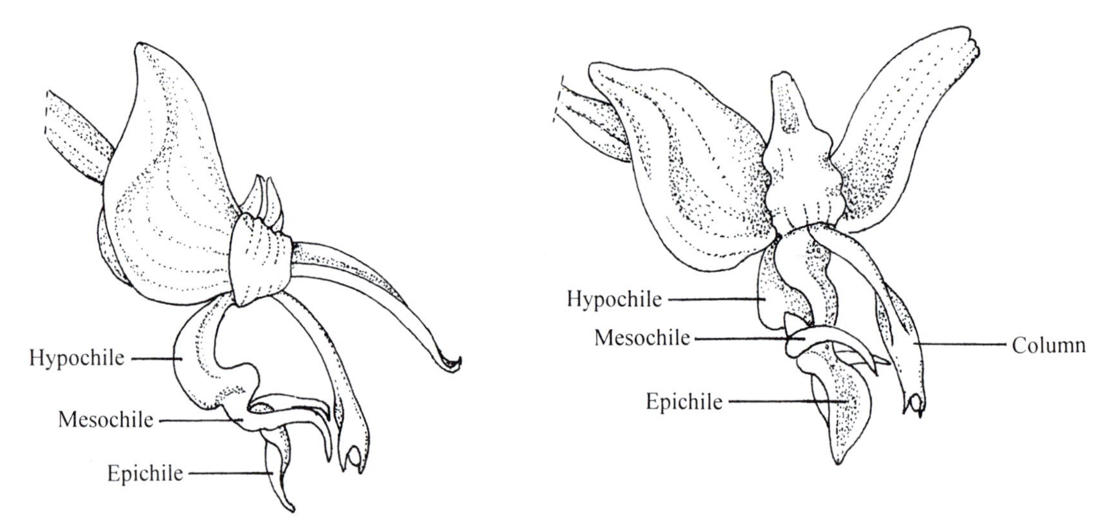

Two species of *Stanhopea* showing details of column (side and front views)

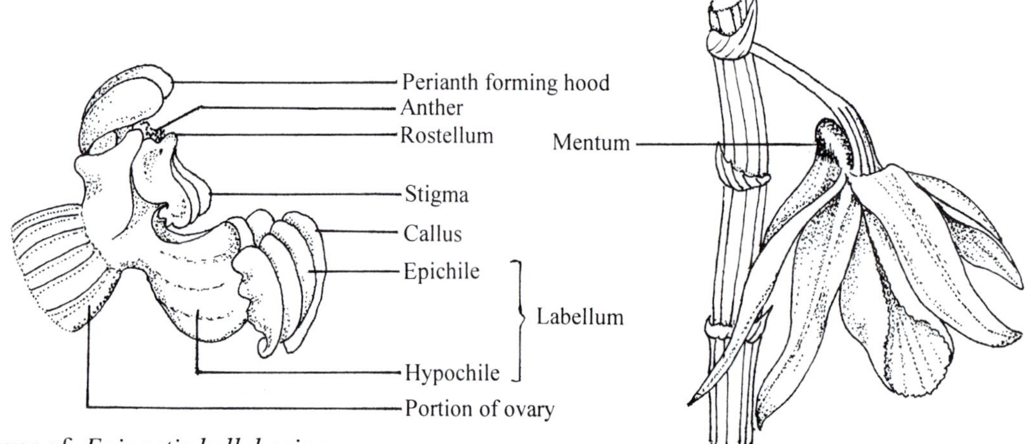

Flower of *Epipactis helleborine*

Dendrobium anosmum showing mentum

12. Fruits

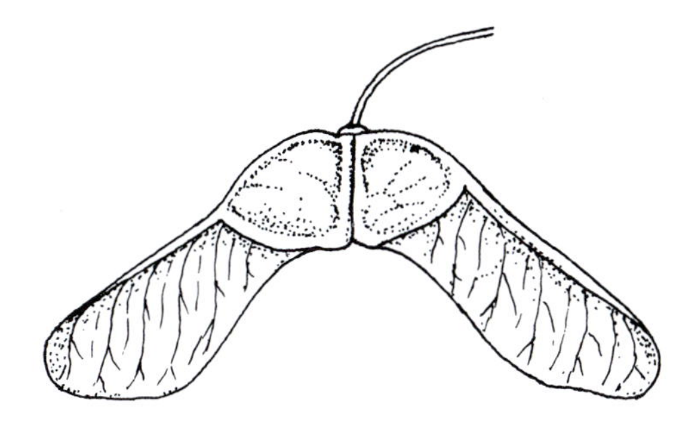

Vexillum Positions and Fruits of Leguminosae (Fabaceae)

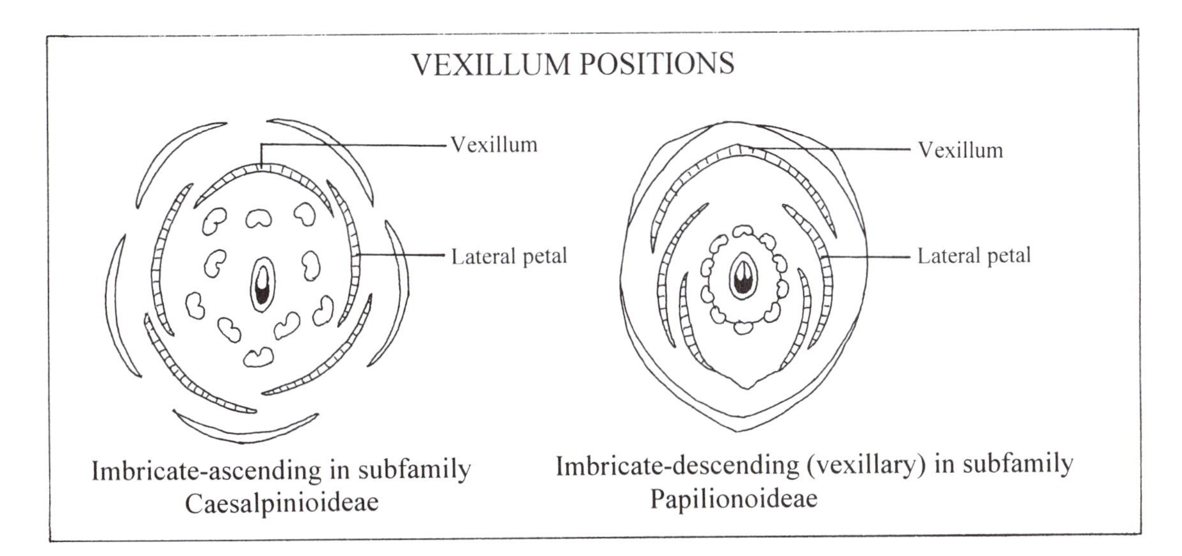

Imbricate-ascending in subfamily Caesalpinioideae

Imbricate-descending (vexillary) in subfamily Papilionoideae

FRUITS 1

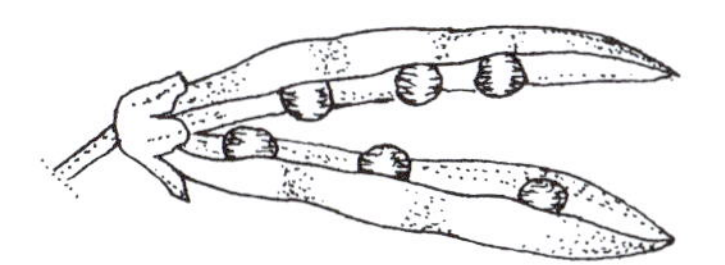

Legume (*Vicia* sp.)

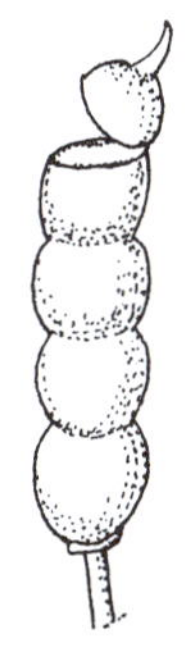

Lomentum (*Ornithopus* sp.)

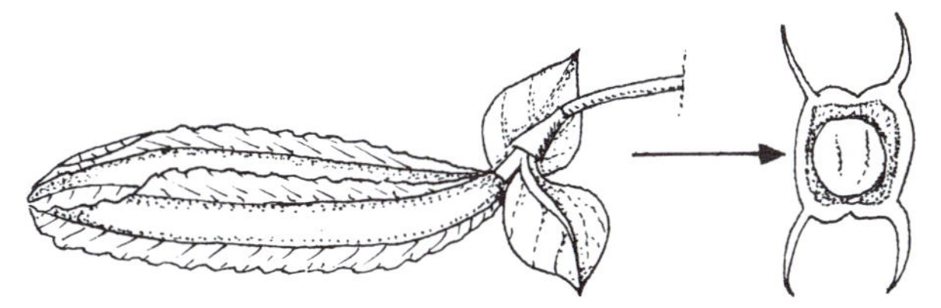

4-winged pod of *Tetragonolobus purpureus*

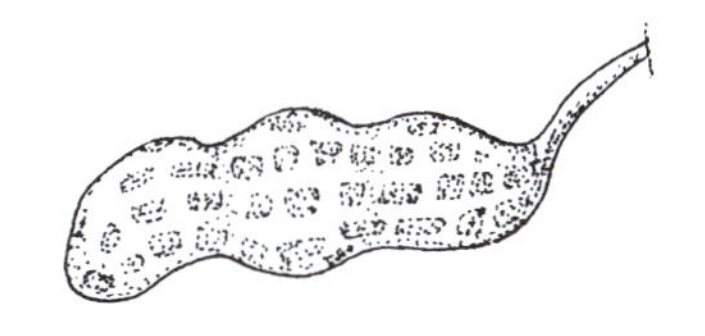

Reticulate pod of *Arachis hypogaea*

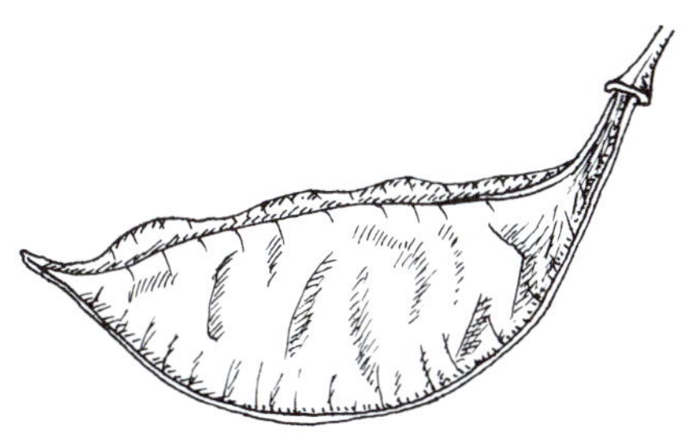

Inflated many-seeded pod with papery walls of *Colutea arborescens*

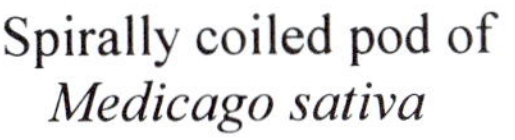

Spirally coiled pod of *Medicago sativa*

1- or 2-seeded pod with beak of *Trigonella caerulea*

Fruits 2

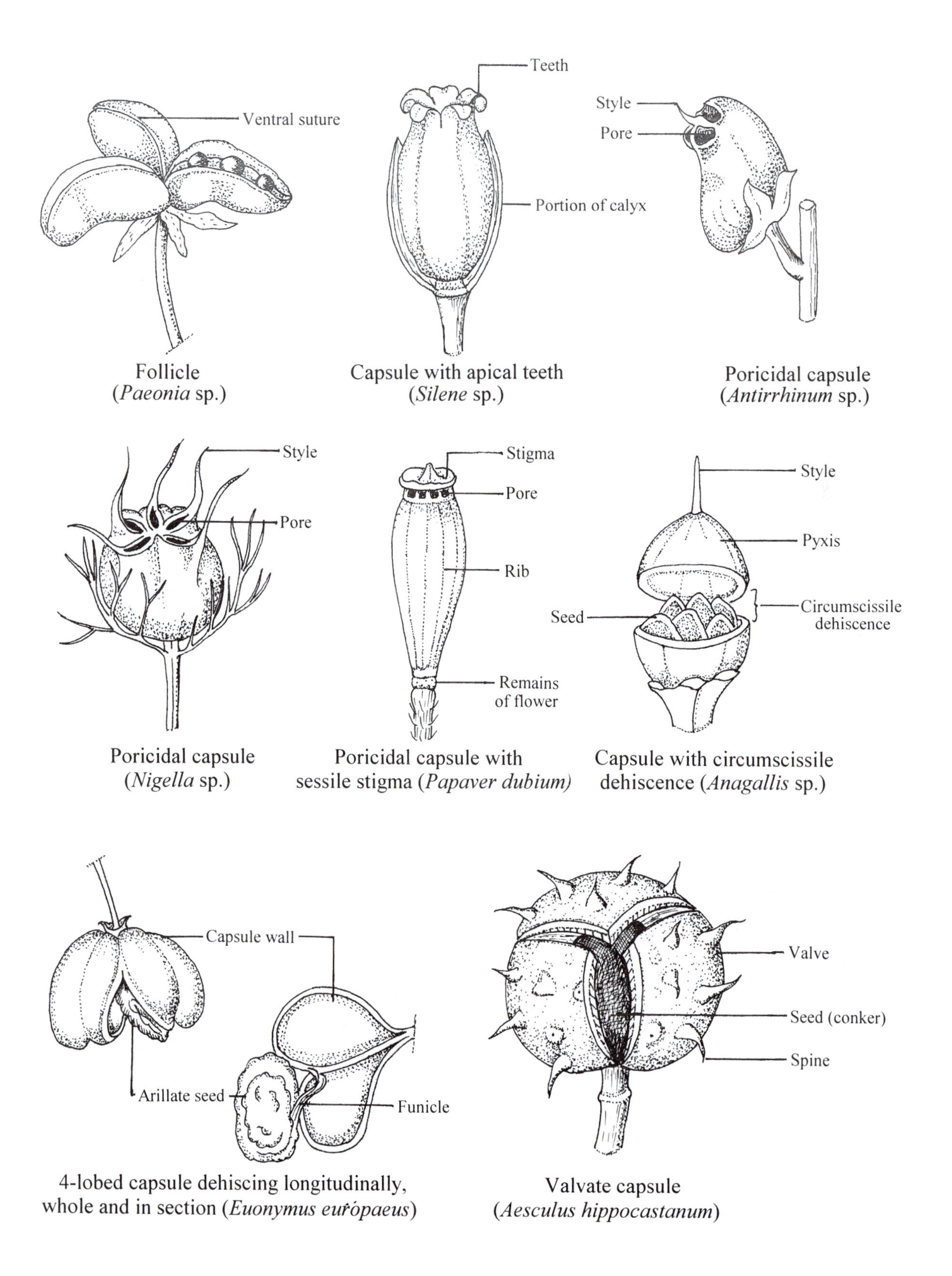

Follicle (*Paeonia* sp.)

Capsule with apical teeth (*Silene* sp.)

Poricidal capsule (*Antirrhinum* sp.)

Poricidal capsule (*Nigella* sp.)

Poricidal capsule with sessile stigma (*Papaver dubium*)

Capsule with circumscissile dehiscence (*Anagallis* sp.)

4-lobed capsule dehiscing longitudinally, whole and in section (*Euonymus európaeus*)

Valvate capsule (*Aesculus hippocastanum*)

Fruits 3

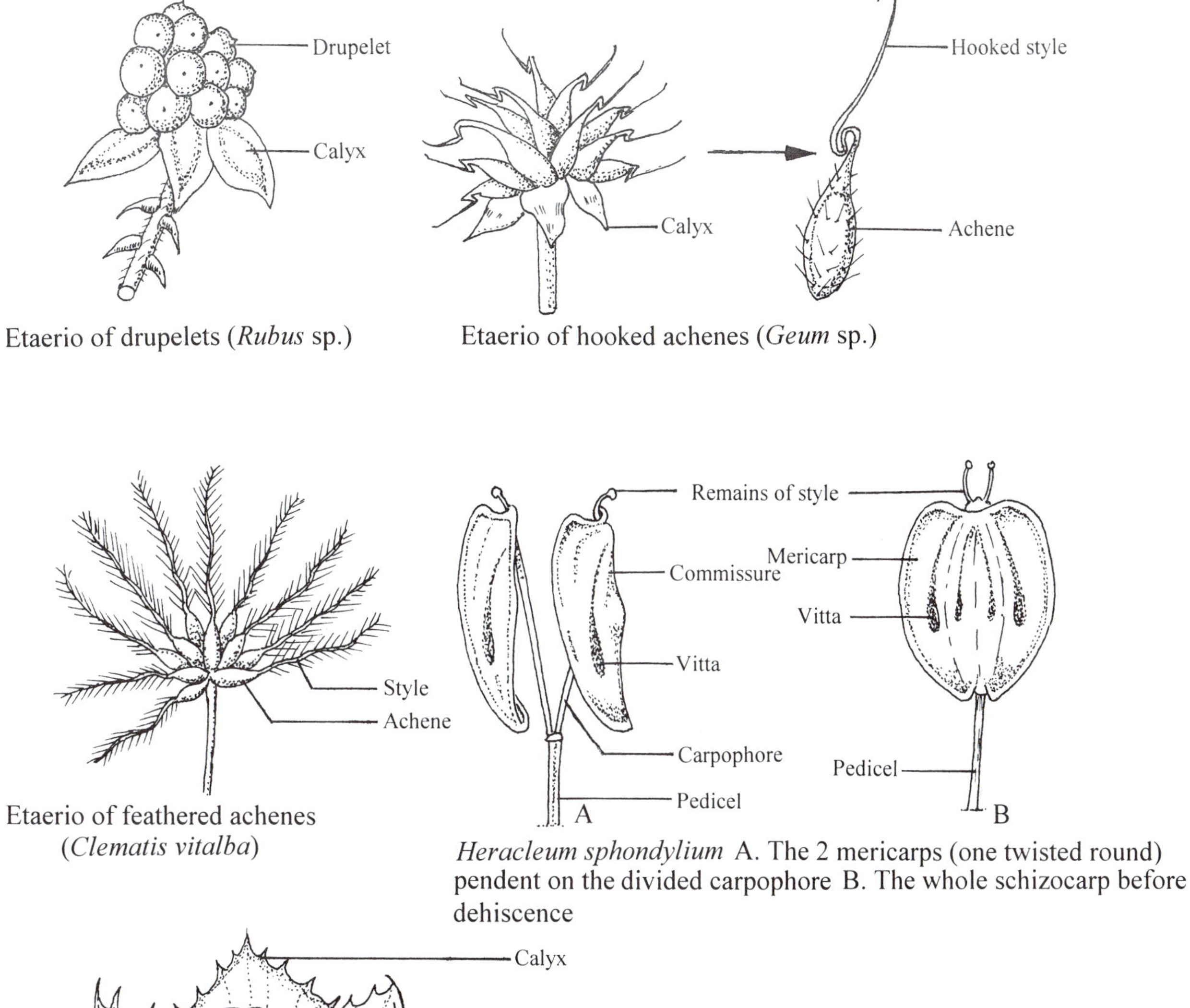

Etaerio of drupelets (*Rubus* sp.)

Etaerio of hooked achenes (*Geum* sp.)

Etaerio of feathered achenes (*Clematis vitalba*)

Heracleum sphondylium A. The 2 mericarps (one twisted round) pendent on the divided carpophore B. The whole schizocarp before dehiscence

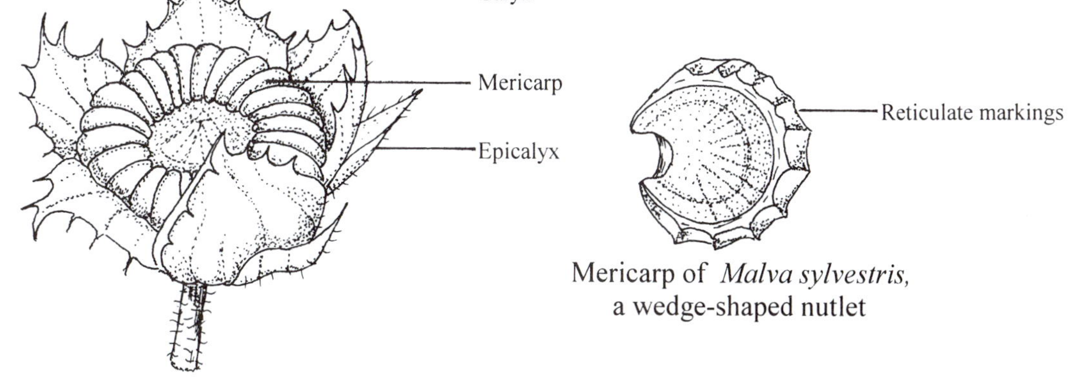

Schizocarp of *Malva sylvestris*

Mericarp of *Malva sylvestris*, a wedge-shaped nutlet

Fruits 4

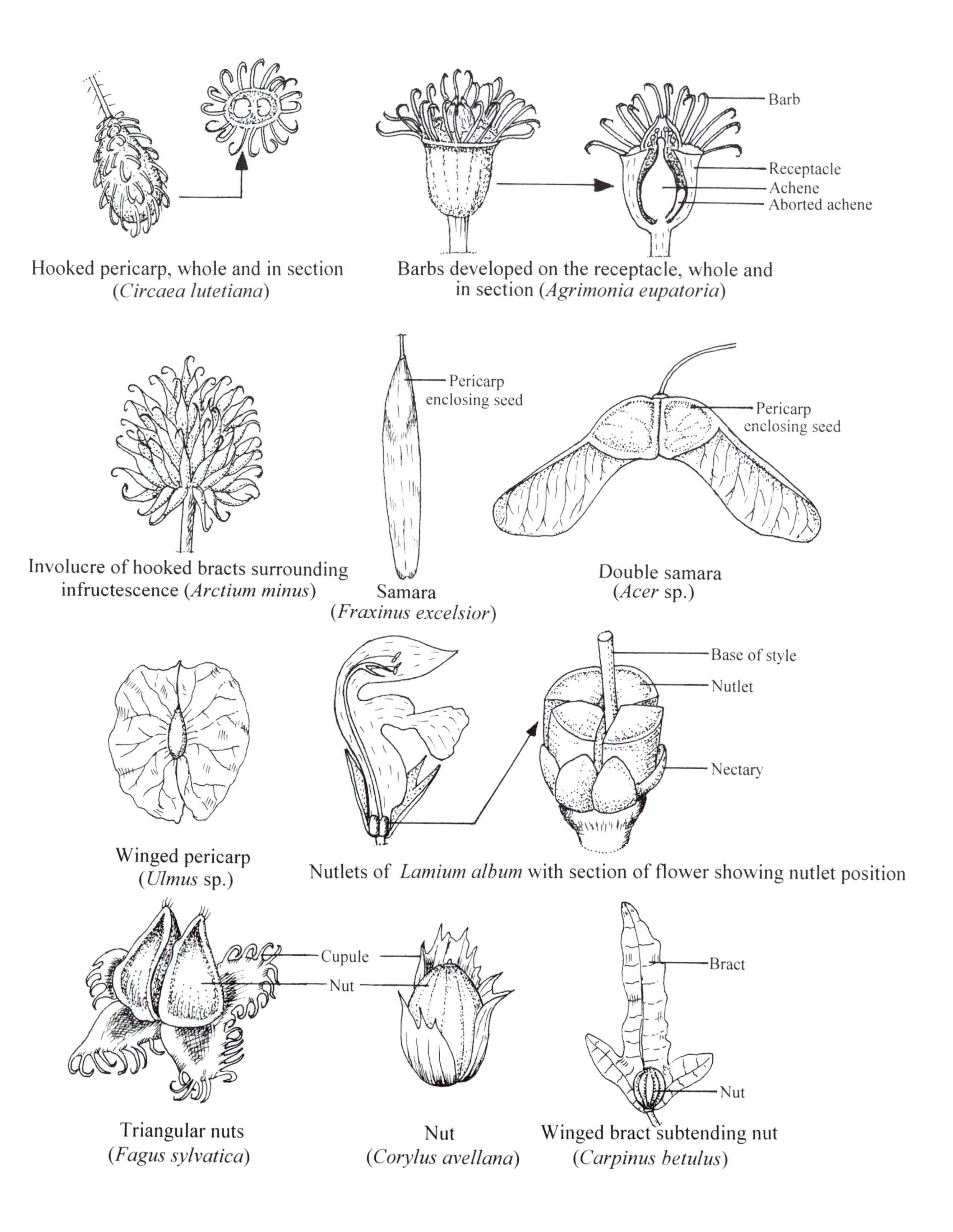

Hooked pericarp, whole and in section (*Circaea lutetiana*)

Barbs developed on the receptacle, whole and in section (*Agrimonia eupatoria*)

Involucre of hooked bracts surrounding infructescence (*Arctium minus*)

Samara (*Fraxinus excelsior*)

Double samara (*Acer* sp.)

Winged pericarp (*Ulmus* sp.)

Nutlets of *Lamium album* with section of flower showing nutlet position

Triangular nuts (*Fagus sylvatica*)

Nut (*Corylus avellana*)

Winged bract subtending nut (*Carpinus betulus*)

Fruits 5

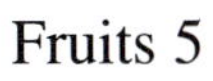

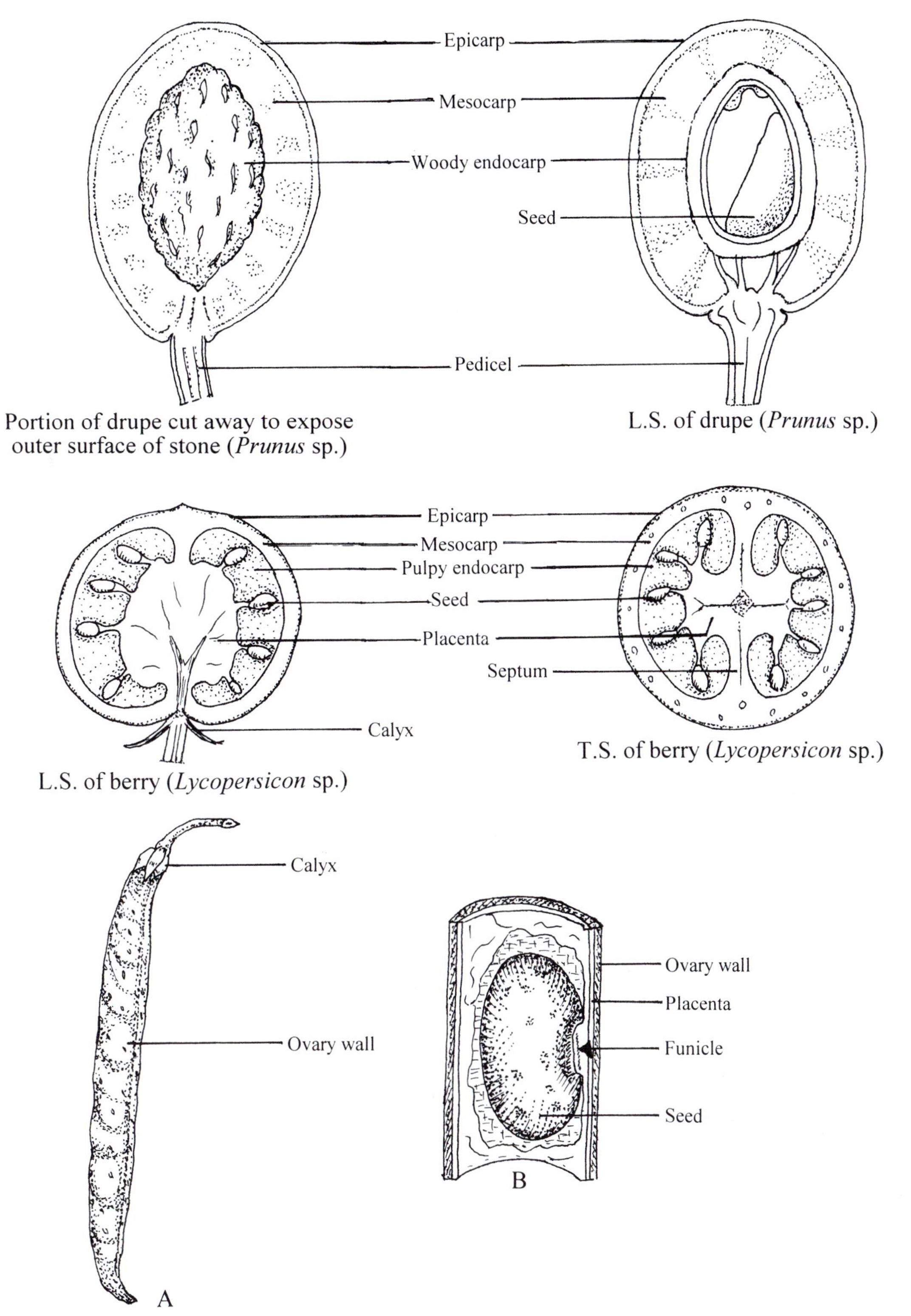

Portion of drupe cut away to expose outer surface of stone (*Prunus* sp.)

L.S. of drupe (*Prunus* sp.)

L.S. of berry (*Lycopersicon* sp.)

T.S. of berry (*Lycopersicon* sp.)

Phaseolus coccineus A. Whole legume
B. Portion of legume opened up to show attachment of seed

Fruits 6

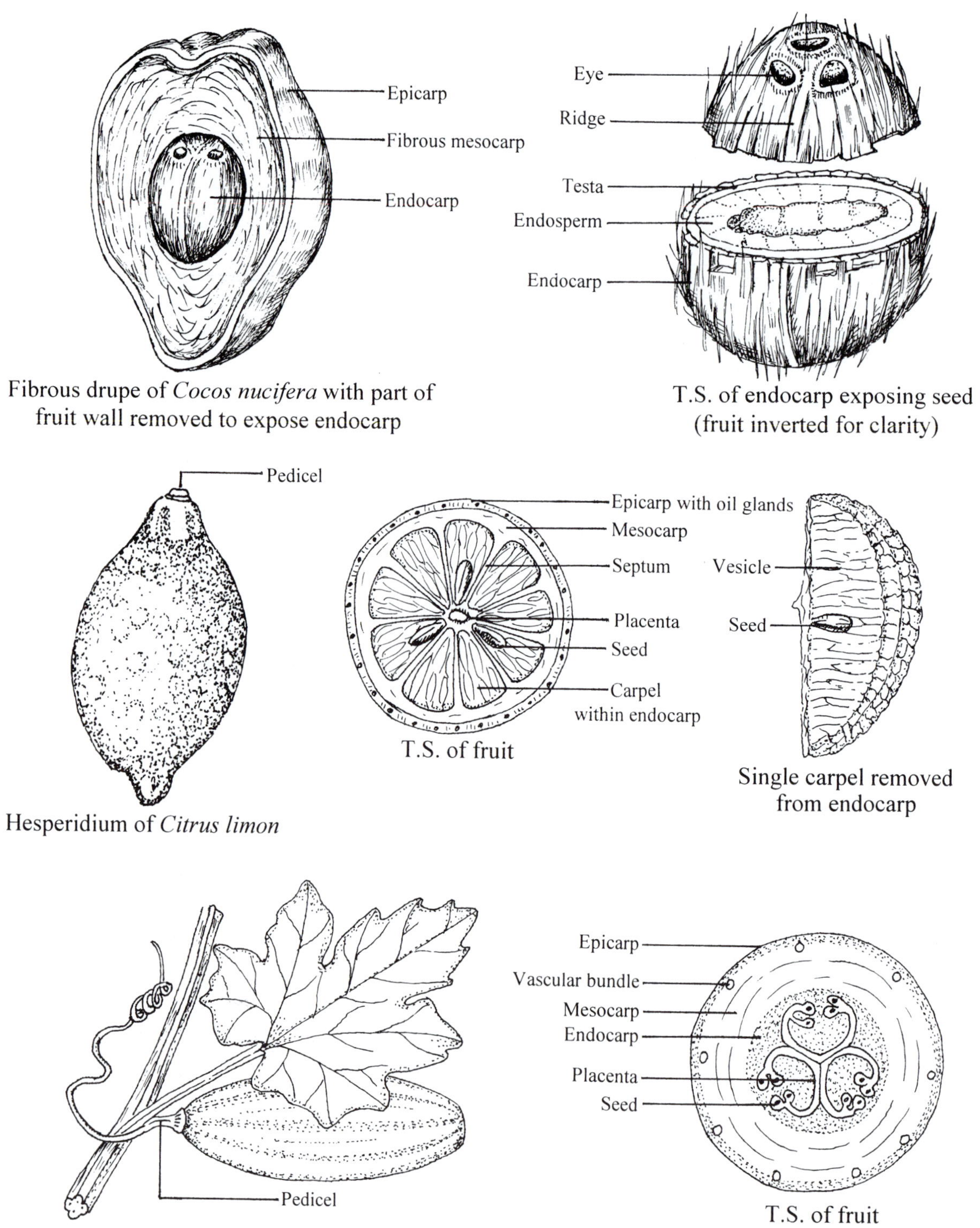

Fibrous drupe of *Cocos nucifera* with part of fruit wall removed to expose endocarp

T.S. of endocarp exposing seed (fruit inverted for clarity)

Hesperidium of *Citrus limon*

T.S. of fruit

Single carpel removed from endocarp

Pepo of *Cucurbita pepo*

T.S. of fruit

Fruits 7

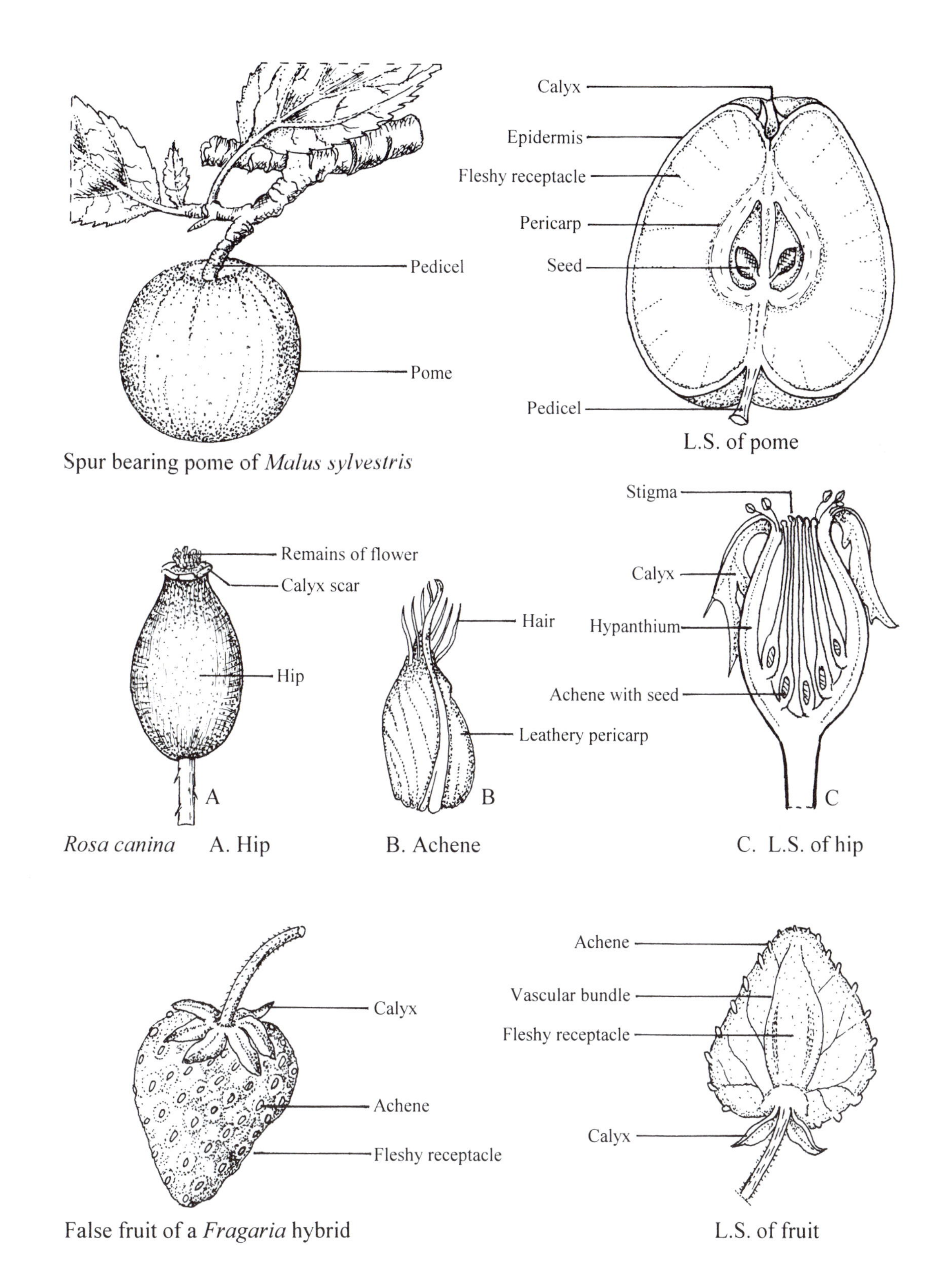

Spur bearing pome of *Malus sylvestris*

L.S. of pome

Rosa canina A. Hip

B. Achene

C. L.S. of hip

False fruit of a *Fragaria* hybrid

L.S. of fruit

Fruits 8

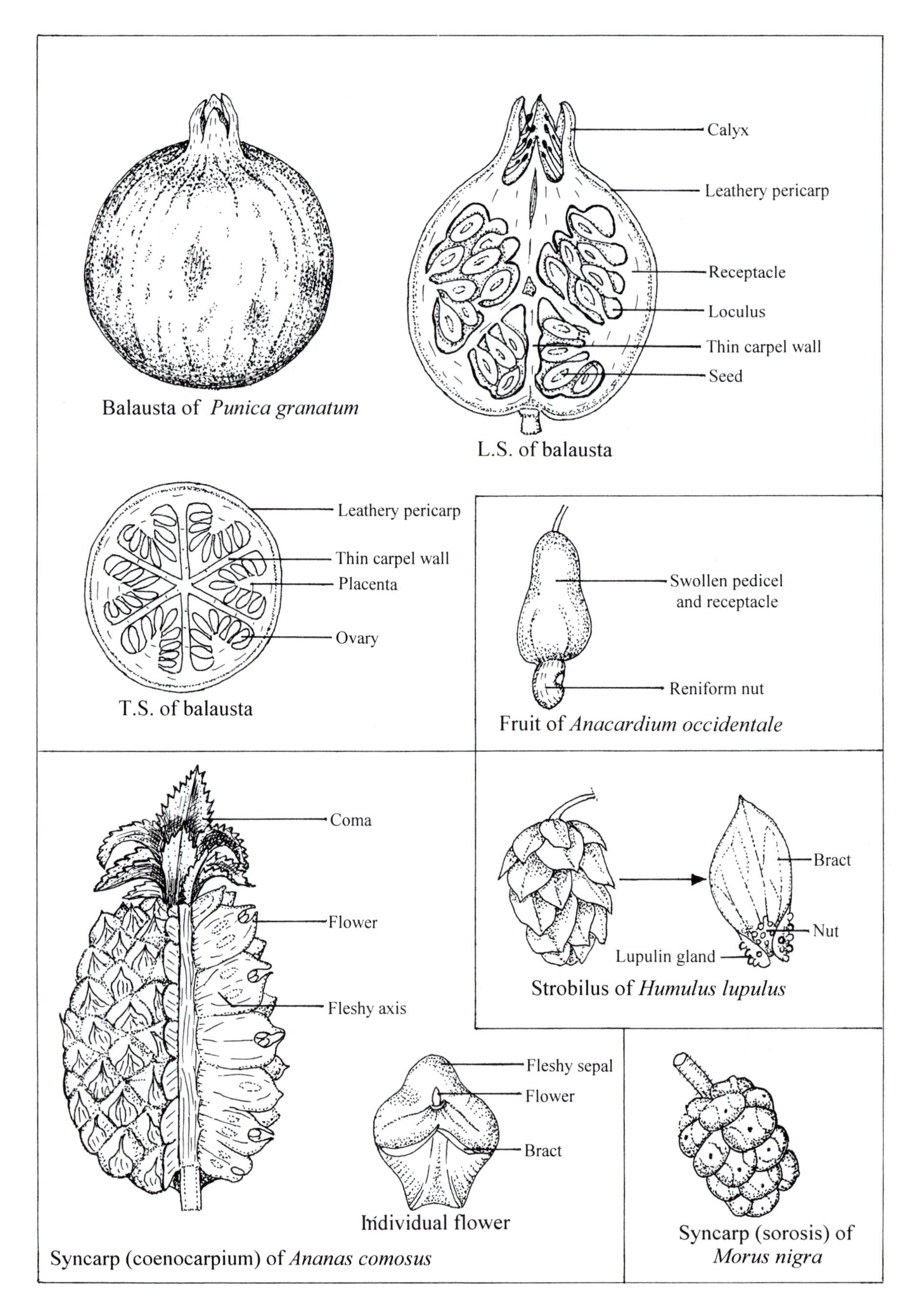

Balausta of *Punica granatum*

L.S. of balausta

T.S. of balausta

Fruit of *Anacardium occidentale*

Strobilus of *Humulus lupulus*

Individual flower

Syncarp (coenocarpium) of *Ananas comosus*

Syncarp (sorosis) of *Morus nigra*

Fruits 9 Syconium

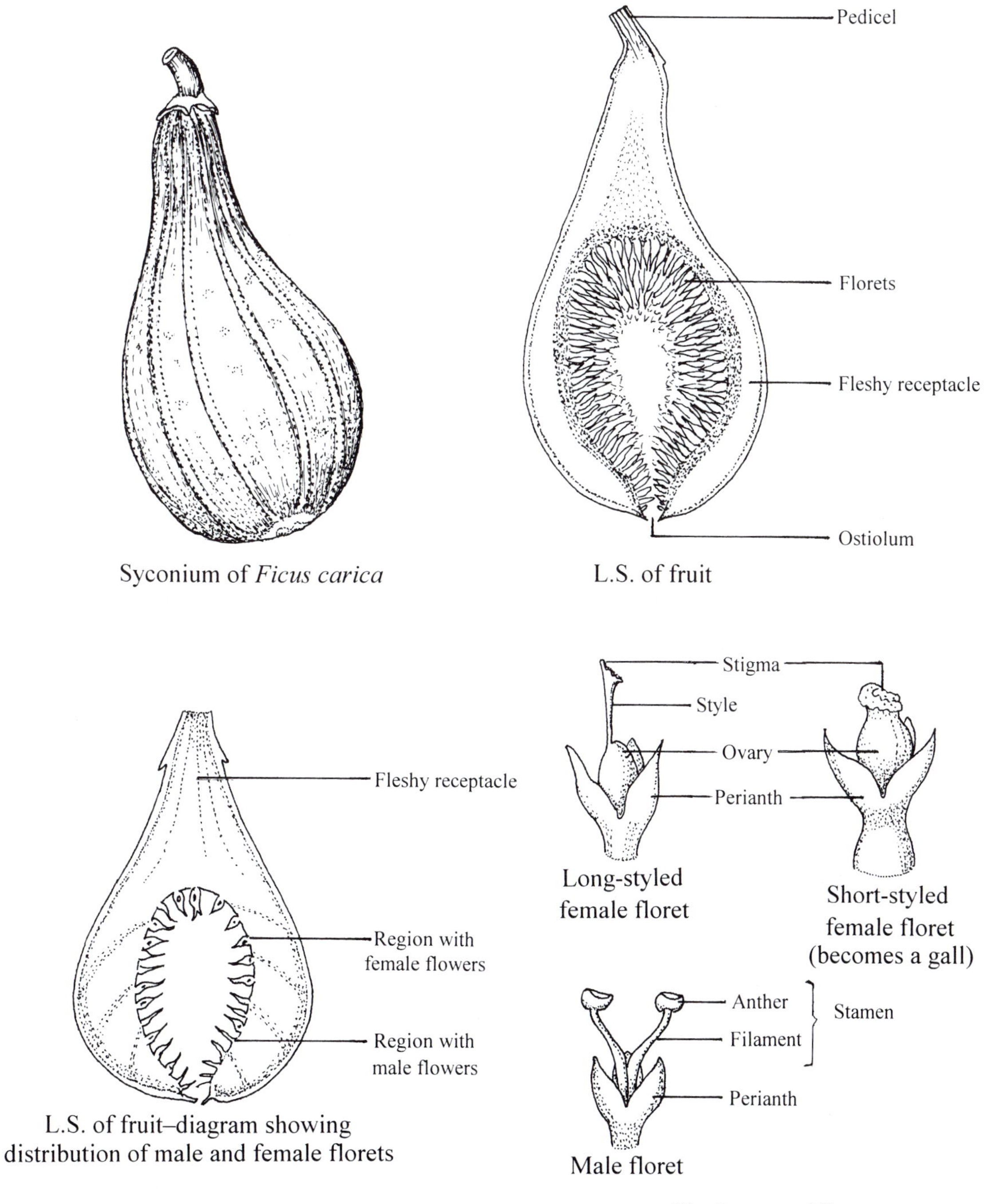

The 3 types of floret

Fruits 10 Cruciferae (Brassicaceae) 1

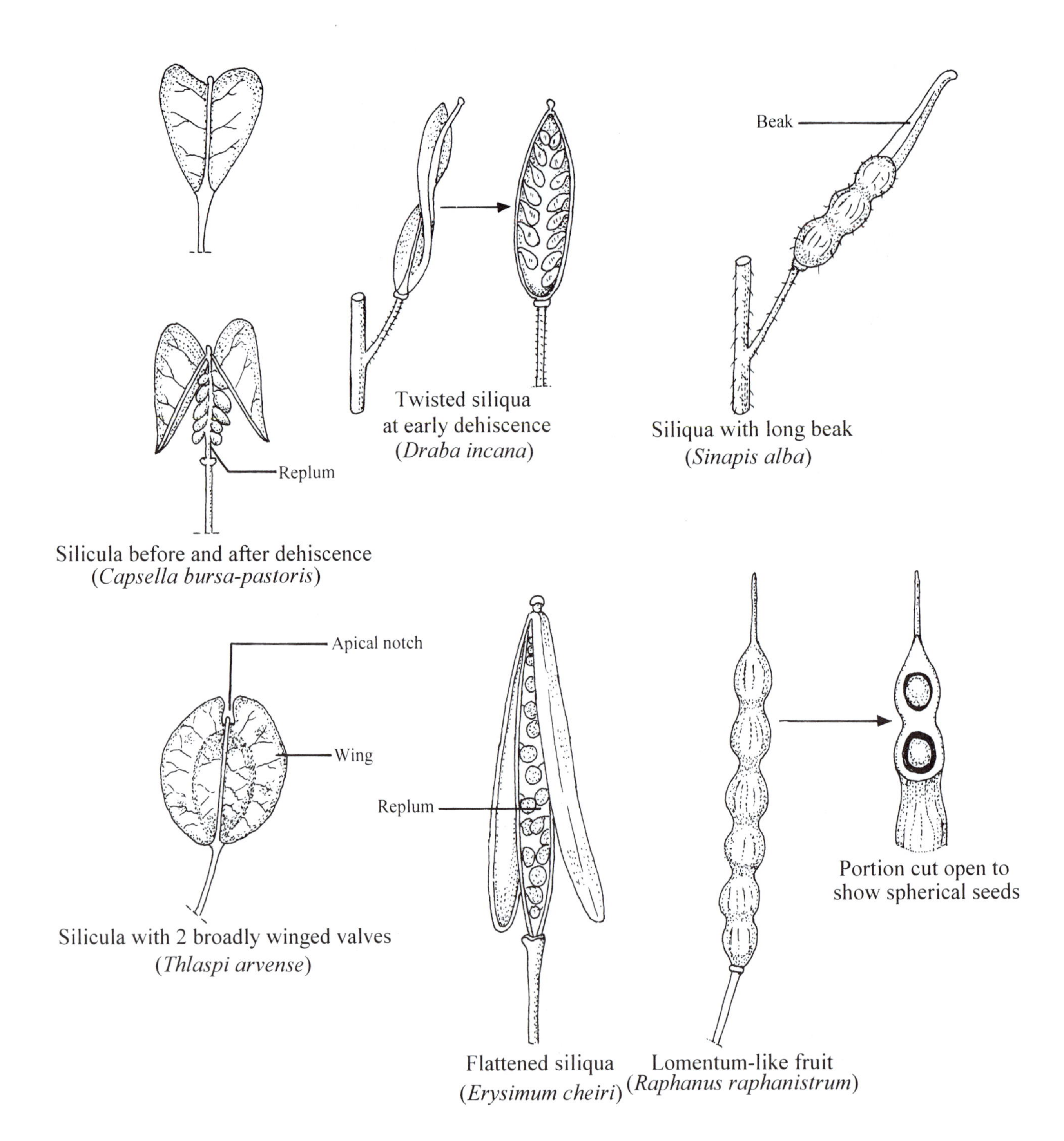

Silicula before and after dehiscence (*Capsella bursa-pastoris*)

Twisted siliqua at early dehiscence (*Draba incana*)

Siliqua with long beak (*Sinapis alba*)

Silicula with 2 broadly winged valves (*Thlaspi arvense*)

Flattened siliqua (*Erysimum cheiri*)

Lomentum-like fruit (*Raphanus raphanistrum*)

Portion cut open to show spherical seeds

Fruits 11 Cruciferae (Brassicaceae) 2

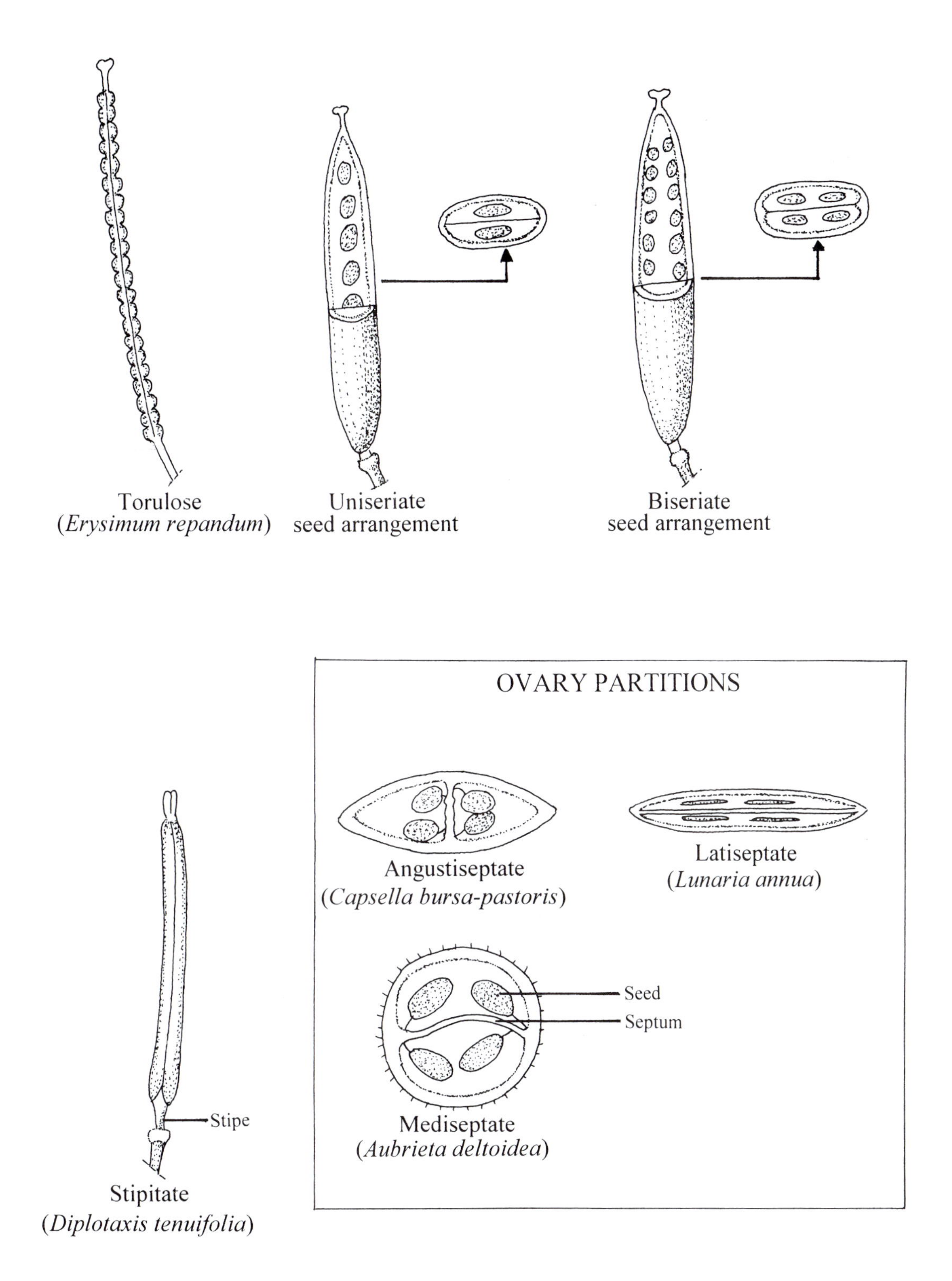

13. Conifers and Conifer Allies

Conifers 1 *Pinus sylvestris*

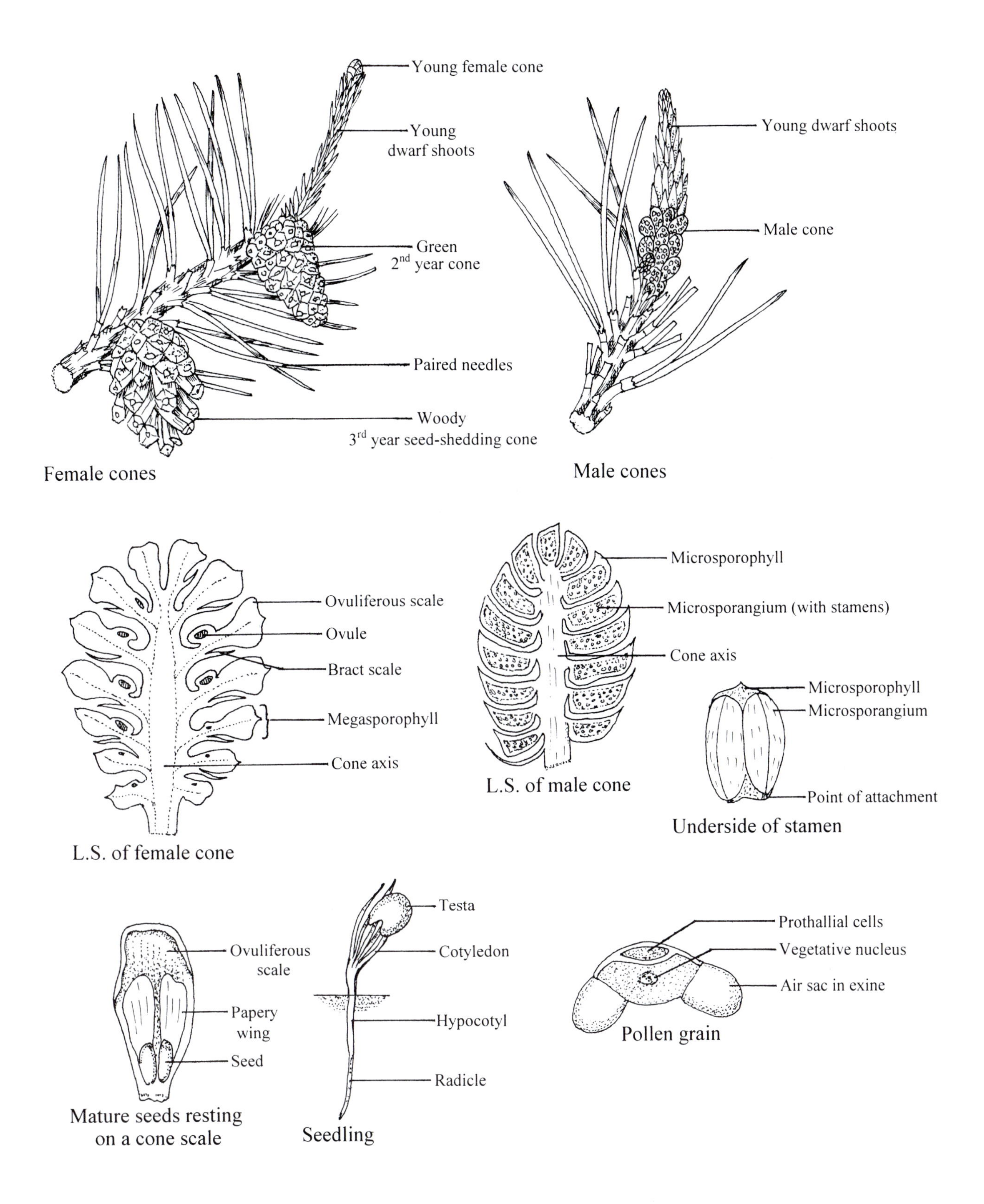

Conifers 2

PINE NEEDLE ARRANGEMENT

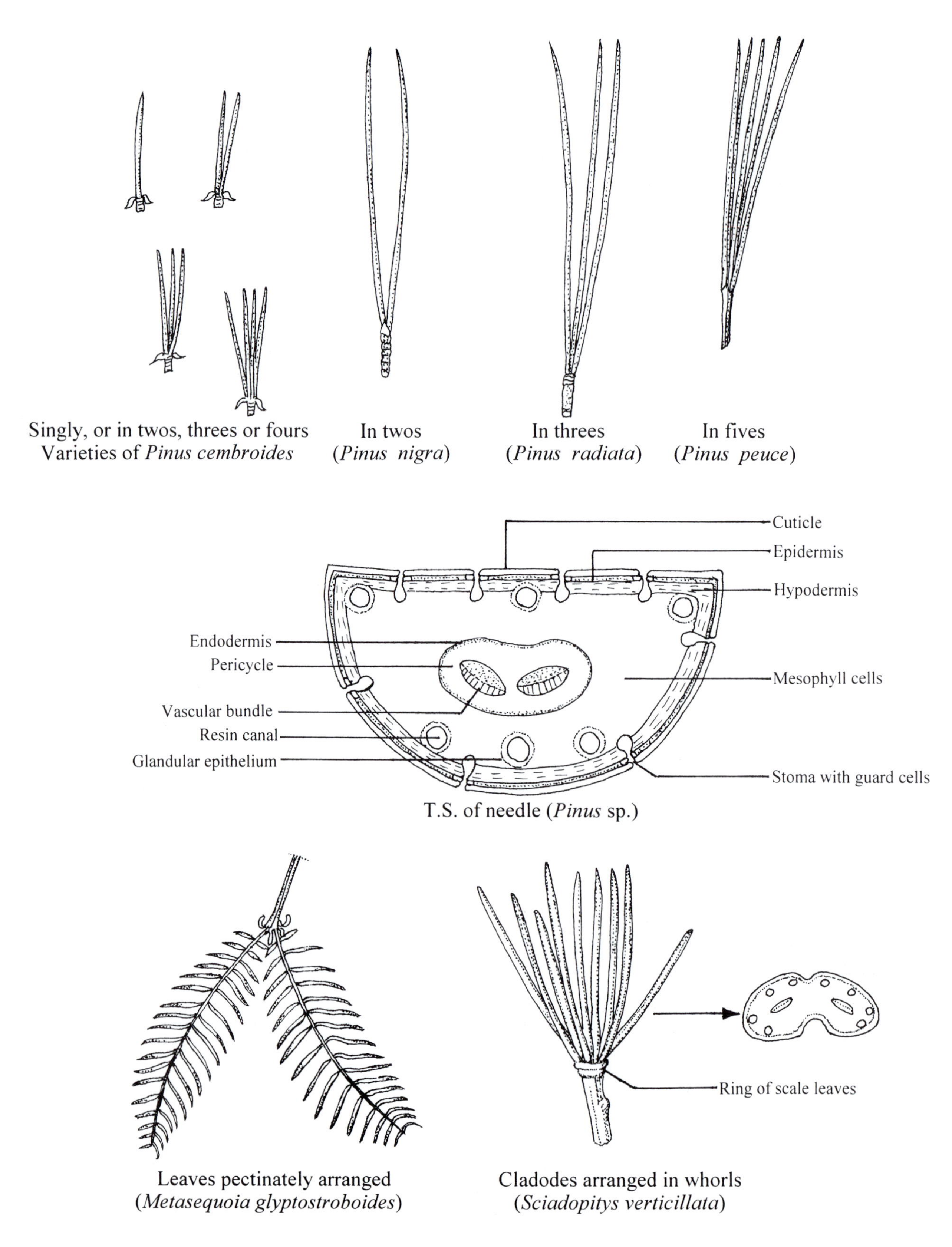

Singly, or in twos, threes or fours Varieties of *Pinus cembroides*

In twos (*Pinus nigra*)

In threes (*Pinus radiata*)

In fives (*Pinus peuce*)

T.S. of needle (*Pinus* sp.)

Leaves pectinately arranged (*Metasequoia glyptostroboides*)

Cladodes arranged in whorls (*Sciadopitys verticillata*)

Conifers 3

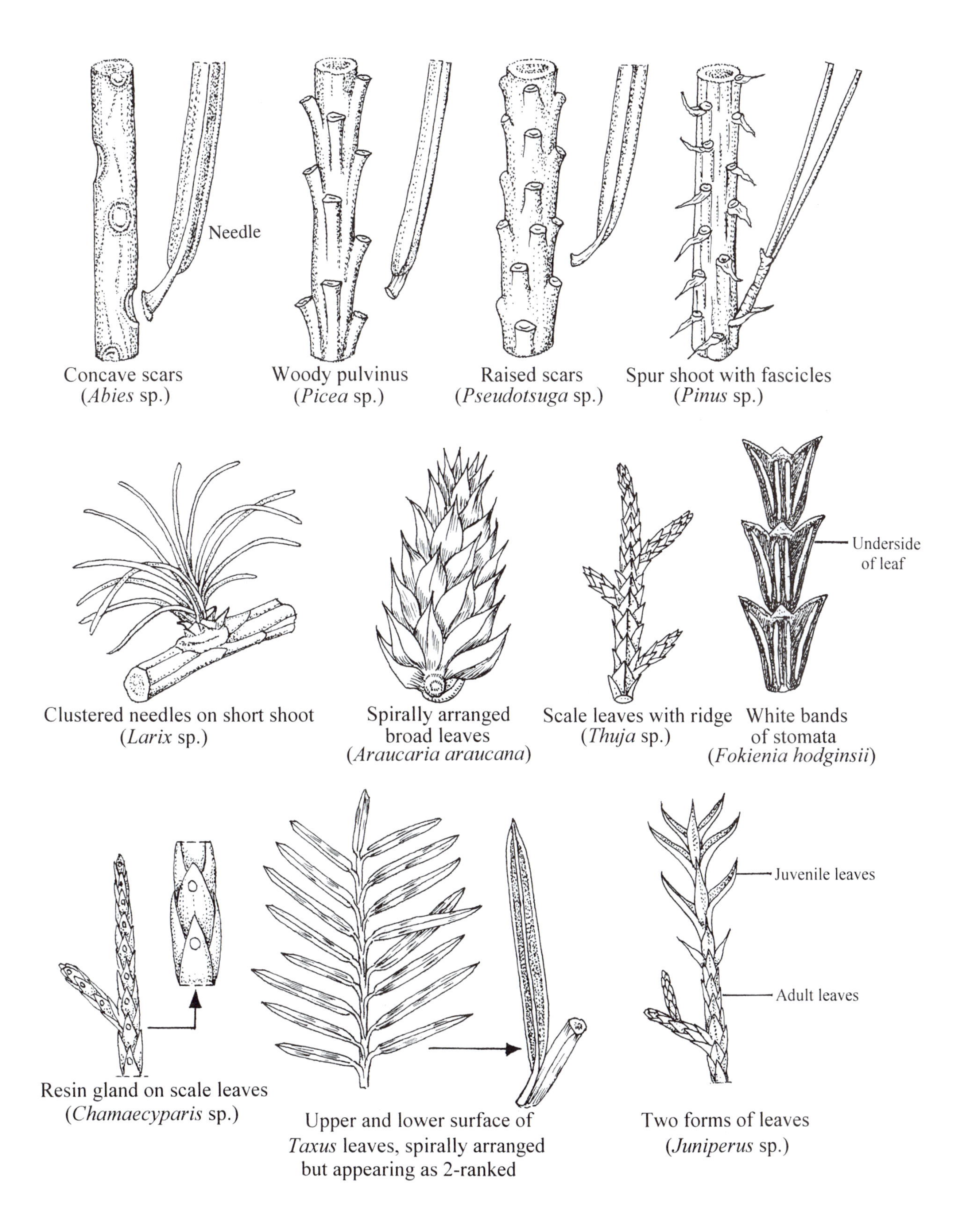

Concave scars (*Abies* sp.)

Woody pulvinus (*Picea* sp.)

Raised scars (*Pseudotsuga* sp.)

Spur shoot with fascicles (*Pinus* sp.)

Clustered needles on short shoot (*Larix* sp.)

Spirally arranged broad leaves (*Araucaria araucana*)

Scale leaves with ridge (*Thuja* sp.)

White bands of stomata (*Fokienia hodginsii*)

Resin gland on scale leaves (*Chamaecyparis* sp.)

Upper and lower surface of *Taxus* leaves, spirally arranged but appearing as 2-ranked

Two forms of leaves (*Juniperus* sp.)

Conifers 4

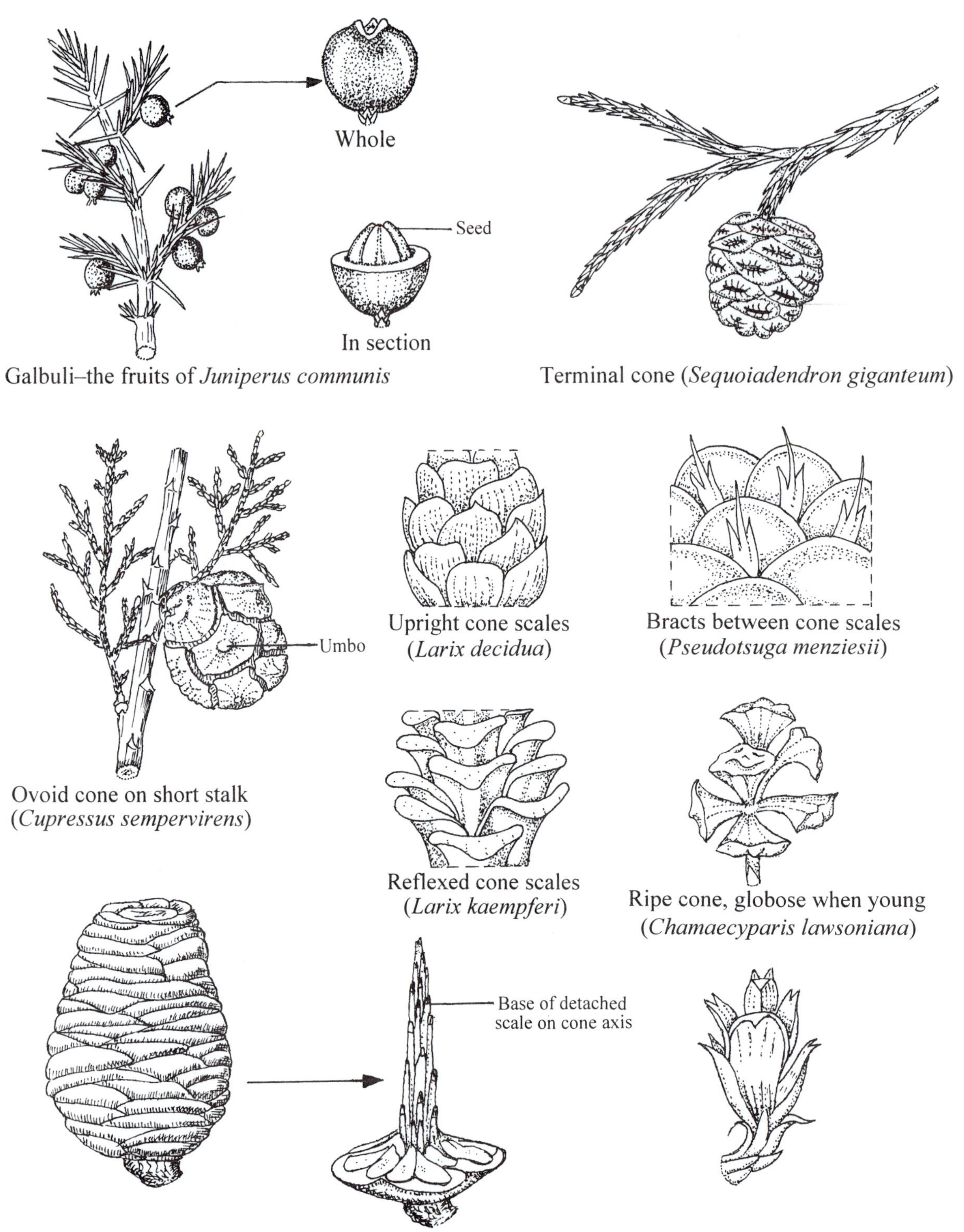

Galbuli–the fruits of *Juniperus communis*

Terminal cone (*Sequoiadendron giganteum*)

Ovoid cone on short stalk (*Cupressus sempervirens*)

Upright cone scales (*Larix decidua*)

Bracts between cone scales (*Pseudotsuga menziesii*)

Reflexed cone scales (*Larix kaempferi*)

Ripe cone, globose when young (*Chamaecyparis lawsoniana*)

Barrel-shaped cone, before and after breaking up (*Cedrus libani*)

Thin flexible cone with reflexed scales (*Thuja plicata*)

Conifer Allies 1 *Ginkgo biloba*

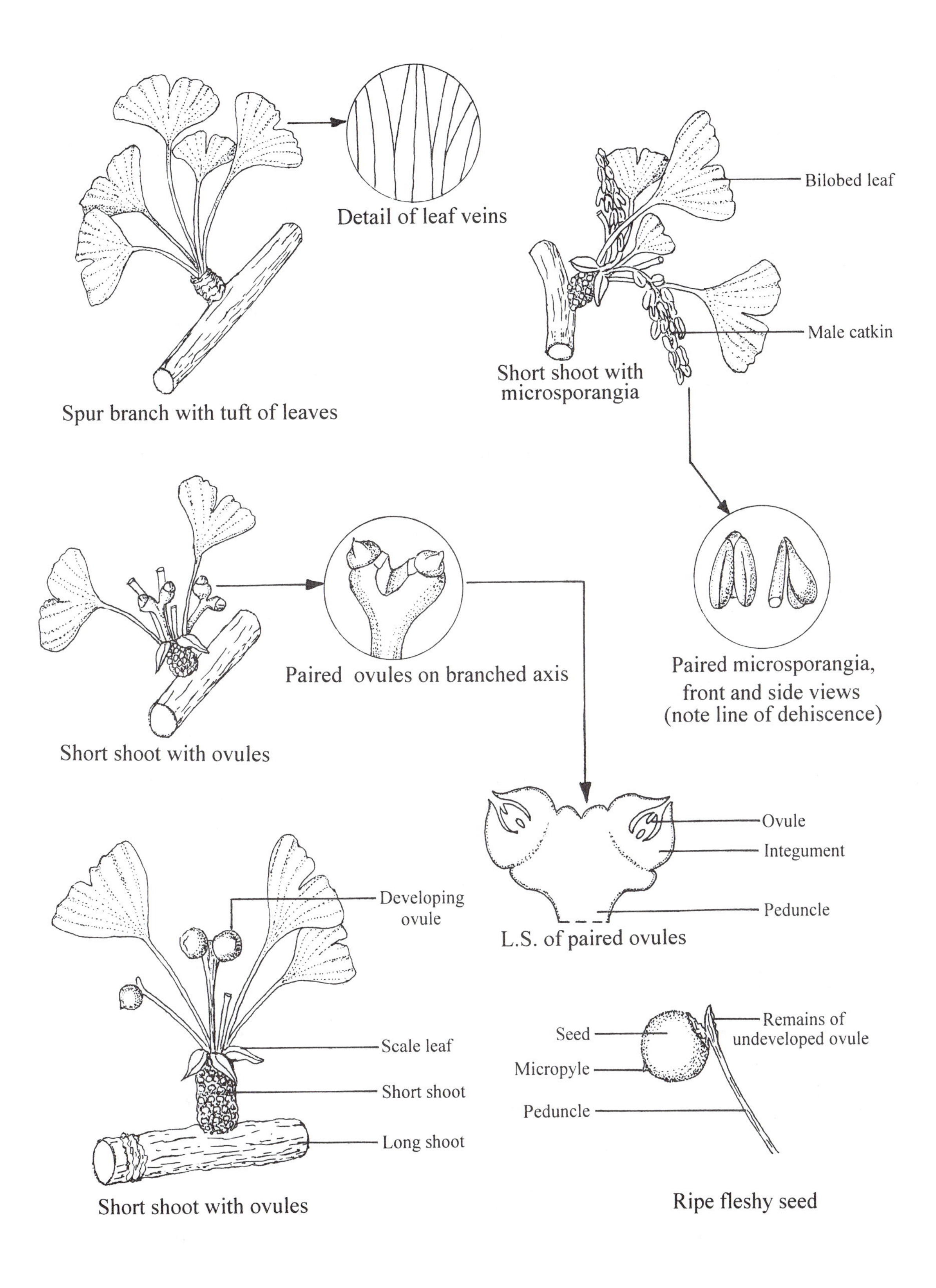

Conifer Allies 2 Cycad

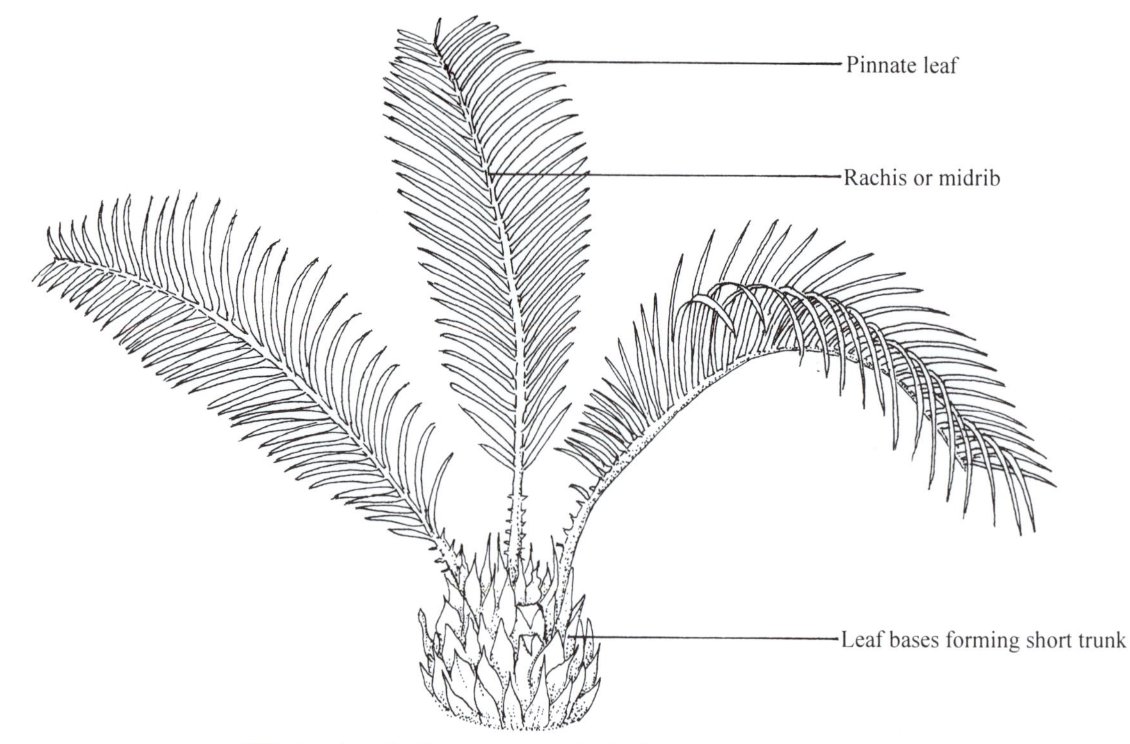

Whole plant before maturity (only 3 leaves shown)

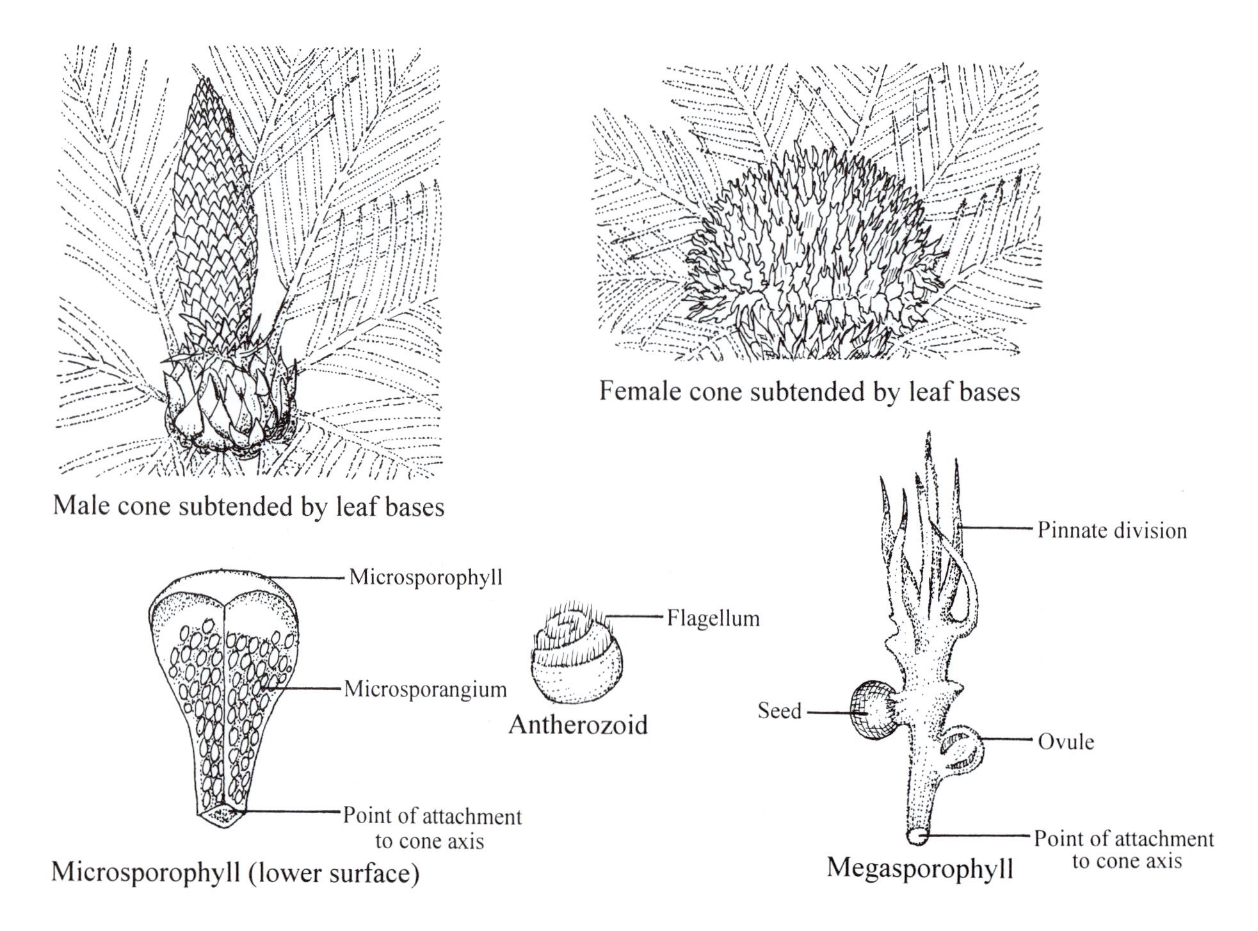

Female cone subtended by leaf bases

Male cone subtended by leaf bases

Antherozoid

Microsporophyll (lower surface)

Megasporophyll

Conifer Allies 3

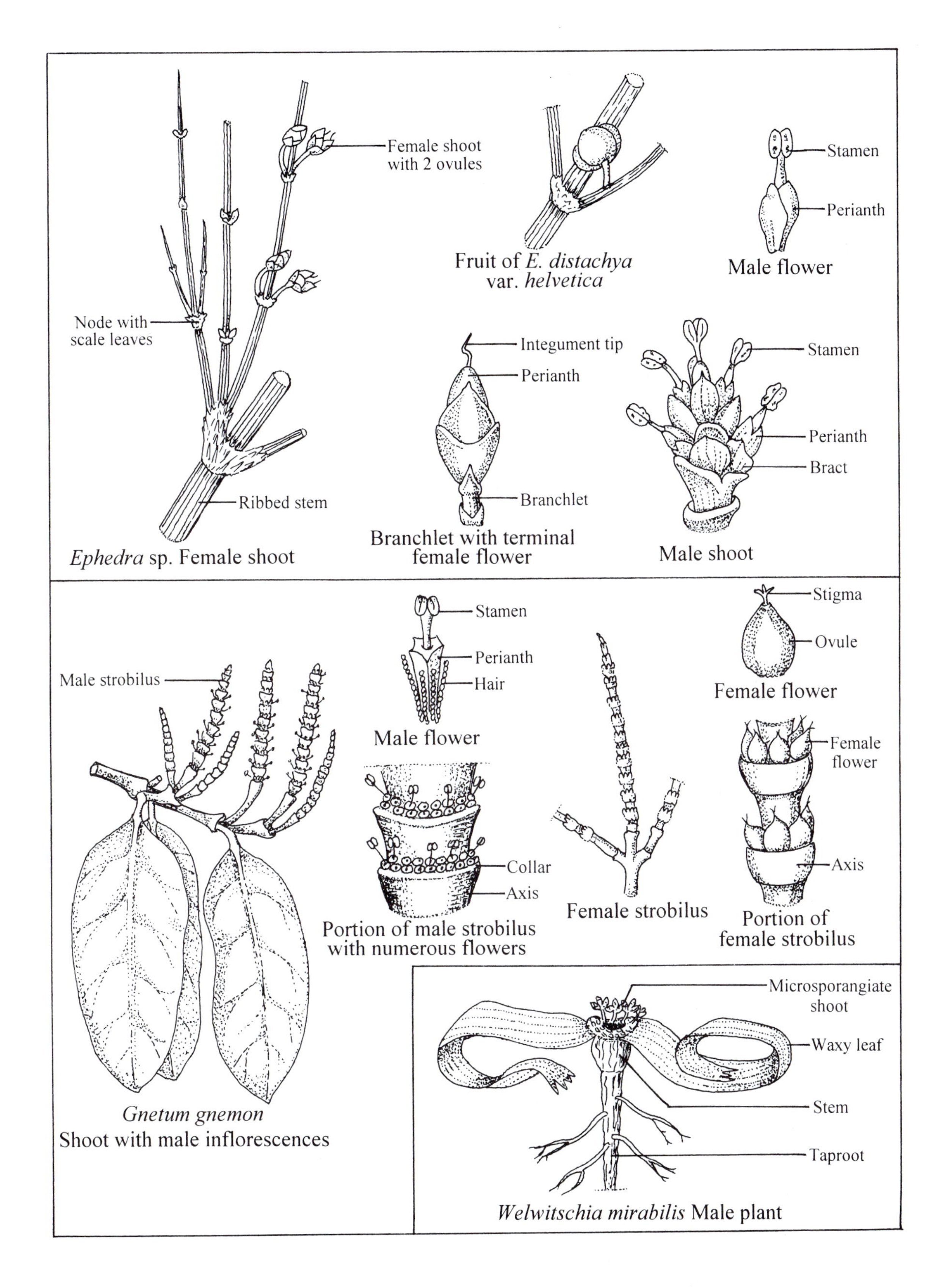

14. Ferns and Fern Allies

Ferns 1 *Dryopteris*

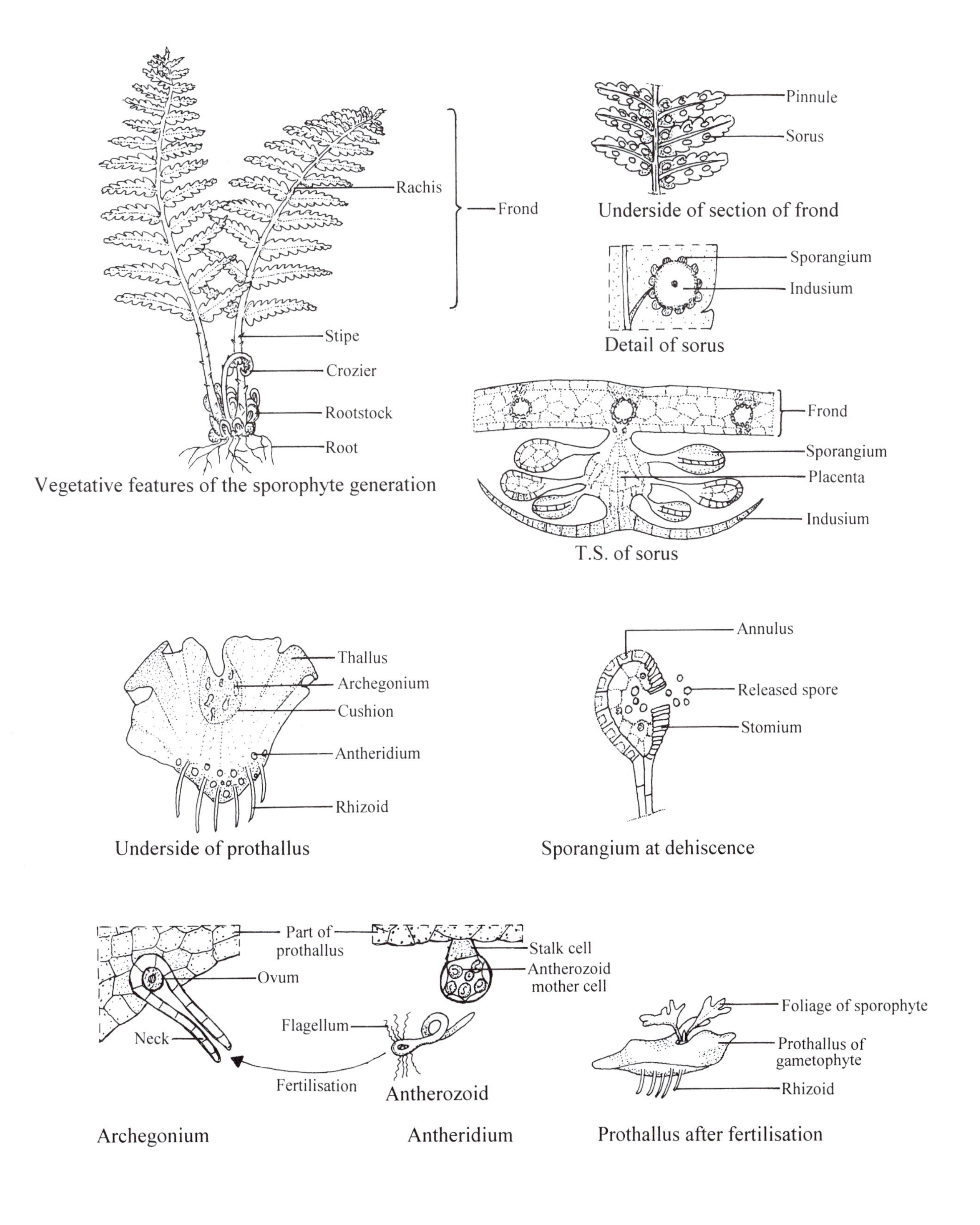

Ferns 2

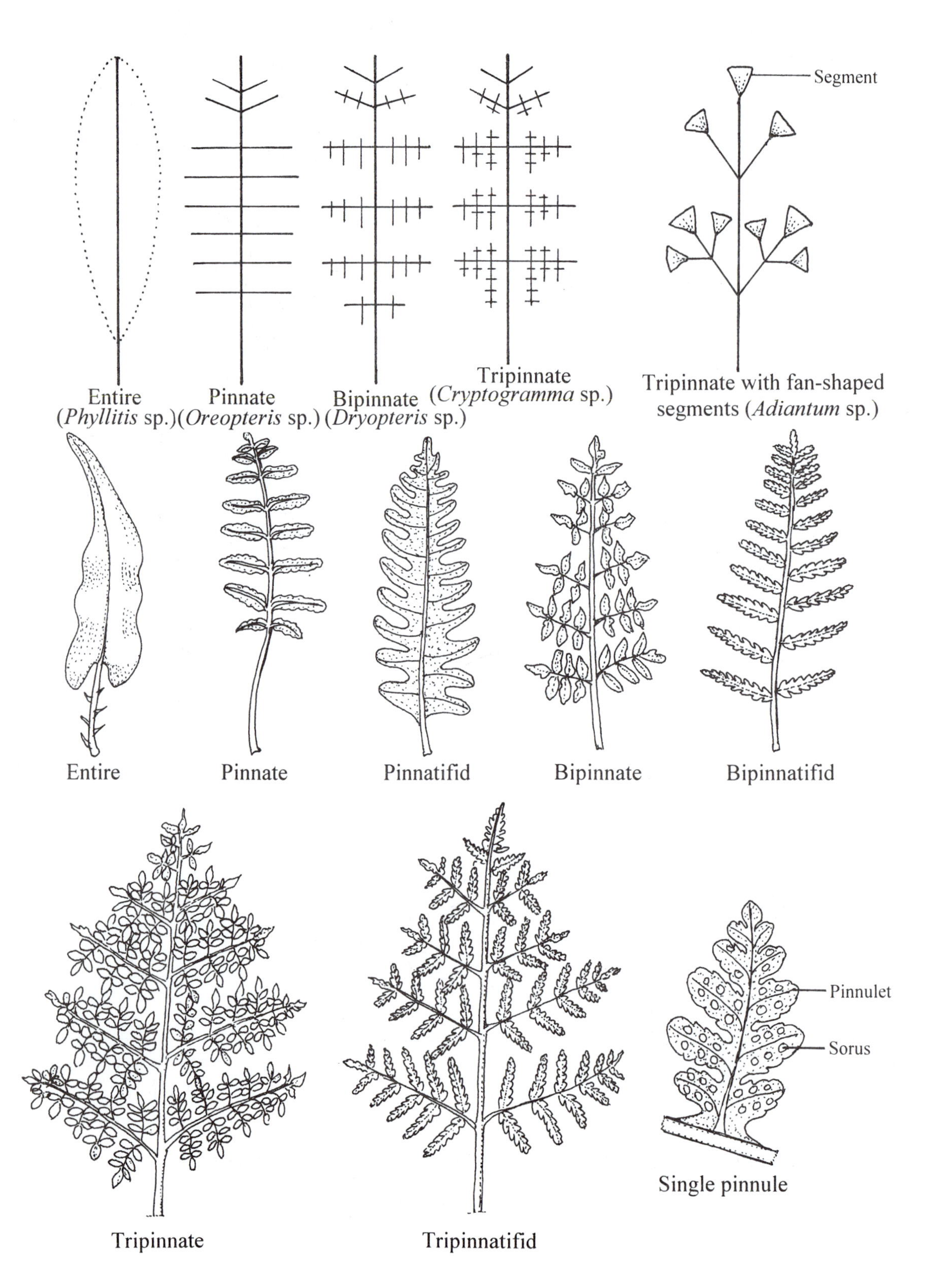
LEAF ARCHITECTURE
Segment
Entire
(Phyllitis sp.)
Pinnate
(Oreopteris sp.)
Bipinnate
(Dryopteris sp.)
Tripinnate
(Cryptogramma sp.)
Tripinnate with fan-shaped
segments (Adiantum sp.)
Entire
Pinnate
Pinnatifid
Bipinnate
Bipinnatifid
Pinnulet
Sorus
Single pinnule
Tripinnate
Tripinnatifid

Ferns 3

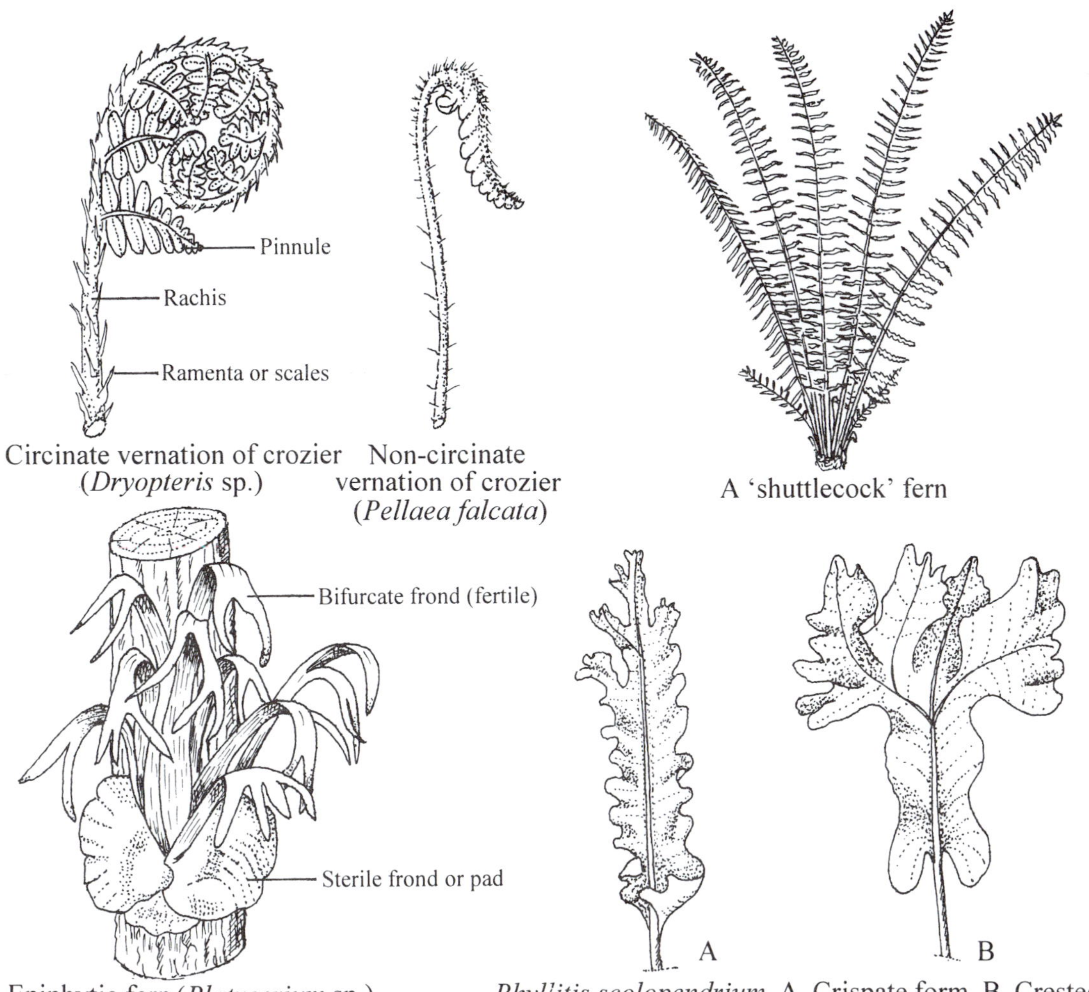

Circinate vernation of crozier (*Dryopteris* sp.)

Non-circinate vernation of crozier (*Pellaea falcata*)

A 'shuttlecock' fern

Epiphytic fern (*Platycerium* sp.)

Phyllitis scolopendrium A. Crispate form B. Crested form

Apospory in *Asplenium bulbiferum*
Plantlet growing on pinna

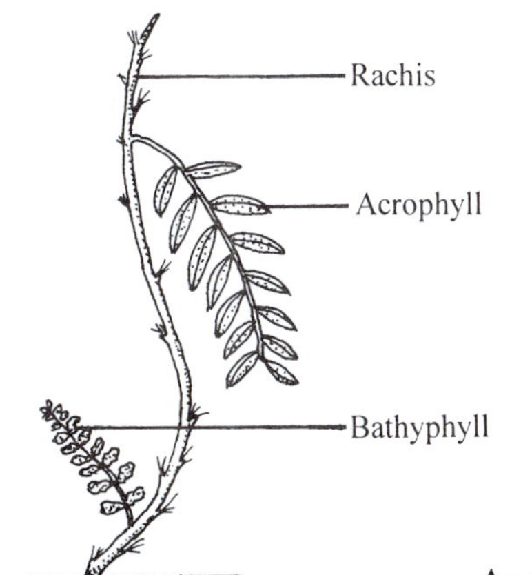

Acrophyll and bathyphyll of a climbing fern

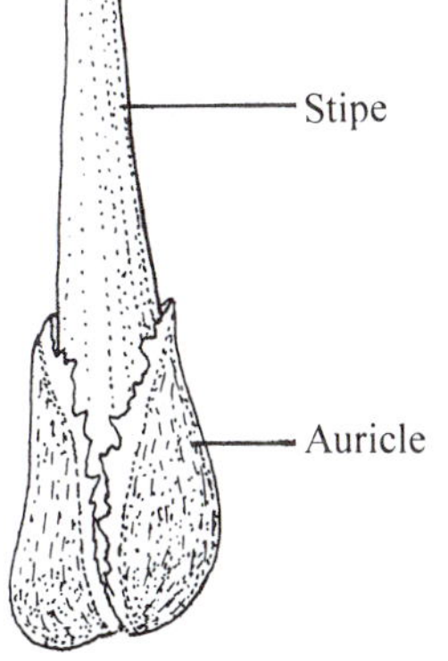

Auricles on *Angiopteris evecta*

Ferns 4

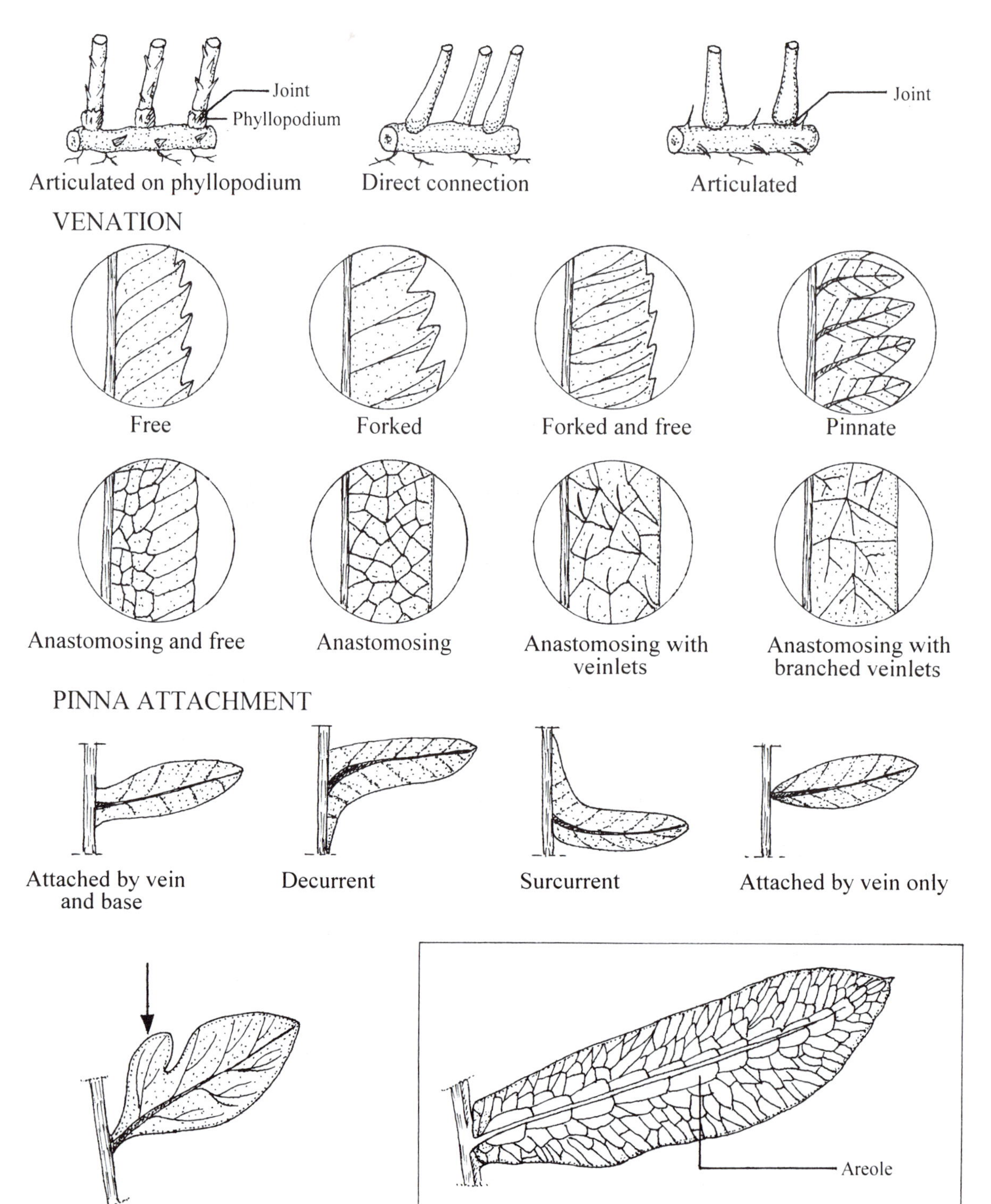

Pinna with acroscopic basal lobe ('thumb')

Enlarged pinna of *Woodwardia areolata* showing venation

Ferns 5

INDUSIA

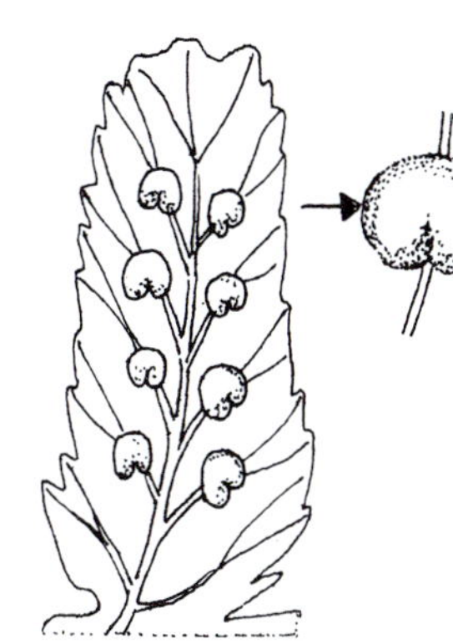
Reniform (*Dryopteris* sp.)

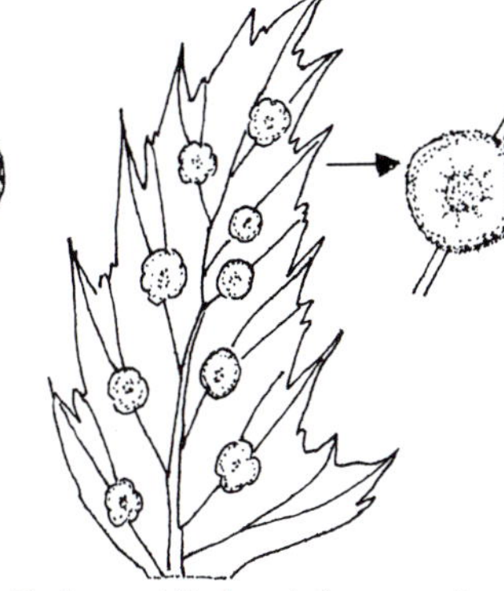
Peltate (*Polystichum* sp.)

Elongate (*Asplenium* sp.)

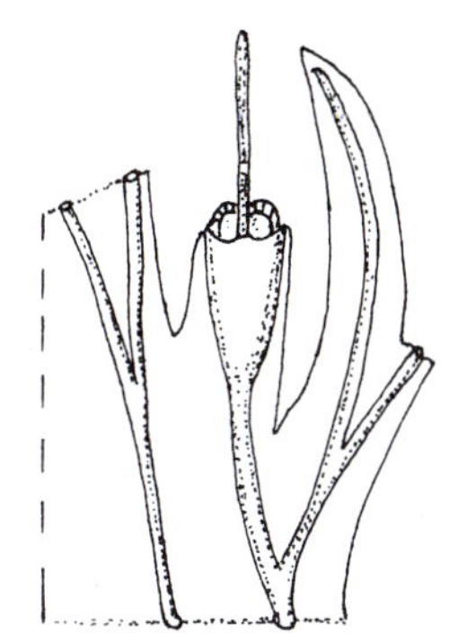
Cup-like or trumpet-shaped (*Trichomanes* sp.)

Valvate (*Hymenophyllum* sp.)

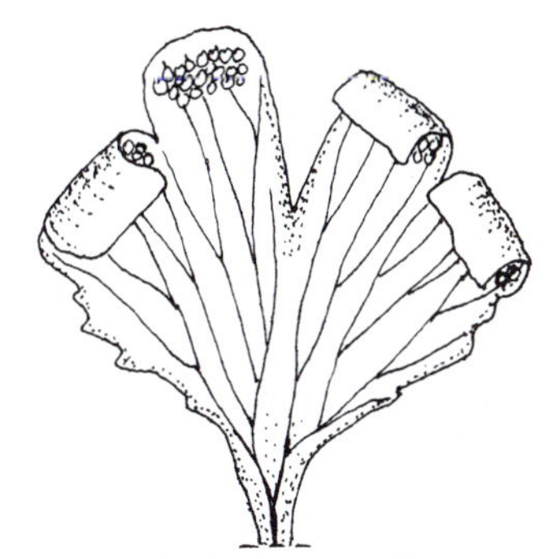
False indusium (*Adiantum* sp.)

Sporangium

Sunken sporangia on the sides of a shaft

Sterile leaf blade

Fertile spike

Whole plant of *Ophioglossum vulgatum*)

Clustered sporangia on a sporophyll (*Osmunda regalis*)

SORUS ARRANGEMENT

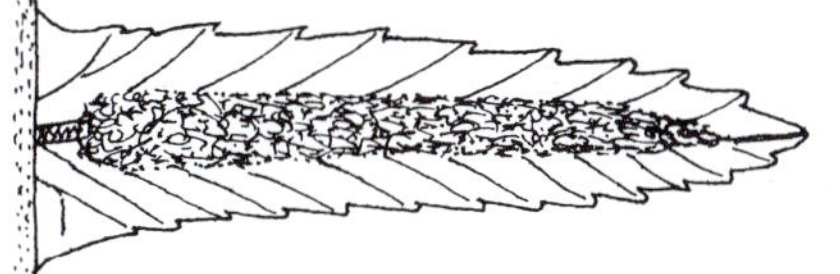
Coensorus along midrib (*Blechnum* sp.)

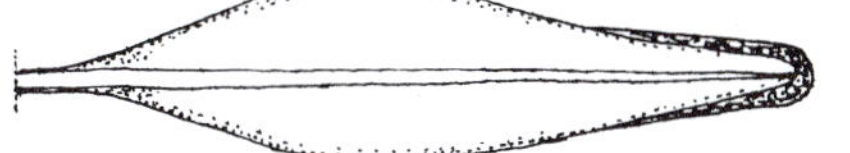
Coensorus on margin (*Pyrrosia* sp.)

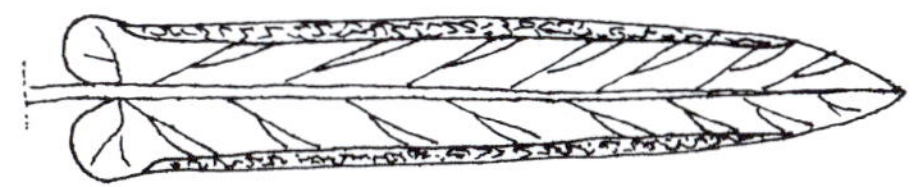
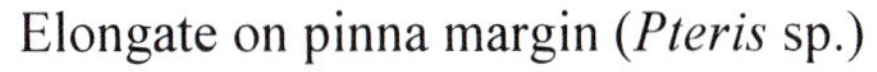
Elongate on pinna margin (*Pteris* sp.)

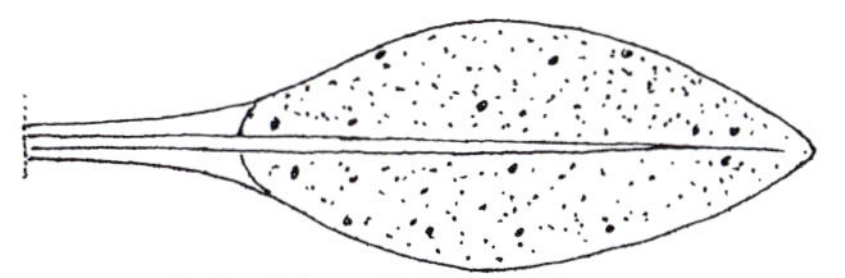
Acrostichoid (*Elaphoglossum* sp.)

Ferns 6, Fern Allies 1

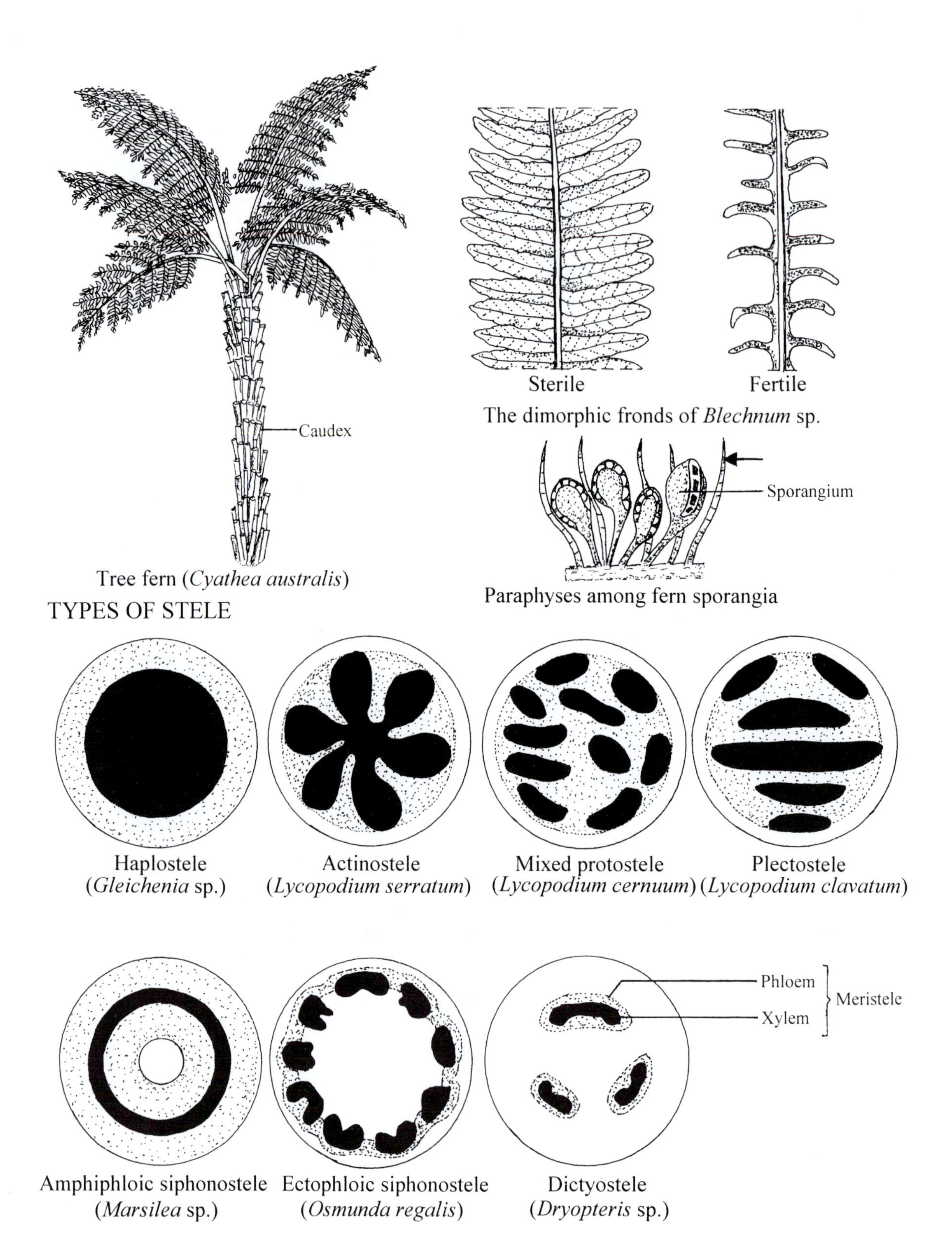

Tree fern (*Cyathea australis*)

The dimorphic fronds of *Blechnum* sp.

Paraphyses among fern sporangia

Fern Allies 2

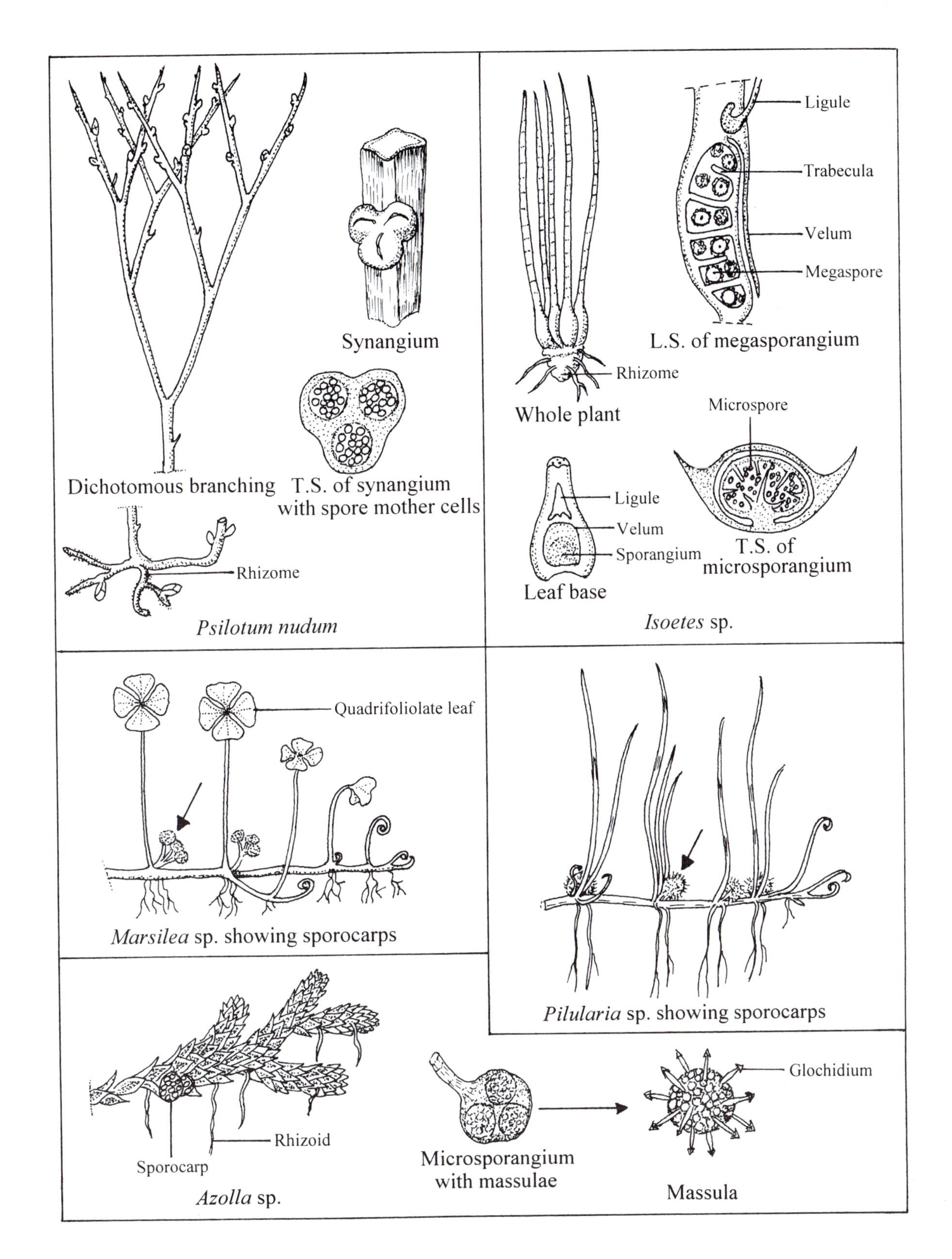

Fern Allies 3 *Lycopodium clavatum*

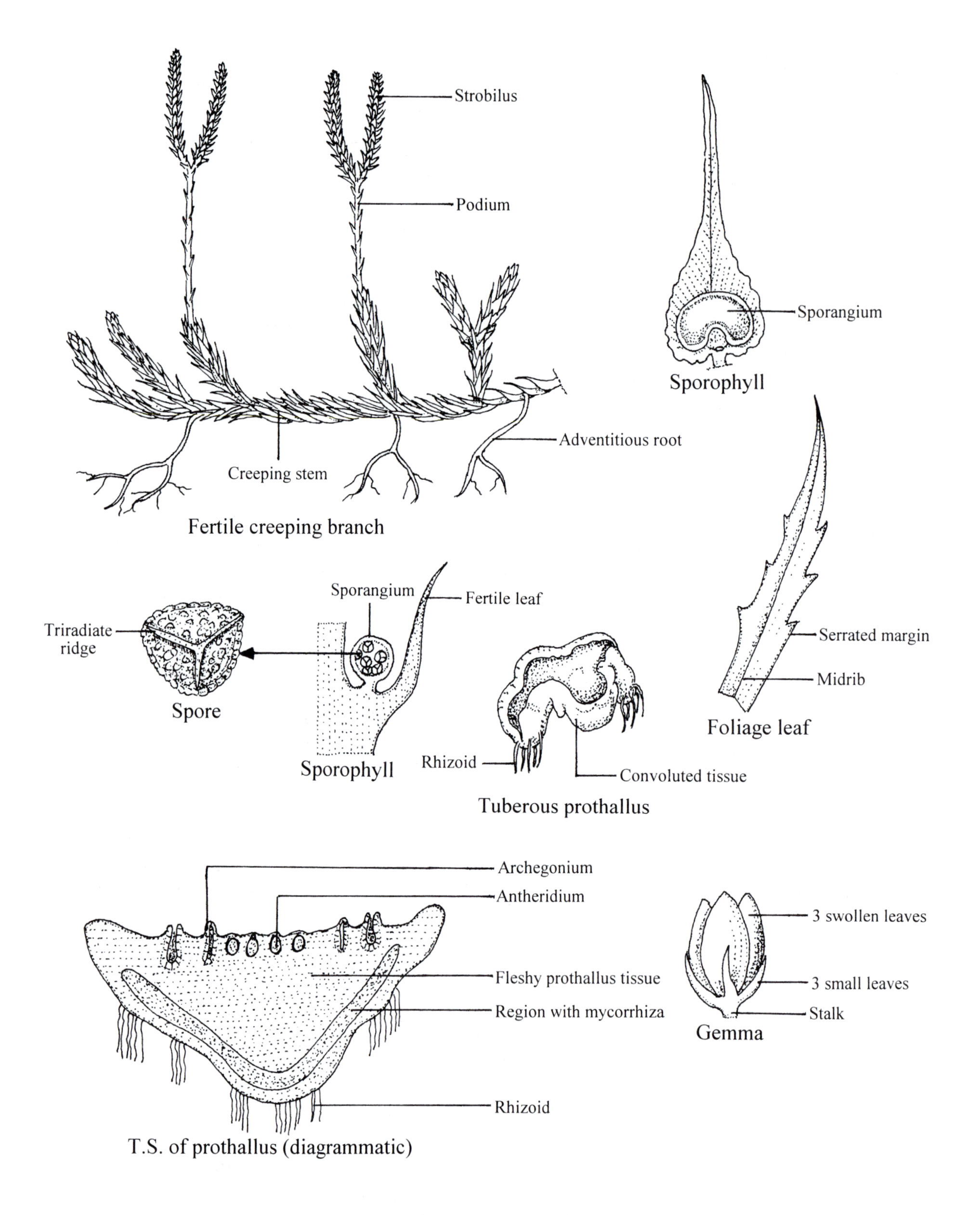

Fern Allies 4 *Equisetum*

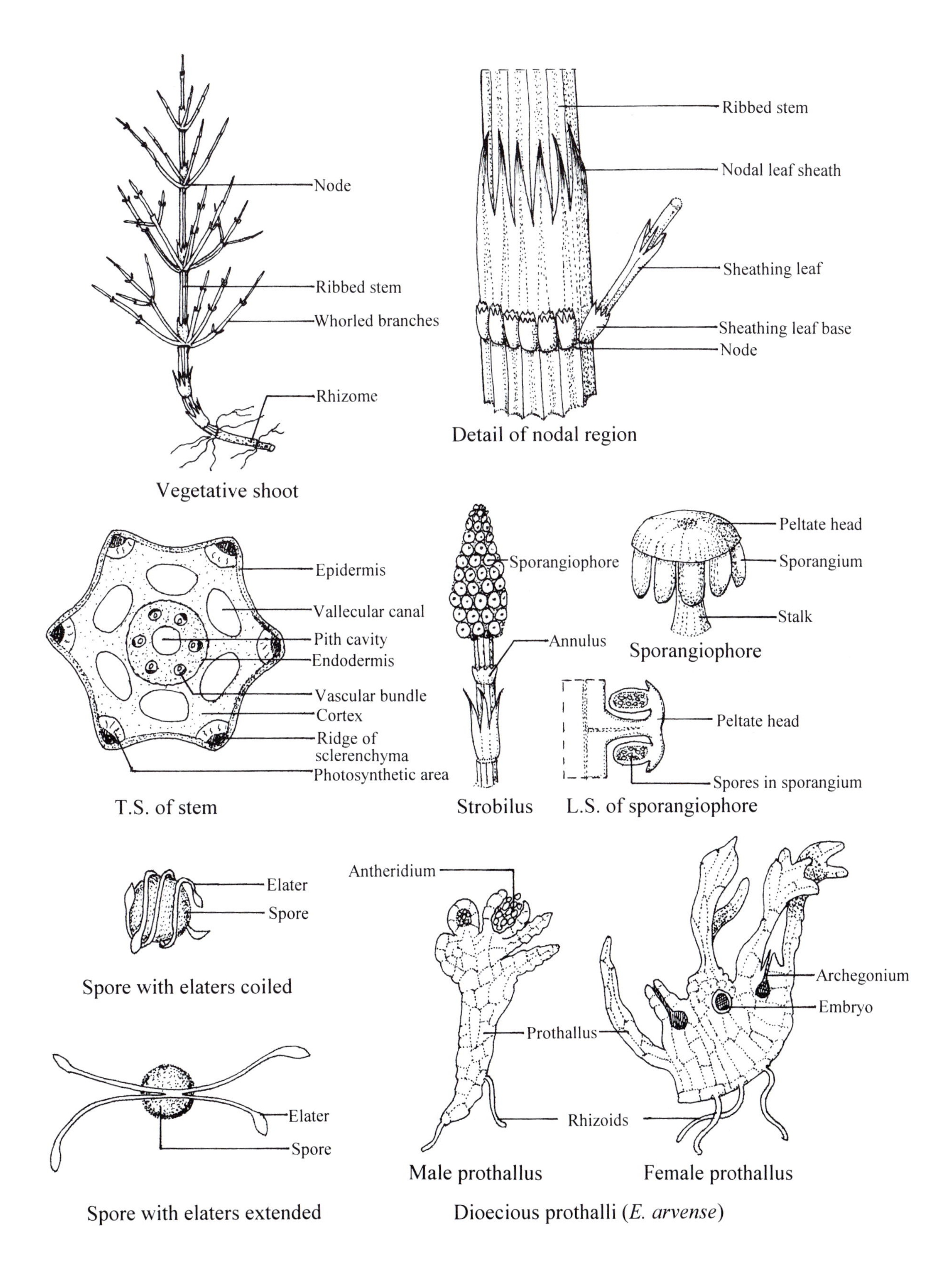

Fern Allies 5 *Selaginella*

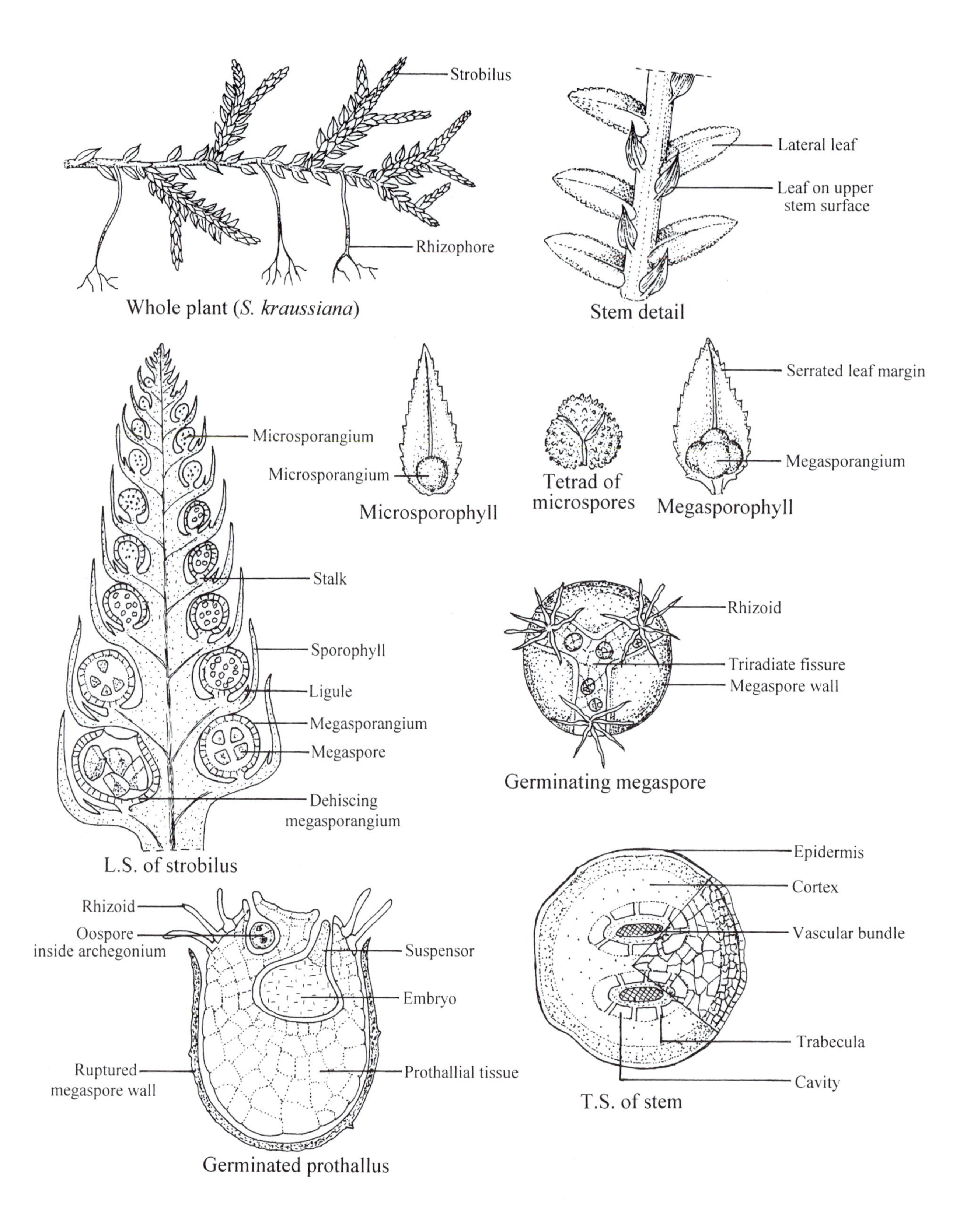